FOREST ECO

FOREST ECONOMICS

Daowei Zhang and Peter H. Pearse

UBCPress · Vancouver · Toronto

20 19 18 17 16 15   5 4 3

Printed in Canada on FSC-certified ancient-forest-free paper
(100% post-consumer recycled) that is processed chlorine- and acid-free.

Library and Archives Canada Cataloguing in Publication

Zhang, Daowei
    Forest economics / Daowei Zhang and Peter H. Pearse.

Revision and expansion of: Introduction to forestry economics.
Includes bibliographical references and index.
Also issued in electronic format.
ISBN 978-0-7748-2152-0 (bound); ISBN 978-0-7748-2153-7 (pbk.)

    1. Forests and forestry – Economic aspects. I. Pearse, Peter H. II. Title.

SD393.Z53 2011   338.1'749   C2011-903547-2

e-book ISBNs: 978-0-7748-2154-4 (pdf); 978-0-7748-2155-1 (epub)

Canadä

UBC Press gratefully acknowledges the financial support for our publishing program of the Government of Canada (through the Canada Book Fund), the Canada Council for the Arts, and the British Columbia Arts Council.

Printed and bound in Canada by Friesens
Set in Syntax and Cambria by Artegraphica Design Co. Ltd.
Text design: Irma Rodriguez
Copy editor: Frank Chow

UBC Press
The University of British Columbia
2029 West Mall
Vancouver, BC V6T 1Z2
www.ubcpress.ca

Contents

Figures and Tables

FIGURES

TABLES

Foreword

For more than 30 years I have taught forest economics to undergraduate and graduate students and to financially sophisticated investors. The latter know much about economics and are interested in forestry but are trying to connect the two. This new textbook will serve these groups well.

Economics is the study of the optimal allocation of resources to achieve society's aims. It has a prescriptive element: how should actors behave if they wish to achieve certain objectives? It also has a normative element: how will key economic factors such as timber prices, harvest levels, and standing inventories evolve if variously incentivized actors (individuals, businesses, governments) interact while seeking to optimize their own positions? This book covers both elements.

While technically a new edition of Peter H. Pearse's classic text, *Forest Economics* is so thorough, comprehensive, and modern a revision that it is really an entirely new book that draws on the individual strengths and collaborations of two of the field's leading practitioners. Pearse has contributed widely to forest policy and natural resource management in Canada, and his seminal paper "The optimum forestry rotation" in *The Forestry Chronicle* has informed many students about the significance of this problem. Because of the large standing inventory of timber in relation to annual harvest, forestry is perhaps the most capital-intensive activity found in a modern economy. The optimal use of that capital has played a seminal role in forest economics for centuries. This book includes the original Pearse formulation and extends it to cover the various elaborations developed in the literature since then. Daowei Zhang's capable analysis of forest sector markets and policy (as exemplified in his book on North American lumber markets, *The Softwood Lumber War:*

Politics, Economics, and the Long US-Canada Trade Dispute) provides a strong normative dimension. Zhang and Pearse have already published together on the impact of forest tenure arrangements on management outcomes, and this important topic is thoughtfully covered in this book.

Both authors have taught forest economics for many years. This book reflects the collected wisdom of that experience. It also includes material drawn from my own courses, which Zhang assisted in presenting at the University of British Columbia. Those who teach forest economics will find this textbook to be a straightforward guide for a semester-long course – there are 13 chapters in the book, a number that does not seem accidental!

Students of forest economics will benefit from the analysis, which links economic principles with both private and public forestry decision making, including resource allocation over time and space, forest sector markets and market intervention (including international trade), valuing unmarketed ecosystem services, and forest taxation. The book extends the traditional discrete-time analysis to the more modern and analytically rigorous continuous-time formulations. The discussion of policy issues (such as the impact of sustained-yield regulation and environmental protection rules, the effect of governmental incentive programs on silvicultural investment, the consequences of alternative forest tenure arrangements, and the role of forestry in economic development) usefully includes contemporary theoretical analysis that is backed up by existing empirical evidence.

Forests in their totality accommodate a wide variety of uses. For billions of us, they are the backdrop to everyday life. Millions of Aboriginal people call forests home. They help regulate key global biogeochemical cycles. Where human interference is absent, or nearly so, they provide important spiritual, ecological, and cultural values. Wood from forests is also one of the most important sources of raw materials and energy in the world today. Sustainable forestry demands that we respect all of these values, and the study of forest economics provides key insights into how forests can be best used and managed. This book provides the tools to begin that task.

Clark S. Binkley
Cambridge, Massachusetts

Preface

This textbook is a thorough revision and expansion of *Introduction to Forestry Economics* by Peter H. Pearse, published in 1990. It is written for undergraduate forestry students taking courses in forest economics and for graduate students with diverse academic backgrounds who are interested in forest management and policy.

This book reflects the two authors' more than 50 years of combined experience in teaching undergraduate and graduate courses in forest economics in the United States and Canada. It differs from the earlier book in its additional and updated content and its more advanced, empirical presentation of materials. Yet the emphasis is still on basic economic concepts, principles, and constructs used to analyze key features of private and public forestry decision making.

Forestry, as we see it, is the applied science of managing land and trees to advance social objectives, which may relate to the production of industrial timber, recreation, or a variety of other goods and services of value to people. Economics is concerned with choices about how resources are allocated and used to create things of value to people. Having begun careers as foresters and later turned to economics, we have found that the two areas of study converge and complement each other. Forestry involves using land, labour, and capital to produce goods and services from forests, while economics helps in understanding how this can be done in ways that will best meet the needs of people.

Moreover, it is increasingly apparent that we cannot isolate forestry from the economic forces that drive other activities. The growing intensity and variety of demands on forests for recreational, aesthetic, and environmental benefits as well as for timber give rise to complicated problems that involve choosing among a wide range of human wants and needs – precisely the subject of economics. And as forestry in North America and elsewhere becomes increasingly concerned with hus-banding and managing forests rather than simply harvesting them, it competes directly with investment needs for other social purposes. Forestry must therefore be understood in its full economic context, and this context can provide a unifying framework for analyzing forestry problems.

Nevertheless, those of us who have taught undergraduate courses in economics for forestry students are often frustrated by the limited refer-ence materials available. Further, there is no introductory forest eco-nomics textbook suitable for a graduate-level course, a complaint often heard from both instructors and graduate students.

Thus we have produced a textbook that will help forestry students understand the economic implications of the work they will do as pro-fessional foresters, and will also help graduate students become forest economists. Whatever undergraduate students learn in one course in forest economics will have to last most of them for many years, while graduate students need more advanced materials. In writing this book, therefore, we have made an effort not only to distill the subject down to the fundamentals – the basic economic principles of forestry and how they bear on forest management and policy decisions – but also to cover more advanced theoretical and empirical issues of forest economics.

Most students of forest economics have already taken a course in eco-nomic principles, and those who have not should be referred to suitable references. This book therefore begins with and builds on the general principles of economics, and elaborates on the particular concepts rel-evant to forest management.

Thus the preparation of this book has been in large part a process of winnowing through economic doctrine on the one hand and forestry problems on the other, in order to focus on the basic connections be-tween the two. Preparation of a textbook that meets the needs of both undergraduate and graduate students is a challenge, however, and runs the risk of satisfying neither. We must rely on instructors to guide stu-dents with varying capabilities towards suitable applications, practical problems, and further readings, some of which are suggested at the end of each chapter. We also rely on instructors using this book for an under-graduate course to identify which materials should be excluded, and those using it for a graduate course to provide additional materials and examples.

We will be content if undergraduate students completing their forest economics course thoroughly understand a few fundamental concepts of economic theory and their relevance to forestry, such as economic efficiency, markets, opportunity cost, incentives, marginal analysis, valuation over time and valuation of non-market goods, optimal rotation age, management intensity, and the implications of property rights, taxes, and international trade. Much of this book deals with the application of these concepts to problems of forestry, and we will be satisfied if graduate students completing their forest economics course understand and gain additional insights into the forest sector, its markets and relevant government functions, and their special challenges for economic analysis.

The book is divided into five parts. Part 1 consists of three chapters. Chapter 1 begins by gathering the threads of economics as they apply to forestry, sketches the scope of the subject, and introduces the issues addressed in the rest of the book. Chapter 2 reviews intermediate microeconomics in the forestry context, and covers important economic concepts and principles, such as economic efficiency, production theory, market failures, and policy failures. Chapter 3 turns to forest investment analysis, which may be seen as an introduction to finance as applied to forestry. This chapter covers the important principles of valuation over time and techniques for assessing forest investments, which are applied to specific problems in subsequent chapters.

Part 2, "The Forest Sector," also has three chapters. Chapter 4 covers market goods in the forest sector, notably timber, and the supply of and demand for these goods. Chapter 5 introduces the variety of non-market goods and services produced through forestry and shows how their values can be assessed. Chapter 6 discusses land allocation among various uses and values.

Part 3 is called "Economics of Forest Management." Chapter 7 deals with the optimal forest rotation age; Chapter 8 considers the regulation of harvests over time; and Chapter 9 presents the temporal dynamics of the forest sector and silvicultural investment.

Part 4, "Economics of Forest Policy," focuses on two of the most important policy issues – property rights and taxation – with a chapter devoted to each. Finally, Part 5 examines some international forestry issues, including trade in forest products and the impact of global forest resources on economic development and the environment.

Forestry involves measurement, but the forestry measurement units used in North America and elsewhere vary considerably. Because foresters and forest economists should be accustomed to working with numbers, in this book we have deliberately used various, often local measurement units in our examples. Students are encouraged to examine the conversion factors listed before Part 1 and to make their own conversions and comparisons.

In preparing this updated book, we have benefited from advice and suggestions from many colleagues, including David N. Laband, Robert Tufts, Yaoqi Zhang, Shaun Tanger, and Gloria M. Umali-Maceina of Auburn University, Michael G. Jacobson of Pennsylvania State University, Janaki Alavalapati of the Virginia Polytechnic Institute and State University (Virginia Tech), Joseph S. Chang of Louisiana State University, Tamara Cushing of Clemson University, and Gary G. Bull and Anthony Scott of the University of British Columbia. Three anonymous reviewers of an early draft were particularly insightful and helpful, so we sought, and received from UBC Press and the reviewers themselves, permission to identify them and acknowledge their contributions. They are Claire A. Montgomery of Oregon State University, Harry Nelson of the University of British Columbia, and Van A. Lantz of the University of New Brunswick.

We must also record our appreciation of the helpful comments we received from William F. Hyde when we began this project, and our thanks to Clark Binkley for contributing the Foreword to the finished product.

Support from our families and the School of Forestry and Wildlife Sciences of Auburn University made this project manageable. Last but not least, we would like to acknowledge the continuing stimulus of our students over many years, which has honed our appreciation of how economics can help to elucidate the challenges of managing forests.

Daowei Zhang
Peter H. Pearse

CONVERSION FACTORS

Length
1 inch = 2.54 centimetres
1 foot = 12 inches = 0.3048 metres
1 mile = 5,280 feet = 1,609.3 metres
1 chain = 66 feet

Area
1 square foot = 0.0929 square metres
1 acre = 43,560 square feet = 0.4047 hectares
1 acre = 10 square chains

Volume
1 cubic foot = 0.028317 cubic metres
1 cubic metre = 35.313378 cubic feet

Weight
1 pound = 0.4536 kilograms
1 short ton = 907.1848 kilograms = 0.90718 metric tons = 2,000 pounds
1 long ton = 1,016 kilograms = 1.016 metric tons = 2,240 pounds

Selected Forest Products
1 cord (80 cubic feet) = 2.27 cubic metres*
1 board foot (logs) = 0.00453 cubic metres**
1 board foot (hardwood lumber) = 0.00236 cubic metres
1 board foot (softwood lumber in US) = 0.00170 cubic metres***
1 board foot (lumber exports and imports) = 0.00236 cubic metres
1,000 square feet (¼ inch panels) = 0.590 cubic metres
1,000 square feet (⅜ inch panels) = 0.885 cubic metres
1,000 square feet (½ inch panels) = 1.180 cubic metres

---

*   A cord is a stack of wood 8 feet long, 4 feet wide, and 4 feet tall (or 128 cubic feet). On average, a cord contains about 80 cubic feet of wood, but a cord of southern pine timber contains 72 cubic feet of wood. Thus, species and log sizes affect the amount of wood in a cord and its conversion factor to cubic metres.

** A board foot is a piece of wood 1 foot × 1 foot × 1 inch, squared in all three dimensions. Because logs are circular and tapered, the quantity of wood they contain is measured more precisely in cubic metres (or tons). Thus, the conversion factor between board feet and cubic metres varies considerably for different species and log sizes.

*** Softwood lumber produced and consumed in the US is measured in nominal, not actual, terms. For example, in the United States, 2 × 4 softwood lumber (historically 2 inches × 4 inches in cross section) is actually 1.5 inches × 3.5 inches, which is why this conversion factor differs from that used in international trade.

PART 1

Markets, Government, and Forest Investment Analysis

Chapter 1   Forestry's Economic Perspective

Forestry calls on a variety of skills and disciplines of study. Foresters must combine knowledge drawn from biology and other natural sciences, applied sciences, and social sciences such as economics. Each discipline brings a different set of tools and methods to the task of managing forests. This book deals with forestry from an economic perspective.

This chapter begins with the economic perspective of valuing forest resources and presents some context for studying forest economics. Those who decide how forests are managed and used must take careful account of the economic and social environment in which they operate, so we review the basic structure of market economies, introduce ideas about society's fundamental economic goals, and sketch the role of governments and private producers. We also explain the scope and substance of forest economics and illustrate how forest economics can help us understand and enhance social values. Finally, we outline a framework for policy development and decision making and indicate where economic analysis fits in.

FORESTRY FROM AN ECONOMIC VIEWPOINT

Forests are *economic* resources because we can use them to help produce goods and services that people want to consume. This is the definition of economic resources (or "factors of production," as they are called in economics textbooks) – things in limited supply that can be combined with other things to produce products and services that consumers want. Thus, we can make use of a forest, combined with some labour and other inputs, to help produce consumer products such as housing, newspapers, fuel wood, outdoor recreation, and environmental services.

It is this usefulness of forests that makes them valuable economic resources. The more value in final goods and services that can be generated from a tract of forest, the more valuable is the forest itself.

Usually there is more than one way in which a forest can be used, and someone must choose from among them. The timber might be harvested and used for making lumber, paper, or fuel wood. It might be kept standing, to support recreational or aesthetic values or environmental services, or it might be saved for industrial use by future generations. Often, a forest can generate two or more kinds of benefits simultaneously or sequentially – such as industrial timber, recreation, livestock forage, wildlife habitat, flood control, and carbon storage – in which case, someone must choose the preferred combination and pattern of uses. In all cases, choices must be made about how a forest will be managed, what goods and services will be produced, how much will be invested in enhancing growth or conservation, and so on.

Economics is the study of such choices, specifically the choices that determine how scarce factors of production are allocated among their alternative possible uses to produce useful goods and services. In other words, it is a science of decision making or the study of how individuals, firms, and societies decide how scarce resources are used.

To illustrate how economics can help individuals make decisions, take the example of college students. Students go to college with one or a few objectives, such as having a good education and getting a good job. On a daily basis, all students have a limited or scarce resource – time. How to allocate time every day is critical to whether students can achieve their goals. Each day, students have several things to do: study, work, engage in sports, relax (for example, by watching TV), and socialize. The economic principle that guides them in allocating their time is that the marginal utility (satisfaction) of allocating the last unit of time to each activity should be equal. In this way, the students get maximum possible utility (satisfaction). Doing this consistently is likely to make them succeed in college. These last few sentences demonstrate the efficiency rule that is developed later.

Forest economics deals more narrowly with choices about how forests are managed and used; how other factors of production, such as labour and capital, are used in forest production, utilization, and conservation; and what and how much forest products are produced and marketed. Forest economics applies the discipline of economics to decision making in forestry and covers the whole forest sector.

Forest economics can be approached from several directions, so it is important to specify at the outset the viewpoint taken in this book and some general assumptions underlying the discussion that follows. First, the focus of attention here is the forestland, the timber, and other goods and services produced directly from forests – the economics of forest resource management. We also deal briefly, however, with the manufacturing and marketing of secondary forest products, such as lumber, plywood, and paper, because the demand for primary forest products such as timber is driven by the value of these secondary products.

Second, our judgments about economic performance are made from the viewpoint of society rather than that of individual forest owners or producers. The criterion we adopt for assessing the economic advantage of one activity over another is a comparison of the net gain, the surplus of benefits over costs, that accrues to society as a whole, taking into account relevant concerns about the distribution of the benefits and costs. This is important because the economic interests of individual entrepreneurs, landowners, workers, or forest users often diverge from that of society collectively. In this respect, this book differs from texts that take a business management approach and analyze problems from the narrower viewpoint of producers or forest owners.

Third, we shall assume throughout that the broad purpose of forest management and the production of forest products is to generate the maximum net gain, or value to society. This assumption is important because much of the literature on forest management assumes, or at least implies, different objectives, such as production of the maximum possible quantity of wood, maximum profits to producers, or stability of harvest rates. Such different objectives have an important place in forestry traditions and, as we shall see, have profoundly influenced forest policies around the world. It is important to recognize, however, that narrower objectives such as these inevitably conflict, to a greater or lesser extent, with the goal of maximizing the forest's economic contribution to society as a whole.

The value that a forest generates for a society can take a variety of forms. Some of these, such as industrial timber, are ordinarily marketed and their value is reflected in their market prices. Others, such as aesthetic benefits and some forms of recreation, are usually provided free, so there is no market indicator of their value. Moreover, while the market values of products like timber are critical to a private company, many non-market benefits of forests are also important to private landowners as well as the public. For example, for many small private forest landowners in North America and Europe, the non-market recreational values of their lands are the primary motivation for owning the land and timber values are secondary. Indeed, a complex range of market and non-market values is a common aspect of forestry almost everywhere in the world. In the forested areas of developed countries, in regions such as British Columbia, Oregon, and Washington, the important values for most landowners may be timber and recreation. In developing countries like India and Nepal, the important values may be construction timber and poles, fruit, and fuel wood; the first two are usually sold in markets, while the latter two are often unmarketed and consumed in homes.

The important point is that in assuming the viewpoint of society as a whole, we must take unbiased account of the full range of social benefits, whether they are priced or not. Even for most private landowners, we must account for a range of both priced and unpriced benefits from forests. Much of our attention in this book will therefore focus on the problems of evaluating environmental and other non-market benefits, trade-offs among uses, and multiple use.

Forest values are classified into two main categories in Figure 1.1. One of these is *extractive values*, which involve physically harvesting and removing resources for use outside the forest; these include not only the familiar timber, poles, and fuel wood but also minor products such as mushrooms, fruit, nuts, livestock fodder, and game. The other main category is *non-extractive values*, which are realized without extracting resources from the forest, and are further divided into two subcategories: *ecosystem services*, such a soil and water protection, biodiversity, and climate modulation; and *preservation values*, which refers to the value that people place on preserving forests in their present state. Recreational or cultural uses of forests can have extractive value (such as hunting and fishing), non-extractive value (such as birdwatching and hiking for relaxation and spiritual renewal), or both.

Although not explained in detail in Figure 1.1, it is useful to distinguish three types of preservation values: *existence value, option value,* and *bequest value.* Existence value is the value that people place on preserving forests from human disturbance for the continuing benefit of future generations; this is the value most often associated with protecting environmentally sensitive forests in their natural state as a park or some other secure form of protection. Option value is the value that forest owners or others may gain from preserving a forest in the present to maintain the option of harvesting it in the future. Bequest value is the value that people derive from bequeathing forests in their present form to future generations. The boundaries between these categories are often unclear and inevitably overlap.

Scholars and others use terms like "ecosystem services" and "preservation values" in various ways. The most broadly encompassing definition

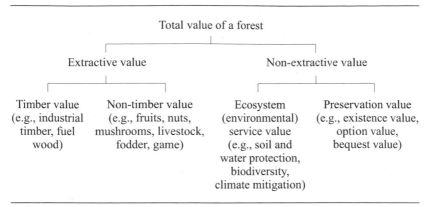

Figure 1.1   A forest's economic value.

of ecosystem services is that of the United Nations 2004 Millennium
Ecosystem Assessment, which embraces all benefits that humans receive
from managed and natural ecosystems – that is, all the values listed in
Figure 1.1. For our purposes, however, it is useful to distinguish these
values to draw attention to the different challenges they present for
measurement and for making trade-offs among them in forest management decision making.

As a final observation on the values derived from forests, it is worth
noting that some biologists and philosophers argue that forests have intrinsic value independent of their instrumental value or usefulness to
human beings, and so an economic valuation is too anthropocentric to
capture their full value. Human values are undoubtedly influenced by
ethical and moral norms, however, especially as these apply to nature,
and whether there are values beyond this – or a practical method of
measuring them – is unclear. In any event, this book focuses on the economic value of forests, defined broadly to include all their identifiable
priced and unpriced values, to guide forest management decisions.

BASIC ECONOMIC QUESTIONS

An economy consists of production, consumption, investment, and other
activities linked by a huge number and variety of transactions taking
place continuously. The bewildering detail and complexity of an economy
can, however, be visualized in terms of a few straightforward processes.

On the one hand, the society being served by the economy has certain
wants. People *want* goods like food, houses, and television sets, and services like medical care and recreation. Their welfare or standard of living
is measured by the extent to which these wants are met: the more

people's wants are satisfied, the better off they are and, since no society has ever been known to be fully satisfied, welfare is always a matter of degree. It is important to note that people's wants extend beyond strictly private desires to collective or public concerns about economic security, equity, and freedom.

On the other hand, any society has a limited capacity to produce the goods and services that will satisfy these wants. The wherewithal to produce these consists of natural resources (or natural capital); human-made capital such as machines, roads, and other infrastructure; labour; and technical knowledge. All of these change over time, but at any point in time they are finite.

The function of the economic process is to determine how these limited resources are used to satisfy some of the unlimited human wants. Thus, economics is the study of how scarce resources are allocated among competing uses.

Every society must deal with three fundamental economic questions. Given its limited endowment of productive resources and the unlimited wants that they must serve, a society must somehow make decisions about:

- *which goods and services,* of the almost infinite variety that it is technically possible to produce with these resources, will actually get produced, and in what quantities
- *which* of the variety of technically possible *ways of producing* each good and service will be adopted in each case
- *how* the goods and services produced will be *distributed* among members of the society.

These basic questions are answered in every economy, but in different ways. Primitive, subsistence societies rely heavily on custom and tradition to make decisions about what to produce, and how. Socialist systems rely primarily on central planning and governmental direction. The capitalist system depends on market forces generated by the independent actions of individual producers, consumers, and owners of productive resources. A mixed capitalist system utilizes market forces as well as government intervention. Nowadays, pure socialist and pure capitalist economies are rare. Rather, most countries have adopted mixed capitalist and socialist systems, which depend on varying degrees of government participation and intervention in markets.

Any study of the economics of forestry must take careful account of the character of the economic system within which forestry is being practised because this governs the issues that need attention. A socialist or subsistence economy raises quite different problems from a capitalist one. In this book, we assume the context of a mixed capitalistic economy, except where we note otherwise.

In a typical form of "mixed capitalism," most production is organized and carried out by private entrepreneurs responding to market incentives. Governments play an important role in regulating economic activity, however – providing a variety of services, manipulating prices and incentives, redistributing income and wealth, and managing the general level of economic activity.

The basic theory developed to explain how mixed capitalist economies operate is thoroughly dealt with in numerous elementary textbooks. This book is intended to build on, rather than duplicate, this general economic theory. Accordingly, the basic principles of economics are reviewed only briefly in the following chapters. Our emphasis is on the particular role that forests play in the economic system and on the economic choices faced by forest managers.

MIXED CAPITALISM AND THE ROLE OF GOVERNMENT

In a market economy, entrepreneurs take responsibility for producing things. They occupy the interface between suppliers of productive resources and purchasers of final goods and services. Entrepreneurs purchase the resources they need in order to produce the goods and services desired by consumers, and the prices they pay for these resources determine the incomes of those who provide the resources. The first of the three basic economic questions referred to earlier – what should be produced – is therefore determined in the first instance by consumer demand, hence the concept of "consumer sovereignty." The second question – about how the output will be produced – is determined by constant competition among individual producers to find the most cost-effective means of production in order to enhance their profits. And the third – concerning the allocation of the fruits of production – is resolved by the distribution of income, which in turn is governed by the market values of the labour, capital, and other productive resources that private suppliers make available to producers.

In the mixed capitalist system typical of most countries in the world today, however, governments intervene in these processes and change the pattern of production in important ways. They provide not only the traditional *public goods* (such as roads, lighthouses, and national defence) that private producers do not normally produce at all but also an increasing variety of goods and services that private producers produce inadequately. Such things as health care, education, and the arts fall into this category of *merit goods,* which have a social value exceeding their value to individual consumers. Some governments even produce industrial timber, a seemingly pure private good. More importantly, governments indirectly influence private production and consumption by means of taxes, subsidies, and regulations. Governmental regulation of activities ranging from marketing to safety procedures for workers affects industrial structure, output, and prices. All these forms of intervention that alter the way in which productive resources are allocated and used comprise the *allocative* role of government.

Governments also substantially affect the distribution of wealth and income. Taxes, government spending programs, transfers, and borrowings of various kinds all redistribute income within and among socioeconomic groups, regions, and generations. Sometimes these redistributional effects are deliberate and obvious, as when pensions are paid to the elderly, but often they are subtle and indirect, requiring complex analysis to trace their full impact. This is the *distributive* role of government.

Finally, modern governments accept responsibility for maintaining a stable level of economic activity. This calls for fiscal policies (spending and revenue-collecting programs) and monetary policies (manipulation of interest rates, exchange rates, and the supply of money) to offset trends towards inflation or unemployment. Related to these stabilization activities are policies for promoting economic growth and regional development. These comprise the *stabilization* role of government.

By intervening in various ways, governments attempt to correct some of the weaknesses and inadequacies of the market system. Expressed in another way, government intervention in the form of *allocative, distributive,* and *stabilization* measures reflects efforts to improve the performance of the economy in terms of achieving the economic objectives that a society sets for itself through the political process.

In studying the economics of forest management, we find ourselves continually confronted with government policies aimed at influencing the way forest resources are developed, managed, and used. The primary objective of some of these policies is to improve efficiency by affecting the rate and pattern of resource use. Other policies are motivated by distributional or equity considerations, or a desire to manipulate community and regional growth. Whatever their primary purpose, all forms of intervention inevitably have implications for all three of the fundamental forms of economic impact, namely, the allocation of resources, the distribution of income and wealth, and economic stability and growth.

ECONOMIC OBJECTIVES: EFFICIENCY AND EQUITY

The allocative, distributive, and stabilization roles of government imply two fundamental economic objectives of society: *efficiency* and *equity*. These objectives provide us with criteria for assessing economic performance.

In any society, there is a presumption, more or less qualified, that the primary objective of economic activity is to satisfy consumer demands to the greatest extent possible. The extent to which these demands are met with the available resources is a measure of the *efficiency* of the economic system.

At the macroeconomic level, if resources were employed in one sector of the economy when they could generate greater value in another, it would be possible through some reallocation to increase the value of total output and hence the efficiency of the total system. In that case, the gross domestic product (GDP), the total value of all goods and services produced in an economy in a year, which is often used as a first approximation of an economy's performance, would be increased. Similarly, at the microeconomic level, an inefficiency exists if a producer fails to employ an available technology that would enable him to produce more with the same amount of inputs.

Thus, *economic efficiency* refers to the allocation of resources that generates maximum value from the resources used. The level of economic efficiency is reflected in the relationship between inputs and outputs: the greater the output relative to input, the greater the efficiency. In economic analysis, efficiency is expressed as the ratio of benefits (outputs) to costs (inputs), both measured in the common denominator of dollar values. Of course, a thorough economic analysis from the viewpoint of society as a whole must account for unpriced benefits and costs as well as those that are more readily observed and measured in market prices.

Efficiency in economic activity is therefore a logical social objective. Unless there are offsetting considerations, the use of any resource in a way that generates less value than it is potentially capable of generating through some other uses is simply a waste, lowering the value that society derives from its resources.

How forests can best be used in light of the variety of demands on them is one of the central questions of economic efficiency in forestry. A second question concerns the intensity of forestry – that is, how much labour and capital can be advantageously devoted to utilizing and managing forests to increase production. There is also an important temporal dimension to economic efficiency in forestry, referring to the pattern of investment and utilization of the resource over time. Because forests take so long to grow and can be harvested over such a broad span of time, this temporal dimension of efficiency is especially important in forestry.

Market economies give producers incentives to operate efficiently and thereby compete successfully. On the other hand, various distortions and *market failures,* or failures of prices and the need for corrective policies to reflect society's real preferences, give scope for governments to improve efficiency through their allocative, stabilization, and growth-stimulating activities.

Equity refers to fairness in the distribution of income and wealth, and therefore the fruits of production, among the population. As noted above, the distribution of income is determined, in the first instance, by payments for the factors of production. It can be altered to a preferred pattern, however, through taxes, subsidies, transfers, and other types of distributive intervention by governments.

Like efficiency, equity invites a comparison of possible alternatives to determine which, of all the possible distributions of income and wealth, best serve society's preferences. Also like efficiency, equity has more than one dimension. *Interpersonal equity* refers to the distribution of income among individuals at any time. Equity among people living in different geographical regions is referred to as *interregional equity.* And *intergenerational equity* refers to the distribution of income among people living at different times. All these dimensions of equity are relevant to forest policy. In forestry, questions of equity often relate to the needs of poorer people or poorer regions, or to conserving forest resources for future generations.

Both efficiency and equity are difficult to measure. Efficiency is usually measured in dollar terms: the value of outputs relative to the cost of inputs, both of which are often reflected in market prices. Market prices are sometimes misleading, however: some benefits are not traded in markets, some costs exceed the amount of compensation paid, and other distortions and market failures make it necessary to supplement market price information with estimates of social values in order to assess efficiency. Equity, which rests on subjective judgments about fairness in the distribution of income and wealth, defies empirical measurement except through political processes and ethical judgments. For example, a policy that makes disadvantaged people worse off is often considered inequitable, but that might not be so if it serves some other social objective, such as a desired regional redistribution of income or benefits to future generations, all of which are difficult to measure and compare.

It is important to note that the objectives of efficiency and equity often come into conflict, and it becomes necessary to sacrifice one for the other. For example, measures that could expand output (increase efficiency) might create unwanted changes in the distribution of income (decrease equity), and vice versa. This illustrates the trade-off between improvement in equity and aggregate production, and the choices that must be made. The relative priority of objectives and the appropriate compromises among them are not matters that can be solved by economic analysis. Political and electoral processes must be depended upon to prescribe the appropriate mixture of allocative, distributive, and stabilization efforts on the part of governments and to reconcile divergent opinions about equity and efficiency. Economic analysis can provide guidance in making these decisions, however.

This book places heavy emphasis on the efficiency of resource allocation, especially the economic efficiency of forest resource development and use. This is not to suggest that concerns about equity and stability are unimportant in forestry; on the contrary, we shall see that some of the most profound issues in forest management, issues that have motivated significant forms of governmental activity, have to do with distribution among groups and regions and economic stability over time. We emphasize the efficiency of resource allocation, however, for two reasons. One is that it provides a necessary starting point for examining the benefits and costs of governmental policies and programs that change the distribution of income, promote economic growth, or affect market outcomes in other ways. The second reason is that from the viewpoint of an economic analyst, much of the uniqueness of forestry, as distinct from other forms of economic activity, centres on problems of efficiency.

FORESTS AS ECONOMIC RESOURCES

In economics, the general term "resources" refers not only to land and natural resources but also to capital, labour, and human skills that are valuable in producing goods and services. The essential characteristic of an economic resource is that it is "scarce," in the sense that there is not enough of it available to satisfy all demands for it. It is this scarcity, or limitation of supply, that raises problems of choice about how resources are to be allocated. It also makes them valuable, even though their value in some uses is not reflected in market prices.

Not all forests are economic resources in this sense. Some are so inaccessible, so remote, or of such poor quality that they are not demanded for any economic purpose, even though they might provide environmental benefits. Having no economic value or alternative uses, such forests do not present the usual problems of choice and allocation among competing uses that are associated with economic resources. Most forests, however, are capable of yielding one or more products or services, and so they constitute part of an economy's total endowment of productive resources. It is this economically valuable part of the total physical stock of forest that we are concerned with in forest economics.

An economy's total endowment of productive resources is commonly divided into four broad categories: *land, labour, capital,* and *entrepreneurship.* Each of these has distinctive economic characteristics, and each generates economic returns of a different kind, namely, *rent, wages, interest,* and *profit,* respectively.

Forest resources fall into two of these categories. The basic resource is the forest *land,* which has the same economic characteristics as agricultural and other land. In any location, it is fixed in supply; it varies in productive quality; and it generates a residual value, or rent, that varies accordingly. The *forest* itself, consisting of trees on the land, falls into the category of capital. It can be built up over time through investment in silviculture and pest control, or it can be depleted through harvesting; it derives its value mainly from the final goods and services that can be produced from it; and it generates returns measured as interest. Standing timber is capital in this economic sense regardless of whether it is a gift of nature or a product of a long period of costly management.

Forestland and timber are economic resources because they are valuable in producing other final goods and services. The demand for land and timber stems from the consumer demand for these final products, and in this sense is a "derived" demand.

Forestland and the capital embodied in timber are part of a society's total endowment of productive resources that can be used in a variety of ways to produce useful goods and services. As with other resources, the extent to which they contribute to social welfare is governed by the efficiency with which they are allocated and used.

Traditionally, forest economics has been concerned with the management of forests for production of wood for industrial manufacture into building materials, pulp and paper, and so on. Forests also yield other goods and services, however, and are often managed to produce fuel wood, livestock, fish and wildlife, recreation, and water supplies. Such benefits are often produced in combination with industrial timber production. Some of these values, especially recreational and environmental benefits, are becoming increasingly important. These increasing and overlapping demands on forest resources complicate the problem of allocating them among alternative uses and combinations of uses.

Moreover, as some forest values are often not priced, they are difficult to evaluate in terms comparable with timber values, but these values are real whether or not they are priced; the absence of price indicators only complicates the problem of economic analysis. Later chapters address these issues in detail.

WHY FOREST ECONOMICS?

Economics deals with all kinds of productive resources, while forest economics focuses specifically on those used in forestry. The latter includes, obviously, the land and forest growth that constitute the forest itself, but it must also consider the labour, capital, and other inputs to forest operations and forest products production. Much of forest economics is concerned with how much of these other resources can be efficiently combined with forestland and timber in producing forest products and services.

This is the subject matter of microeconomics – the half of economic science that deals with how prices and incomes are determined, how producers find the most efficient scale and form of production, how consumers behave, and so on. Forest economics builds on this basic theory as it applies to forests and land. Forest economics is therefore in large part a study in applied microeconomics.

Like other special fields of applied economics, forest economics draws on the particular threads of economic theory that are relevant to the unique or especially important problems of the field. For forest economics, the theory of production, especially the theory of capital and rent, is fundamental. And, as a relatively narrow area of applied economics, it draws on broader applied fields such as the long-established specializations in land and agricultural economics and the newer branch of natural resource and environmental economics.

The special characteristics of forest resources that justify considering forest economics as a special field of study can be summarized as follows:

- Forests can produce a wide variety of goods and services and combinations of them, some of which are not priced in markets. This gives rise to special problems relating to the allocation of resources among uses.
- With the exception of some tropical and temperate species, forests typically take a long time to grow, often involving investment periods of decades. This gives rise to special problems in investment analysis, harvesting schedules, risks in carrying forest crops over long periods,

and market uncertainty. It also means that forests can be altered only slowly in response to changed economic and natural conditions.

- Forestry usually involves very high capital and carrying costs relative to production because the slow rate of forest growth means that large forest inventories must be carried to sustain a modest harvest. As a result, the costs of forest production are often dominated by the burden of carrying land and capital over time.

- Forests valued for industrial timber are both productive capital and product. This fact distinguishes forests from other forms of capital and gives rise to special analytical problems in selecting the best age to harvest and in designing taxes and regulations.

- Governments often wield a heavy hand in forest management and utilization. Partly because forest production is a long process, partly because forests produce many goods and services that are not sold in markets, and partly because timber harvesting and intensive forest management often have adverse side effects, governments in various countries have usually had more involvement in forestry than other sectors of the economy, through public ownership, timber-harvesting and forest practices regulations, taxation, and subsidies.

These features are not unique to forestry, but forestry illustrates them to a unique degree. They are also issues that underlie most of the analytical problems addressed in this book.

The economic choices in forest management are constrained by the biological capacity of the resource. Those limits, and the scope for manipulating them, are the subject of the natural science of forestry, or silviculture. Silviculture is a specialized field of biology, just as forest economics is a specialized field of applied economics. Whereas silviculture is concerned with all the things that can be done to manipulate the structure and growth of forests, forest economics deals with decisions and choices within that range of possibilities, focusing attention on their social rather than biological implications.

Forest economics is concerned not only with silviculture but with all aspects of forest management and forest products markets – protection, consumption, development, harvesting, and utilization of the full range of goods and services associated with forests. The natural science of forestry identifies the limits of natural systems and the range of choices available to forest managers; this range provides the framework of natural constraints within which economic analysis can help in identifying the social implications of alternative courses of action.

ECONOMIC DECISION MAKING

Forest resource management ranges from the design and implementation of high policy to the execution of everyday field tasks. Broad policy objectives are determined by governments and corporate boards of directors; how particular forests are to be used is usually the decision of their private or public owners; for detailed matters, it is often foresters, superintendents, or foremen employed by the owners who make the decisions.

Whatever the level, the process of decision making can be viewed as involving at least the following steps: (1) identification of goals or objectives, (2) identification of the alternative possible means of pursuing those objectives, (3) evaluation of the alternatives, (4) choice of the preferred alternative, and (5) implementation of the decision. In practice, decision making seldom follows the orderly sequence implied by this list of steps. The objectives of the parties involved are often unclear or conflicting, their motives may range from self-interest to altruism, and their time perspective may range from the immediate to the distant future. The processes of investigation and evaluation often bring out new information that causes those involved to change positions and shift alliances. As a result of this ongoing process, decision making often appears confused and disorderly, especially in matters of public policy. Nevertheless, it is helpful to the understanding of decision making to identify these separate components of the process.

Identifying Goals or Objectives

To make appropriate decisions, the decision maker needs a clear purpose to serve as a frame of reference for judging the desirability of one course of action compared with another. Thus, forest managers facing a decision about how to plan a harvesting program, or how much provision should be made for wildlife, or where to direct silvicultural effort must assess their alternatives in light of the objectives they are striving to achieve. There are two fundamental economic objectives: the efficiency and equity objectives discussed earlier. The relative importance of these must also be considered in decision making. Other social objectives, such as public safety, cultural awareness, and social harmony, although not normally economic objectives, have economic implications as well.

Several points about objectives deserve mention. First, objectives, even at the same level of decision making, vary depending upon who is responsible for defining them. A government, for example, is likely to have different and broader objectives for the management of public forests than a corporation does in managing its private forestland. A major objective guiding the forest operations of industrial corporations is the corporations' responsibility to shareholders to generate profits, and corporate decisions are consequently aimed largely at increasing profits. They may also be influenced by other goals, however, such as corporate growth, security of markets or resource supplies, protection of dependent manufacturing activities, or social responsibility, and firms may be prepared to compromise their current profits to advance these other goals. Similarly, small private landowners may be guided by a desire not only for profit but also for financial security, amenity, or other values that they derive from their forests. The management of public forests in democratic countries reflects the perceived wishes of the populace, which nowadays typically puts considerable emphasis on the non-market and environmental benefits of forests, on distributional considerations, and on regional development. In short, those who make the decisions about how forests are to be managed have varying frames of reference, and hence differing objectives that lead to differing decisions.

Second, the objectives of decision makers depend on the hierarchical structure of the organizations within which they work. As one moves down through the organizational structure of a government or corporation, the relevant objectives of decision makers become more narrowly defined. For example, at the highest policy-making level in a government, the goal might be to promote regional economic stability. Towards this end, the government's forest management agency might adopt a sustainable yield objective in regulating timber harvests in each region. That objective would provide regional administrators with production objectives, the official in charge of silviculture with reforestation objectives, and the foreman of the planting crew with daily planting targets. This illustrates that at each subsidiary level of decision making the objective is different and narrower, but derives from and is consistent with the next higher objective and ultimately with the general goal of advancing regional economic stability.

Third, it is important to distinguish ends from means in this context, because they are often confused. Sustained yield (examined in Chapter 8) is an example. Sustained yield is a principle that has become so enshrined in the forestry administration of some jurisdictions that it has become institutionalized as an end in itself. It is, however, merely a formula for meeting a higher purpose, such as regional industrial stability, and unless it is clearly seen as a means to such an end, its limitations for that purpose and the implications of alternatives to it cannot be properly assessed.

There are many examples of such confusion between goals and means. Concepts such as conservation and sustainable development have garnered enormous popularity as goals for natural resources management, but on closer examination both imply means to higher goals, primarily intergenerational equity. The latter (sustainable development) is defined by the World Commission on Environment and Development (or the Brundtland Commission) as "development that meets the needs of present without compromising the ability of future generations to meet their own needs."

Fourth, forest managers are often expected to pursue several objectives simultaneously. As has been suggested, a corporation might be concerned with such things as security of raw material supply or avoidance of risk as well as profit maximization, and a government may seek to provide stable regional employment or environmental benefits as well as revenues from a public forest. These various objectives are rarely perfectly complementary, and to pursue them simultaneously requires compromises among them. Economic analysis can assist in identifying and evaluating possible trade-offs, but the ultimate choice usually requires some weighting of the competing values, which are often not easily quantifiable.

Similarly, the objectives of individuals or governments vary in the short term and long term, and short-term objectives may conflict with long-term goals, again calling for trade-offs and compromise. To return to an earlier example, college students can readily attest to multiple objectives in a dynamic or inter-temporal fashion. Either by conscious choices or by default, all students have broadly defined, long-term goals. Their intermediate goals in college may be to get a good education, to find a job, to find a life partner, and to enjoy sports and a social life. In each semester or term, they may want to have good grades in various classes and learn certain skills. In everyday life, they must balance several immediate goals (or demands), such as attending classes, reading class materials, playing some sports, partying, and relaxing.

Fifth, although orderly decision making calls for explicit objectives, the objectives that forest managers are expected to pursue are sometimes vague. This is a difficulty faced most frequently in government forest agencies, where guidance about the broad purposes to be served in managing public forests is often ambiguous or even conflicting in the legislation, regulations, and administrative arrangements that articulate public policy. In such circumstances, managers are forced to infer or guess their intended objectives, and this can lead to inconsistencies and inefficiencies.

Finally, it is worth emphasizing that specification of objectives is not usually the responsibility of an economist. The special expertise of the economist is not in prescribing corporate or governmental goals but, as discussed below, in analyzing and evaluating the means of achieving them. There is, however, a normative aspect of economics in which economists incorporate their value or normative judgment about what the economy *ought to be* like or what particular policy actions *ought to be recommended* to achieve a desirable goal. In this sense, economists, along with other decision makers, can prescribe what these goals are and the ways to achieve them.

Identifying Alternative Means

Most corporate or public goals can be pursued in a variety of ways. At the level of high economic policy, a goal such as increased regional employment might be served by promoting industrial development, for example, which can be done in various ways through adjustments of taxes, subsidies, infrastructural improvements, or direct governmental enterprise. Forestry may be only one of several alternative means. Or, at the level of forest management planning, a goal of increasing yields might be accomplished by such varied means as improved protection, reforestation, spacing and fertilization of stands, or more complete utilization of harvested trees. Thus, once the decision maker's objectives are identified, the next step is to identify the range of alternative strategies that can feasibly be adopted to achieve most of those objectives.

This step involves assessing the technical alternatives and the inputs required to achieve a particular level of output, which in economic jargon is referred to as determining the production function. Sometimes it is also important to assess the risk or uncertainty associated with the alternative strategies.

Identification of the feasible means of pursuing an objective, and of their technical production functions, is typically the responsibility of engineers, foresters, and other technical experts. For large ventures, this task sometimes becomes highly formalized in feasibility studies that involve economists and financial analysts, while at the operational field level it typically depends on continuing subjective assessments by working supervisors.

Evaluation of the Alternatives

The next step is to evaluate the technical alternatives in order to provide guidance for choosing among them. It is at this stage that economic analysis is brought to bear on the decision-making process. It involves assessing the extent to which the goals of the decision maker would be advanced by particular actions, and their costs.

The relationship between the value of the outputs and the cost of the inputs associated with a particular activity provides a measure of the potential net gain it can generate. Economic efficiency calls for maximizing the surplus of benefits over the cost of resources utilized, so the greater the value of output relative to the cost of inputs, the more efficient the activity is. Chapter 3 describes how alternative courses of action can be assessed according to the efficiency criterion.

The task of identifying the relative advantage of alternative courses of action is often complicated by incomplete information, distorted or unpriced costs and benefits, and uncertainty about future circumstances and outcomes. Notwithstanding these difficulties (which are examined in subsequent chapters), economic evaluations offer a means of ranking alternative courses of action according to consistent criteria for the guidance of decision makers in selecting among the alternative strategies available to them.

Economic decisions are never made with complete certainty, of course. Information about the resources and markets, and about the range of possible actions and their outcomes, is always more or less uncertain. Most decision makers are averse to risk, and so the degree of uncertainty surrounding alternative courses of action has significant influence on their choice. Attitudes towards risk taking vary considerably, however.

The degree of uncertainty is therefore an important influence on decision making, and a later chapter considers how it can be taken into account in economic analysis. It is particularly important in forestry because knowledge about the biological character of forests and their potential responses to treatments is usually limited. In addition, the long planning periods involved in forest production exacerbate the difficulty of predicting the long-term benefits and costs of actions taken today. Thus, the risks of losses from fire and other causes also contribute to the uncertainty in forestry decision making.

Traditionally, economists have approached their subject in two ways. *Positive economics* is concerned with describing and explaining economic behaviour, without judgments about its desirability; in contrast, *normative economics* assesses behaviour in terms of given criteria or objectives, and is therefore more concerned with how economies *should* be organized and regulated. This forest economics book does not follow either of these schools exclusively. We try to avoid prescribing the objectives that decision makers should adopt, or the best distribution of income; rather, our emphasis is on using economic analysis to assist those who make decisions about managing forests in identifying the most beneficial courses of action in light of their objectives.

Choice, Implementation, and Evaluation

The economist's role in evaluating alternative courses of action is to provide guidance in decision making; it is not to make the final choice. Decision makers may, for a variety of reasons, reject the activity that appears most advantageous on purely economic grounds because of considerations of corporate strategy, political sensitivities, or other concerns not accounted for in the analysis. Nevertheless, economic analysis will help decision makers better understand the economic implications of their choices.

Once a decision maker has chosen a preferred course of action, the final step of initiating and implementing the action is taken. Sometimes an additional step is added, that of review and assessment of the action taken, drawing attention to the dynamic and continuous character of decision making and the need for evaluation.

The process of decision making sketched here is a central concern of management science. The growing literature in this field of study presents a variety of decision models and models of strategic behaviour to help understand the relationships among decision makers, the problems of multiple and conflicting objectives, means of minimizing and coping with uncertainty, and so on.

The following chapter reviews the forces that guide decision making in the context of a market economy, introduces the reasons that most countries today regulate and modify the activities of businesses, individuals, and markets, leading to what we have identified as mixed capitalist economies, and explains why government interventions in markets may fail to achieve their purpose. Subsequent chapters examine the special problems encountered in forest management and how economic analysis can contribute to their solution.

REVIEW QUESTIONS

1 Explain why economics is concerned only with the allocation of "scarce" resources. In what sense are forest resources "scarce"? Where are they not scarce?

2 Compare how management decisions are made for (a) a privately owned forest in a capitalist economy, (b) a government-owned forest in a planned socialist economy, and (c) a tribal forest in a primitive subsistence economy.

3 What are private goods? What are public goods? What are merit goods?

4 Explain how innovations in mechanized forestry can affect (a) economic efficiency in timber production, and (b) the distribution of income.

5 Describe the importance of objectives in evaluating forest management decisions.

6 If the fundamental objectives of society are either economic efficiency or equity, how would you classify the following? (In other words, which would follow the criteria of the equity objective, and which would follow the criteria of the efficiency objective?)

a) A forestry firm's immediate profits
b) A forestry firm's long-term profits
c) Employment in a forested region
d) Selective employment of indigenous populations
e) Recreation in a forest park
f) Long-term security of timber supply for a pulp mill
g) Long-term preservation of biodiversity on public forestland.

7 Compare the approaches of a silviculturalist and an economist in considering how best to manage a forest. What are the main concerns of each likely to be? Can their approaches be reconciled?

FURTHER READING

Clawson, Marion. 1975. *Forests for Whom and for What?* Baltimore: The Johns Hopkins University Press, for Resources for the Future. Chapters 3 and 7.

Duerr, William A. 1988. *Forestry Economics as Problem Solving.* Blacksburg, VA: Author. Part 1.

Gregory, G. Robinson. 1987. *Resource Economics for Foresters.* New York: John Wiley and Sons. Chapter 1.

Klemperer, W. David. 2003. *Forest Economics and Finance.* New York: McGraw-Hill. Chapter 1.

Quade, Edward S. 1989. *Analysis for Public Decisions.* 3rd ed. New York: North-Holland. Chapters 4-7.

Samuelson, Paul A., and William D. Nordhaus. 2004. *Economics.* 17th ed. New York: McGraw-Hill/Irwin. Chapters 1-3 and 32.

Chapter 2   Market Economies and the Role of Government

Decisions made by forest managers can be viewed as economic decisions involving the allocation of resources in the broad sense. The focus of immediate interest is typically land and timber, but decisions about these also involve decisions about the use of other resources, such as labour, capital, and entrepreneurial skills. In effect, forest management decisions determine how, and how much of, a wide variety of economic resources will be directed towards forest production. Thus, forestry decision making is part of the much larger mosaic of resource allocation activity that goes on continuously in an economic system.

This chapter deals with the efficiency of economic decision making and the deviations from efficient results that we frequently observe in forestry. We begin with a brief outline of the economic theory of production. We then review the conditions that must be met in a market economy to ensure that all resources will be used with maximum efficiency. This provides a framework for examining how the markets affecting forestry decision making often fall short of the perfectly competitive market conditions that promote economic efficiency. These "market imperfections" or "market failures" give rise to the special economic problems that provide much of the substance of forest economics as a field of study and that often lead to government interventions. Finally, we examine the various forms of market failures, the ways to correct them, and the "policy failures" sometimes associated with government efforts to correct them.

Pulling together the relevant elements of economic theory is a necessary precursor to later chapters, but it must be said at the outset that it is a rather abstract and mechanical task. Readers already familiar with microeconomics may want to pass over this chapter quickly. Readers who are not so familiar with basic microeconomics may want to keep a basic textbook handy for more complete reference.

ECONOMIC EFFICIENCY AND OPPORTUNITY COSTS

Economic analysis is based on the plausible presumption that the objective of all economic activity is to satisfy human wants, and, as noted in Chapter 1, the greater the total satisfaction achieved with the limited resources available, the greater the efficiency of the economic system.

To determine whether productive resources are being used efficiently, we must compare their productivity in their actual use with the alternatives. For example, if a parcel of land – or a unit of labour or capital – could generate more value in timber production than in agriculture, its next most productive use, the cost to society of employing it in forestry is the value of agricultural output forgone. This is the *opportunity cost* of a factor of production – the value of output it could generate in its best alternative use.

The most efficient use of factors of production is the use in which they generate the most value, which means that the value it generates is at least equal to, if not greater than, its opportunity cost. If all factors of production were used in their most efficient use, the whole economy could operate with maximum efficiency, thus maximizing the total value of production and satisfaction of human wants. In a competitive economy, factors of production tend to be directed to their most efficient use because producers get the most value from those uses and can therefore pay the most for them. This is the basis for the efficiency of market economies.

Thus, in a competitive market economy, the opportunity costs of inputs are accurately reflected in their prices. The wages paid to truck drivers, for example, will tend to reflect their marginal productivity in the various industries that employ them. Sometimes, however, prices are unreliable guides to opportunity costs. For example, if some of the labour employed in a forestry operation would otherwise be unemployed, then there is no sacrifice in other production by employing it in a forestry operation; its opportunity cost is zero even though its price (or wage) is positive.

In their pursuit of profit, however, private producers respond mainly to the monetary cost or price they must pay for their inputs. As a result, whenever the market prices of inputs differ from their opportunity costs, an assessment of the efficiency of an economic process by a private corporation will diverge from that by an analyst taking the viewpoint of society as a whole (as is done here). Both compare benefits with costs, but the strictly monetary benefits and costs that private producers use to assess profitability must often be supplemented or adjusted to estimate more accurately the net gains to society as a whole.

Opportunity cost is one of the most important concepts in economics. It is everywhere in our daily life. Whatever we do has an opportunity cost, and we do things when the benefits or satisfaction (broadly defined) derived from doing them outweigh, or at least equal, our (opportunity) costs – the benefits or satisfaction that we forgo by not choosing our next-best alternative. For example, why do students go to college? To answer this question, we need to know the opportunity costs of students who spend four years getting their bachelor's degrees and the added benefits of having college degrees. If they don't attend college, most students would end up in different job markets after graduating from high school. If working after high school is the best alternative for college students, their opportunity costs are the income and experiences that they sacrifice by not working during these years. The added benefits, in terms of income, are the (appropriately discounted) income that college graduates can earn over a lifetime over and above what they could earn without college degrees. Studies show that the added benefits, on the basis of income, exceed the opportunity costs plus college tuition. Assuming that their living expenses are similar whether they work or attend college, it is not surprising that most high school graduates who can go to college do so, despite recent tuition hikes.

THE THEORY OF PRODUCTION

There are always alternative ways of producing things; the task is to find the most efficient, which means that every resource is employed in its most productive use. For example, timber can be harvested using different combinations of machinery and manpower, each having a different opportunity cost. The combination that can harvest the timber at the lowest cost per unit harvested is the most efficient. And if the market prices of these inputs accurately reflect their opportunity costs, then the combination chosen by private producers will converge with the most efficient combination from the viewpoint of society as a whole. The concept of economic efficiency thus focuses attention on the relationship between production and the cost of the resources used in the production process – the relationship between inputs and outputs.

This section summarizes some of the basic concepts of the economic theory of production. It does not cover all the relevant theories, but rather ties together the economic relationships most critical in understanding forest production, which are referred to in subsequent chapters. A thorough discussion of the conventional theory of production can be found in most textbooks on microeconomics.

Here, we start with technical relationships in production – that is, how technology enables products to be produced by combining inputs in various ways. We then show how these technical relationships, coupled with information about the costs of the inputs, can be used to identify the most efficient combination of inputs for producing a particular product and the optimal level of output. Finally, we address economies of scale, output supply, and input demand before turning in the next section to the efficient allocation of resources across the range of their potential uses.

Production Functions

The theory of production begins with the relationships between inputs and outputs, referred to in economics as a *production function.* A production function for timber, for example, might indicate the relationship between the output of timber and varying inputs of land and labour.

In its most general form, the quantity, Q, of a product produced per period depends on the quantities of the inputs, $X1, ..., X_n$, employed in the productive process. This is represented algebraically as:

$$Q = f(X1, X_2, ..., X_n) \tag{2.1}$$

Production functions can take complicated mathematical forms to account for the substitutability of one input for another and for multiple outputs, but here we are concerned only with the general relationships between inputs and one output, and among inputs. Although we will use two inputs in our production function above, the principles we learn from it apply to a production function with multiple inputs. For example, in empirical studies, production functions for forest products such as lumber and paper use at least four conveniently defined inputs: wood (or timber, logs), capital, labour, and energy.

Substitutability of Inputs

It is almost always possible to substitute one input in a production process for another. The degree of substitutability among inputs depends on the technology of production. Because technology advances over time, so does the scope for substitution among inputs in production.

For example, a particular volume of timber can be produced on a tract of land of particular size and quality either with a small input of capital equipment and a large quantity of labour, or with less labour and a larger quantity of capital. With the tract of land fixed in this example, it does not formally enter into the production function. Thus, the production function for timber, Q, produced with labour, L, and capital, K, can be written as a function:

$$Q = f(L, K) \tag{2.2}$$

The substitutability among inputs is illustrated graphically in the upper quadrant of Figure 2.1. The curve *ab*, known as an *isoquant*, connects all the possible combinations of labour and capital that are capable of producing a particular quantity of output, say, 100 cubic metres of timber. Each point along *ab* indicates a different combination of labour and capital that can produce 100 cubic metres of timber on the fixed tract of land. One such combination on the curve is $0K$ capital and $0L$ labour. Points further to the right on the curve indicate less capital and more labour, and vice versa as we move to the left of the first point.

The curvature of an isoquant reflects the substitutability of the two inputs or, more precisely, the *marginal rate of technical substitution* of one input for the other. If they are highly substitutable, the curves approach straight sloping lines; at the other extreme, if the inputs cannot be substituted for each other at all, the curves become right angles, with straight vertical and horizontal sides. In most productive activities, the relevant isoquants fall between these extremes, as shown in the upper quadrant of Figure 2.1. Advances in technology tend to increase the substitutability of inputs in production processes. For example, technological improvements in specialized equipment have significantly increased opportunities to substitute capital for labour in silviculture, timber harvesting, and forest products manufacturing.

Typically, the marginal rate of substitution of one input for another is not constant. As more labour is substituted for capital, it will take progressively larger increments of labour to maintain the same level of output, which explains the decreasing slope of the isoquant. This illustrates the *law of diminishing marginal substitutability*, or the general rule that it becomes increasingly difficult to substitute one input for another in producing a given level of output.

The isoquants above *ab* in Figure 2.1 represent higher levels of output; those below apply to lower levels of output.

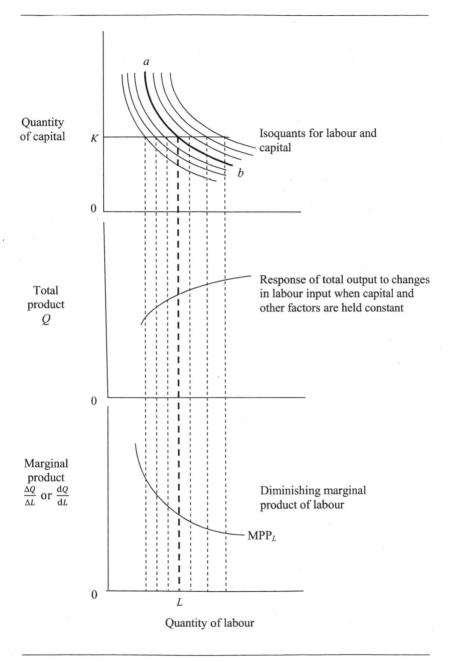

Figure 2.1   Relationship between output and inputs.

To examine the effect on total output of varying one input while holding the other constant, assume that each successive isoquant in the upper quadrant of Figure 2.1 represents an equal increment of output above the isoquant below it. If the amount of capital is fixed at $0K$, it can be seen that successively higher isoquants, or levels of output, can be reached only with progressively larger increments of labour. This is reflected in the middle quadrant of Figure 2.1, which shows how total output increases with larger inputs of labour when other factors are fixed. The dashed vertical lines represent equal increments of output (from the top quadrant), so the vertical rise of the total product curve between each increment is equal. Because progressively larger increments of labour are needed to produce successive equal increments of output, however, the slope of the curve diminishes to the right while the gap between the dashed lines increases.

This illustrates the *law of diminishing returns* or, more specifically in this context, the *law of diminishing marginal product* of one input when others are fixed, illustrated graphically in the bottom quadrant of Figure 2.1. This law states, generally, that if one input in a production process is increased by successive equal increments while other inputs are held constant, output will increase, but the increments of output will become smaller and smaller. In our example, the marginal physical product for labour (MPP_L or $\Delta Q/\Delta L$) declines once capital is held constant.

A forest harvesting operation, for example, may employ a crew of a certain number and some capital equipment. If another worker were added, output would increase by a certain amount. If yet another worker were added, output would increase further, but by a lesser amount, and by progressively diminishing amounts for additional workers.

Efficient Factor Proportions (Least-Cost Combination of Inputs)

Isoquants and production functions illustrate the variety of input combinations that can be used to produce a given level of output, but they do not indicate which combinations are economically efficient. This requires information about the cost of the inputs.

For a given cost, it is possible to acquire varying quantities of capital and labour. In Figure 2.2, the cost of an amount of capital, $0A$, alternatively would be used to hire a certain amount of labour, shown as $0B$. Any point along the line AB represents a combination of capital and labour, such as $0K$ capital and $0L$ labour, that can be obtained for the same total cost. This line is referred to as an *isocost curve*.

It follows that the ratio of $0A$ to $0B$, and hence the slope of the isocost curve, is equal to the ratio of the price of capital to the price of labour.

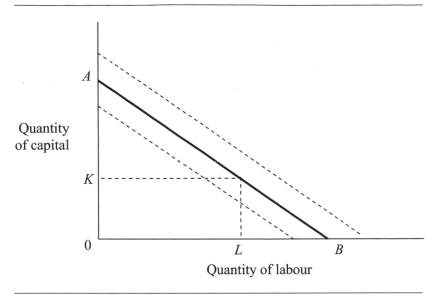

Figure 2.2   Isocost curve for capital and labour.

The isocost curve will be a straight line as long as the prices of the inputs are constant and unaffected by how much of each is employed.

Isocost curves reflecting higher total costs can be drawn parallel and to the right of AB in Figure 2.2, and lower total costs to the left. A set of such isocost curves can be represented by the total cost expression

$$TC = P_L \cdot L + P_K \cdot K \tag{2.3}$$

Thus the total cost, TC, of a production process depends on the quantities of labour, L, and capital, K, employed and their respective prices, P_L and P_K.

The least-cost combination of inputs can be identified by superimposing isocost curves on isoquants, as shown in Figure 2.3.

For any level of output, such as that represented by the isoquant ab in Figure 2.3, there is a particular combination of capital and labour that will minimize the total cost of these inputs. This is shown geometrically by the point at which the isoquant is just tangent to an isocost curve. In Figure 2.3, this point indicates that the combination of $0K$ capital and $0L$ labour is the most efficient means of producing the level of output represented by ab. The tangency of an isoquant and an isocost curve implies that the substitution ratio between the two inputs, or the ratio of

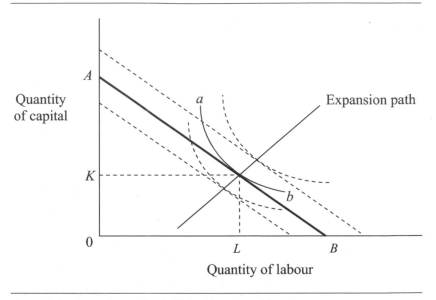

Figure 2.3   Expansion path of efficient input combinations.

their marginal physical products (MPP_L/MPP_K, reflected in the slope of the isoquant) is just equal to the ratio of their input prices (P_L/P_K, the slope of the isocost curve):

$$\frac{MPP_L}{MPP_K} = \frac{P_L}{P_K} \tag{2.4}$$

where MPP_L and MPP_K are the marginal physical products of labour and capital, respectively, and P_L and P_K are the prices of labour and capital, respectively. In other words, the least-cost combination of inputs to produce a given output is precisely defined as that at which the marginal rate of substitution of one input for another is equal to the ratio of the prices of the two inputs.

Equation 2.4 characterizes an efficient production process. Let us consider a manager's response to an inequality in equation 2.4 – for example, the left-hand side is greater than the right-hand side. Rearranging, we get:

$$\frac{MPP_L}{P_L} > \frac{MPP_K}{P_K}$$

This means that the marginal physical product per dollar spent on labour exceeds that per dollar spent on capital. The manager will substitute labour for capital in order to achieve the least-cost combinations of inputs until the equality in equation 2.4 is restored.

Such least-cost combinations of inputs exist for any possible level of production. A line drawn through these successive optima for increasing levels of production, as shown in Figure 2.3, is called an *expansion path.* It shows how the most efficient combination of inputs varies over different levels of output.

Cost Minimization, Profit Maximization, and Duality

The points of tangency between isoquants and isocost lines illustrate how producers, in pursuit of profit, choose the cost-minimizing combinations of inputs for producing given quantities of output. We now need to examine how they choose their profit-maximizing level of output.

As we shall see, the conditions for profit maximization are the same as those for cost minimization at a given level of output. Thus, the theory of production can be viewed from either a profit-maximization or a cost-minimization perspective. This *duality* helps economists in empirical studies whenever the profit function is not readily observable but the cost function and level of output are known.

In order to maximize their profits (π) or net returns, producers must find the level of production that generates the greatest surplus of total revenue over total costs. This is determined with reference to marginal cost and marginal revenue.

The *marginal cost* of production is defined as the additional cost of inputs required to produce an additional unit of output. The *marginal revenue* from a unit of output sold in a perfectly competitive market is reflected in its price. To maximize profits, producers must expand their output as long as their marginal revenue exceeds their marginal cost. As long as the marginal cost rises with output, however, there will be a point beyond which the marginal cost exceeds marginal revenue, generating a loss at the margin of production. The profit-maximizing level of output is therefore that at which marginal cost, MC, just equals marginal revenue, MR:

$$MC = MR \tag{2.5}$$

Mathematically, the profit-maximization problem for a firm in a competitive market can be written as

$$\text{Max } \pi = P \cdot q - C(q) \tag{2.6}$$

where P is the price of output, q is the firm's quantity of output, and $C(q)$ is the firm's total cost function that is dependent on the level of output, q.

Taking a derivative of π in equation 2.6 with respective to q, and setting the resulting equation to zero (which is a condition for maximization or minimization), yields the following:

$$\frac{dC(q)}{dq} = P$$

which indicates that marginal cost equals marginal benefit (P) for the firm in a competitive market.

This profit-maximizing rule can also be cast in terms of the equality of the marginal cost and revenue associated with an additional unit of factor of production. To see how this is done, we write the profit-maximization problem for a firm in a competitive market with one output and two inputs (L and K) as

$$\text{Max } \pi = PQ - (P_L L + P_K K) \tag{2.7}$$

where P is the price of output, $Q = f(L, K)$ is the firm's production function, and the term in parentheses is the firm's cost function.

Taking a partial derivative of π in equation 2.7 with respect to L and K, and setting the resulting equations to zero (which, again, is a condition for maximization or minimization), yields the following:

$$P\frac{df(L,K)}{dL} = P_L \tag{2.8}$$

$$P\frac{df(L,K)}{dK} = P_K \tag{2.9}$$

The left side of equation 2.8 is the marginal revenue product for input L, and the right side is the marginal cost for input L. The firm will continue to add input L until this condition (marginal benefit = marginal cost) is met. Similarly, the firm will continue to add input K until the condition in equation 2.9 is met.

Dividing equation 2.8 by equation 2.9 yields the following:

$$\frac{\frac{df(L,K)}{dL}}{\frac{df(L,K)}{dK}} = \frac{dK}{dL} = \frac{P_L}{P_K} \tag{2.10}$$

This is identical to equation 2.4. The left-hand side (and the middle part) of equation 2.10 is the ratio of marginal physical products (MPP$_L$/ MPP$_K$), which is, again, called the *marginal rate of technical substitution*, whereas the right-hand side is the ratio of input prices (or the *rate of economic substitution*). Graphically, the point of tangency between the isoquant curve and isocost curve satisfies this condition.

Another way to look at the resource allocation faced by this firm is to minimize the cost (C) subject to output constraint – say, when $Q = f(L, K)$ $= q$, q is a constant. This cost minimization yields the same conditions as profit maximization:

Min $C = P_L L + P_K K$ s.t. $f(L,K) = q$

where *s.t.* stands for "subject to."

This is a conditional minimization problem, or a minimization problem subject to a constraint. In order to find a minimum, we need to combine these two functions in a new objective function:

$$\begin{aligned} M &= C - \lambda[f(L,K) - q] \\ &= P_L L + P_K K - \lambda[f(L,K) - q] \end{aligned} \tag{2.11}$$

where M is the new objective function and λ is a Lagrange multiplier.

Taking a partial derivative of M with respect to L and K generates the following:

$$\frac{dM}{dL} = P_L - \lambda\frac{df(L,K)}{dL} \tag{2.12}$$

$$\frac{dM}{dK} = P_K - \lambda\frac{df(L,K)}{dK} \tag{2.13}$$

Setting these two equations to zero and dividing them yield the following:

$$\frac{P_L}{P_K} = \frac{dK}{dL} = \frac{\frac{df(L,K)}{dL}}{\frac{df(L,K)}{dK}} \tag{2.14}$$

Note that equations 2.10 (equation 2.4) and 2.14 are identical. Thus, profit maximization and cost minimization at a given level of output achieve identical results, giving rise to *duality*, which means that one can study producers' behaviour based on either a cost-minimization approach or a profit-maximization approach.

Returns to Scale

In making production decisions, producers must pay careful attention to how their costs are affected by the level of output or scale of operations. In some cases, output is proportional to all of the inputs used, which is referred to as *constant returns to scale*. Successive equal increments of all inputs will generate equal increments of output, so the isoquants will be equally spaced.

In other cases, the additional output generated by an increment of all of the inputs decreases as the scale of operations is expanded; this is referred to as *decreasing returns to scale*. In these circumstances, where output increases less than proportionally to all of the inputs, the isoquants for successive equal increments of output become more widely spaced at higher levels of output. The opposite case is that of *increasing returns to scale*. The expansion path will form a straight line only under constant returns to scale. Decreasing, constant, and increasing returns to scale can all be found in the production of forest products.

The precise relationship between outputs and inputs is a critical influence on the choice of producers about their level of production because it determines their marginal cost – that is, the additional inputs required to increase output by a small amount multiplied by their input prices. Many production processes, including the production of most forest products, demonstrate increasing returns to scale over a range of output, causing marginal cost to decline, and decreasing returns over higher levels of production. Producers have strong incentives to expand their operations until they exhaust increasing returns to scale. Beyond that point, decreasing returns and rising marginal cost limit the scale of efficient operations. Constant and increasing returns to scale over all levels of production are rare; they imply no limit to the efficient scale of operations, and invite all production to fall to a single producer.

Note that the law of diminishing returns is true when all but one input is held constant. On the other hand, economy of scale, whether it is increasing, decreasing, or constant, considers the relationship between outputs and all of the inputs.

With a given market demand and possible existence of other firms, even if there are increasing returns to scale in an industry, firms in the industry can produce only so much before the output prices start to fall. Eventually the fall in output prices will cause the firms' profits to fall. How do firms choose their level of output?

Output Supply and Input Demand
The profit-maximizing equality of marginal revenue and marginal cost (MR = MC) explains why producers are willing to produce and sell more at higher prices, resulting in upward-sloping market supply curves for products. Profit-maximizing producers employ more of any factor of production as long as its marginal revenue product exceeds its cost or price of input (that is, the left-hand sides of equations 2.8 and 2.9 are greater than their respective right-hand sides), because this enhances profits.

If a firm that is operating at the profit-maximizing level of production experiences an increase in the price of its product, its marginal revenue will rise above its marginal cost. To maximize profit under the new price, the firm will expand output, moving upward along its marginal cost curve until equality between the two marginal variables is restored. Thus the firm's supply curve slopes upward, following its marginal cost curve.

The market supply curve for a product is simply the sum of the supply curves of all the producing firms. Conceptually, aggregating all firms' upward-sloping supply curves into a market supply curve is a horizontal summation of the quantities supplied by all firms at various prices. As a general rule, therefore, when the price of a forest product rises, more of that product is produced and offered for sale, and less is produced at lower prices.

Since we are talking about output supply curve here, any such higher prices are assumed to be purely driven by increases in demand. Similarly, when we focus on demand, as we will in the next paragraph, price changes are assumed to be purely supply-driven. Supply and demand are discussed in greater detail in Chapter 4.

These relationships also explain why producers demand more inputs when input prices fall. A fall in the prices of its inputs lowers a firm's marginal cost, and to restore the equality between its marginal cost and marginal revenue, the firm must expand production, using more inputs. Thus, the demand curves for inputs slope downward in the usual fashion, consistent with the observation that producers employ more labour, capital, and other inputs in production when the prices of these inputs are lower.

Time as an Input in Forest Production

The relationships outlined above help to explain how a profit-maximizing producer chooses his level of output and his use of inputs at a point in time. The behaviour of forest products–producing enterprises can be analyzed satisfactorily within this theoretical framework. In producing forest crops, however, producers must pay special attention to time in addition to the usual kinds of inputs considered in manufacturing and other forms of production. Because forest growth is a biological process, once a forest is established, the time it is left to grow is often the most important determinant of the output produced.

Thus, time, because it influences the value of capital embodied in forests, can be seen as an input in forest production, and it has the essential characteristics of other inputs. Generally, the more time a managed forest crop is left to grow, the greater the output will be; the quantity of output (or growth) usually shows increasing returns over an initial period, then diminishing returns to time; and time involves a cost, measured by the opportunity cost of tying up the capital and land as long as harvesting is postponed. Moreover, because the same output can be produced in less time with the use of more labour and other inputs of intensive silviculture, the optimal combination of inputs must take account of this substitutability between time and the other factors of production. Chapters 3 and 7 deal specifically with time as a special dimension in the economics of forest production.

ALLOCATING INPUTS AMONG ALTERNATIVE USES

Most factors of production in an economy can be used for a variety of purposes. An additional unit of one of the inputs employed in a production process will increase output by an amount referred to as its marginal product or, more precisely, its *marginal physical product* – MPP, or $df(L, K)/dL$ and $df(L, K)/dK$ in equations 2.4, 2.10, and 2.14. The *value* of the additional output generated by adding the additional unit of an input – the marginal physical product of the input multiplied by the marginal revenue of the extra product – is called its *marginal revenue product* – MRP, or $Pdf(L, K)/dL$ and $Pdf(L, K)/dK$ in equations 2.8 and 2.9 – of the input. The marginal revenue product of an input declines as more of it is employed because diminishing returns cause the marginal product to decline and, in imperfectly competitive product markets, the marginal revenue declines as well.

A profit-maximizing producer will employ more of any input as long as its marginal revenue product exceeds its cost because this will contribute to profit. The same is true for other producers who use the same input. As a result, all the producers purchasing the input in a competitive market will bid its price up to a single level that will reflect its marginal revenue product in all uses. When the marginal revenue product of an input in production is equal in all its uses in an economy, it can be said to be allocated efficiently, because as long as it generates a value, at the margin, at least equal to its opportunity cost, there can be no possible reallocation that would increase the total value of the input's contribution to economic production.

Similarly, in deciding how to allocate an input among different forms of production, a firm producing several different products will endeavour to balance its marginal revenue product in each line of production. For example, manufacturers of timber often face the task of allocating logs among several uses, such as lumber, panels, and paper production. Again, maximum efficiency can be gained by allocating the logs in such a way that their marginal revenue products are equal in each use. This is the *equi-marginal principle:* the maximum benefit from any factor of production will be realized when its marginal revenue product is equal in all its uses. The same principle was applied in our previous example of college students trying to allocate their time on a daily basis.

These conclusions can be generalized to describe the conditions that must hold throughout the economy in order to achieve maximum economic efficiency, or the greatest possible benefit to society from all the resources available for production. If the revenue that producers earn from employing each factor of production at the margin accurately reflects the value of the output to society, then the marginal revenue product measures its *marginal social benefit* (MSB). And if all producers employ each factor to the point at which its marginal revenue product equals its price, that price will reflect its true opportunity cost, or its *marginal social cost* (MSC). Resources will be allocated to achieve maximum social efficiency if MSB = MSC in all forms of production. No reallocation could then yield a higher social value in the aggregate.

This rule for economic efficiency holds for all sectors of the economy and all forms of production, including production of goods and services that are not priced and sold in markets, such as some forms of forest recreation. Thus, maximum economic efficiency requires that land, labour, and capital inputs used to provide outdoor recreation be allocated such that their marginal social costs equal the marginal gains that society realizes from them. Of course, the unpriced nature of recreational benefits makes estimates of its marginal social benefit difficult. This will be a topic for Chapter 5.

MARKET FAILURES AND EXTERNALITIES IN FORESTY

Private producers and market processes can achieve this socially optimal result only under the stringent conditions of a "perfect" market economy. These conditions are never fully met, of course; in reality, market processes are always more or less imperfect. Nevertheless, the theoretical model of a perfect market system helps us identify the sources and effects of imperfections in the real markets that we have to deal with, and the opportunities for improving market performance. The appendix to this chapter summarizes the conditions that must prevail in a market economy in order for it to achieve the socially optimal allocation of resources identified above, and some of the ways in which these conditions are not met in forestry.

Obstacles to perfectly competitive markets for both products and inputs are common in the forest industry and other forest-related activities, and they take a variety of forms. In some markets, the number of producers may be too small to ensure effective competition, or the size and market power of suppliers may distort prices. Other impediments to perfect competition include ill-defined property rights, lack of market information and knowledge, and divergences between market interest rates and the social rate of time preference. In later chapters, we will examine in detail these obstacles to efficient markets and how governments can offset or mitigate their adverse effects. Here we note that economists use the term *market failure* to refer to a situation in which the market on its own fails to allocate resources in a socially optimal fashion. Further, we want to draw attention to a particular class of market failures because of their pervasive importance in forestry, namely, *externalities.*

Externalities are the costs or benefits of economic activity that are external to the firm or individual that creates them; they accrue to others instead. For example, a forest managed for timber production might incidentally improve the habitat and abundance of game birds, benefiting hunters who have free access to hunting in the area. This unpaid-for benefit is an *external benefit* or *positive externality* of timber production. Correspondingly, a cost or loss inflicted on others for which no compensation is paid, such as the loss that might be suffered by fishermen if the forest enterprise reduced the productivity of a fish stream, is an *external cost* or *negative externality* of forestry.

It is useful to distinguish different types of externalities. *Ownership* or *technological externalities* are of the kind noted in the examples above, where the producer of the external benefit or cost is not connected through a market to those who bear the result. They are technological market failures insofar as producers cannot gain by producing positive externalities or by reducing negative ones, and so they have no incentive to take these side effects into account in making their production decisions. As a result, producers will produce less of the positive externalities (such as enhanced wildlife in the example above) than economic efficiency would dictate, and more than the efficient amount of negative externalities (such as pollution of streams). These effects are becoming increasingly important in managing forests, with growing concern about non-market impacts on the forest environment, biodiversity, climate change, and aesthetic and recreational values.

These technological failures frequently induce governments to regulate forest practices to mitigate their effects. This involves restricting practices that produce negative externalities, such as logging unstable banks of fish streams, and requiring management actions that generate positive externalities, such as enhancing bird habitat. As we explain later, however, government regulation of economic activity is not costless, and not always effective.

Another type of externality is *public goods,* which refers to benefits that can be consumed without diminishing their availability to others. Classic examples are lighthouses and national defence, which, once in place, benefit all, with no one beneficiary diminishing their availability to others. Pricing such goods and services is often impractical, because no one can be excluded from access to them. Moreover, because any individual's consumption or enjoyment of such public goods imposes no marginal cost on the economy, the efficient price is also zero, limiting the ability of private producers to supply them. Instead, they must often be produced, if at all, by governments and provided to consumers without direct charge.

Forestry examples of such public goods include aesthetically attractive landscapes, clean air, and protected endangered species. Many are not pure public goods, however, because they have a private element as well. For example, a national park provides public goods in the form of attractive landscapes, scientific values, and clean air, but it may also provide camping and recreational opportunities that are "private goods" because they are quite capable of being priced through entry fees.

Some economists identify other types of externalities. *Technical externalities* are impediments to competition, such as natural monopolies, oligopolies, and barriers to entry. *Pecuniary externalities* are the impacts, beneficial or adverse, that one firm may have on another through their markets for inputs or products, such as the adverse effect of a new competitor on existing firms. These are not concerned with unmarketed benefits and costs that accrue to other parties, however, and in this book we refer to externalities only in the form of technological externalities and public goods.

GOVERNMENT INTERVENTION AND POLICY FAILURE

Because weaknesses and failures in market processes impair the efficient production and use of resources in market economies, governments often intervene. Government intervention may also be called for because of an inequitable distribution of income, excessively high costs of marketing, or ill-defined property rights.

Intervening in economic activity to make it work more efficiently or to make it conform more closely to the public interest is a major role of modern governments, but correcting market failures is not always easy or costless. The corrections themselves impose costs in public administration and in private adjustment and compliance. Because of this, the mere existence of a market failure is insufficient reason for intervention, even on efficiency grounds. A gain in efficiency will result only if the resulting benefits exceed all the costs of intervention. When government intervention is not justified or fails to achieve its stated goals of efficiency in resource allocation and desirable level of income distribution, it is called *government failure* or *policy failure.*

Furthermore, measures taken to correct market failure may cause inefficiencies elsewhere, and the best solution may be to tolerate the market failure. In an economy with some unavoidable market failure in one sector, actions to correct market failures in another sector with the intent of increasing overall economic efficiency may often actually decrease economic efficiency. In theory, at least, it may be better to let two market imperfections cancel each other out than to make an effort to fix either one. This is the so-called *theory of a second-best solution,* or the *theory of the second-best.*

For example, consider a mining monopoly that is also a polluter: mining leads to tailings being dumped in the river and deadly dust being deposited in the workers' lungs. We have two market failures here: a monopoly and an externality (side effect). If the government wants competition, it needs to break up the monopoly. Competitors can be expected to produce more than a monopolistic producer, however, and because pollution is associated with production, pollution would likely increase, and the situation may end up being worse than without government intervention.

It is also important to recognize that, whatever the objectives, interventions used by governments are not always well chosen or effective.

Like private enterprises, governments sometimes suffer from imperfect information, make mistakes, function inefficiently, or find it difficult to adjust to changing circumstances. As a result, public policies sometimes do not succeed in advancing their apparent purposes, or do so less effectively than alternative measures would, or cause unintended side effects. When the interventions of government are badly designed and poorly implemented, *policy failure* may cause even greater misallocation of resources than the market process itself. It has been argued, for example, that efforts to assist marginal agriculture through subsidies have hindered the transfer of marginal farmland to more productive use in forestry in the US and elsewhere, thereby aggravating rural poverty.

Moreover, governments may not always be motivated by the public interest. A standard theory of democratic governance, the *public interest theory*, is based on the assumption that government decisions are always aimed at advancing the public interest and that any policy failure is simply an honest mistake. A contrary view, the *interest group theory*, assumes that all those involved in government policy making, including elected officials, bureaucrats, lobbyists, private firms, and individuals, act in their own self-interest in seeking votes, influence and power, business advantage, personal gain, and favour, so that the interest of the public at large is often sacrificed for the interest of special groups. This is exemplified by *pork barrelling* (or simply "pork"), where legislators steer government funding and projects to their own constituencies regardless of more deserving alternatives. In Chapter 12, we explain how interest group theory works in restricting forest products trade, but the theory applies to many forestry and non-forestry examples, including tax relief, subsidies, and other government assistance programs.

Finally, it must be noted that many shortcomings in economic performance are not due to market imperfections in the usual sense but rather to institutional failures such as ill-defined property rights, ineffective laws, and poor governance. *Institutions* are a society's rules and constraints that have been devised to structure and guide human interaction. They are composed of formal rules, informal constraints, and the enforcement characteristics of both. Poorly designed institutions or weak enforcement distort the economic incentives and behaviour of producers and consumers and generally weaken economic performance.

The significant role played by governments in the management of forest resources can be explained in large part as their response to market imperfections. In later chapters, we explain how economic analysis can help in understanding the implications of their actions.

APPENDIX: MARGINAL CONDITIONS FOR EFFICIENCY

This appendix summarizes the conditions that must prevail in a market economy in order for it to achieve the socially optimal allocation of resources, and some of the ways in which those conditions are not met in forestry.

Profit maximization. A fundamental precept of the market system is that producers seek to maximize their profits, and the efficiency of the system depends on their doing so. As we have seen, this incentive will lead them to choose the most efficient or least-cost method of production and to select the level of output at which their marginal revenue equals their marginal cost (MR = MC).

We have also seen that this profit-maximizing rule can be cast in terms of the equality between the marginal cost and revenue associated with production of another unit of output, or the marginal cost and revenue associated with employing another unit of a factor of production. In the latter case, the rule is precisely expressed in terms of the marginal revenue product of the factor (MRP, which we defined in the main part of this chapter as the factor's marginal physical product multiplied by the producer's marginal revenue) being equal to the cost to the producer of employing the additional factor, the marginal factor cost (MFC); that is, MRP = MFC.

If the firm fails to maximize its profit by adjusting all its inputs in this way, it will misallocate resources in the sense that they could have been used more efficiently in some other way. For example, if a firm producing timber produces beyond this point of equality of marginal revenue and marginal cost, the cost of producing the last cubic metre of timber will be greater than the revenue it generates. This not only reduces the firm's profits but also means that the resources used in producing the timber, at the margin, would be capable of generating more value in other things.

There are many ways in which forest production and use fail to be guided by profit maximization. Some forests are owned not by private firms but by governments that often pursue other objectives. Even private owners are constrained in their profit-maximizing behaviour by governmental regulations. Some forest products and services, such as recreation, amenity, and other environmental values, are not marketed and their value cannot be realized by private forest enterprises. Such circumstances prevent forest production from being consistently guided by the marginal costs and revenues incurred by private firms, and thereby impede efficient allocation of resources through market processes.

Perfectly competitive product markets. To provide producers with incentives to produce at the socially most advantageous level, the markets for the goods and services they produce must be perfectly competitive. Perfect competition is characterized by a large number of sellers of a homogeneous product, each seller being too small relative to the total market supply to influence the product's market price.

Such market conditions imply that the firm's marginal revenue, the extra revenue it gains from selling another unit of product, must be equal to the price of the product. It follows that if the firm alters the amount it sells, the change in its revenue will be equal to the market price multiplied by the change in quantity sold, and that this will be identical with the market's evaluation of the change in output. In other words, the market value of the marginal product (VMP) must equal the firm's marginal revenue product (MRP), that is, VMP = MRP.

This condition will not hold if, as is often the case in forest product markets, competition is imperfect. For example, when a firm has a *monopoly* in a market, or is such a dominant seller that it can influence the market price of its product by adjusting how much it sells, its marginal revenue will be less than the price of its product. That is, monopolies, oligopolies, or other dominant firms can usually sell more of their products only if they lower their prices, and the lower price must apply to all that they sell, not just the increased increment. If a monopoly lowers the price for one unit of homogeneous product, then it must sell all other units also at the new lower price. Therefore, firms with strong market position find that the marginal revenue they obtain from the last unit they sell is less than the price they obtain for it. When they equate marginal cost with marginal revenue to maximize profits, their marginal cost will also be less than the price of the product, giving rise to another inefficiency.

Perfectly competitive factor markets. Correspondingly, the efficient operation of the market system requires that the markets for all factors of production be perfectly competitive, so that no single producer or factor owner can influence factor prices. Thus each producer must pay the going market price for all of the factors purchased. The producer is a price taker, facing horizontal factor supply curves; correspondingly, suppliers face horizontal demand curves for their factors at the going market prices. This condition is more general in forestry, as many forest landowners are too small relative to the market to have any significant effect on the going wage rate for woodworkers or on the prices of trucks or other capital equipment.

Under these circumstances, the producer's cost of producing an additional product, the marginal cost, is equal to the value of the factors needed to produce it. Put another way, the cost to the producer of employing another factor, the marginal factor cost (MFC), must be equal to the price that the producer pays for it, the value of the marginal factor (VMF) – in other words, MFC = VMF.

This condition will not hold if those who supply factors have monopoly power or if those who purchase them can exercise monopsony power to influence factor prices. A monopsony in local land and timber markets, bilateral bargaining in organized labour markets, and restricted access to capital markets for small landowners are examples of impediments to competition in the markets for factors of production.

Other requirements for optimal allocation of resources and an efficient market system include:

Divisible inputs and outputs. The optimal allocation of resources requires fine marginal adjustments in production decisions, which implies that all inputs and outputs are highly divisible. In natural resource activities, however, the scale at which inputs are used is often constrained by natural circumstances. The "natural" management unit for an industrial forest may be a drainage basin or area served by a transportation network that does not readily lend itself to division. A wilderness usually requires a significant area to maintain its natural character. So while the theory of market economies rests on discrete, divisible, homogeneous, and mobile inputs, the natural resources used in forest production are often heterogeneous, immobile, multipurpose, and sometimes indivisible, restricting producers' flexibility in making adjustments at the margin.

Control of inputs. An efficient market system must provide each producer with complete control over inputs and outputs. Producers must be able to acquire the inputs they need, to use them in the most efficient way, and to dispose of the goods and services they produce in the most beneficial manner, incurring the full benefits and costs of their actions.

Enterprises that use land and natural resources rarely have full control of their beneficial use. An obvious impediment, more common in fisheries and water management than in forestry, is common property regimes where producers are never free of interference from other users of the same natural resources. In these circumstances, producers lack appropriate incentives to manage and conserve because the benefits will accrue, in large part, to others. Even where producers acquire exclusive rights to resources, government taxes, royalties, and other charges drive a wedge between the private and social benefits of using them. As Chapter 11 will show, these levies may create incentives to use resources in ways that reduce the social value they generate.

A wide variety of leases and licences provide forest enterprises with only qualified property rights in land and timber, described in Chapter 10. They typically limit the holder's rights to a particular resource, such as timber, while excluding rights to others, such as the water or wildlife on the same land; they often involve charges that distort incentives governing the management and use of the resources; and they are of limited duration, which may lead holders to ignore the long-term impact of their actions.

In short, although the theory of the market system rests on well-defined private ownership of all factors of production, private property in forest resources is often absent or constrained. Limitations on private ownership and control may serve important social objectives, which we will consider in later chapters. Here it is important to note that they distort the incentives of operators and owners and thereby interfere with the efficient performance of the market.

Optimal time preference. Like many other economic activities, forest management involves decisions about timing: when to harvest a stand, when to apply silvicultural treatments, when to invest in roads, and so on. Forest planning also calls for ways of evaluating investments that yield returns only over long periods. The key to comparing values that occur at different times, discussed in Chapter 3, is the rate of interest, which measures the degree by which present values are preferred over future values.

In a market economy, decisions about the relative value of a dollar today and a dollar at some future date are made by private demanders and suppliers of capital with reference to the market rate of interest. Thus, in order for the market to allocate resources over time in the best interest of society as a whole, the rate of interest used by private producers in making their decisions must accurately reflect the preference of present over future values for the society as a whole. In other words, market rates of interest must reflect the *social rate of time preference.*

For many reasons, this may not be the case. All producers rarely have equal access to capital markets, so that even if some market rates correspond to the social rate of time preference, others do not. For example, if small forest owners have to pay higher interest rates on loans than large corporate owners of similar forests, they will be less willing to invest and more likely to discount future values at a higher rate than society as a whole. Moreover, as we shall see later, the broader interests of society may mean that market rates of interest are an unreliable guide to social time preference.

No externalities. As noted in the main body of this chapter, the market system will work efficiently only if those who make production decisions account for all the costs and benefits of their actions. Many forest activities, however, involve *external economies* and *diseconomies* that have beneficial or adverse impacts that do not enter into the economic decisions of private producers because they do not involve market transactions.

Perfect knowledge. In order that the market system can efficiently allocate all resources, consumers must have perfect knowledge about the goods and services available to them, their prices, and the satisfaction to be derived from them. Producers, for their part, must have perfect information about factor markets and the technology of production. In reality, of course, knowledge is always more or less faulty, and ignorance in these matters will lead to faulty allocation of resources.

Even more stringent is the requirement that efficient utilization and production over time calls for perfect knowledge of product and factor markets not only in the present but also in all future periods. That is, there must be no market or technical uncertainty facing producers or consumers.

In forestry, however, uncertainty is a particularly important influence because decisions must be based on long-term plans and projects. Uncertainties about future resource supplies, rates of growth, natural losses, and technological change bear heavily on decision making. Some of this ignorance can be overcome by increased effort in data collection and research, but these efforts to improve knowledge all involve a cost, and they can rarely eliminate uncertainty altogether.

Equitable distribution of income. We have already emphasized that economic efficiency and equity in the distribution of income and wealth are two separate considerations, although changes in one have implications for the other. Equity is a subjective concept, resting on social and political attitudes towards fairness. It cannot be defined by economic analysis, but rather must be resolved collectively through political processes.

Nevertheless, for market processes to allocate resources efficiently, the distribution of income must be optimal in the sense that it is equitable. This is because the distribution of income governs the pattern of demand for goods and services, and hence also the efficient allocation of productive resources. The most efficient allocation of resources therefore depends on the distribution of income. A market economy can respond reliably to social values only if the distribution of income corresponds to what society considers equitable.

If some group in society – such as farmers or retired people – receive too small a share of the national income in this sense, then the demand for the goods and services they buy will be less, and fewer of those particular goods and services will be produced than if the group's income were higher. Correspondingly, production of other things will be greater. It follows that if their incomes were raised, resources would be reallocated as a result of the changed pattern of demand. So, if the distribution of income is faulty, resources will be misallocated. Only when income and wealth are equitably distributed will the amount that individual consumers are willing to pay for goods and services provide an accurate measure of the social benefits derived from producing and consuming them.

With an optimal income distribution, the market price that consumers are willing to pay for the marginal product will equal its marginal social benefit – that is, VMP = MSB. If the distribution of income is not optimal, the willingness of consumers to pay for different products will not correspond to social priorities, and this equality will not hold.

Summary of Marginal Conditions for Efficiency

We can now summarize the conditions for social efficiency in a market economy in a series of equalities:

$$MSB = VMP = MRP = MFC = VMF = MSC$$

where

MSB = marginal social benefit
VMP = value of the marginal product
MRP = marginal revenue product
MFC = marginal factor cost
VMF = value of the marginal factor
MSC = marginal social cost

All these variables refer to one unit of a representative factor employed at the margin of production. The first equality on the left-hand side implies a prior social decision about the distribution of income. The value of the marginal product produced by employing another unit of the factor, as indicated by the amount that consumers are willing to pay for it, accurately measures its marginal social benefit (MSB = VMP).

The second equality will hold as long as perfect competition prevails in the product market, so that consumers' value of the marginal product equals the producer's marginal revenue product from employing the additional factor (VMP = MRP). The next equality is the condition for profit maximization, indicating that the producer's additional revenue is equal to the marginal factor cost incurred to generate it (MRP = MFC). In conditions of a perfectly competitive factor market, this marginal factor cost equals the price that the producer must pay for an additional unit of the factor, or the value of the marginal factor (MFC = VMF). Finally, the value of the marginal factor will be equal to its marginal social cost if all the other equalities hold, so that the factor earns its opportunity cost at the margin in all its uses (VMF = MSC).

All of these equalities must hold if resources are to be optimally allocated in a market economy. We noted earlier in this chapter that the ultimate criterion for maximizing social welfare is the equality between marginal social benefit and marginal social cost, the first and last terms in the series, which implies that all factors of production earn their opportunity cost and no reallocation of any factor can increase the social benefit because it would involve a cost equal to the benefit it could generate. Perfect markets, meeting all the conditions outlined above, would ensure that all other terms were made equal.

This formulation of efficiency conditions follows that of the economic theorist Abba Lerner, who attempted more than 60 years ago to summarize the rigorous conditions that must be met in a market economy in order for it to succeed in maximizing social welfare. Such an economy could achieve this result only if certain preconditions, noted above, hold as well, including perfect knowledge among producers and consumers, institutional arrangements that give producers complete control over their inputs and outputs, no externalities, convergence of market and social rates of time preference, and an equitable distribution of income. It also rests on certain technical relationships in production, such as the divisibility of inputs and outputs, decreasing returns to scale, and diminishing returns to all factors of production, which are examined in this chapter.

The focus on *marginal* conditions is critical. It should be noted that criteria for efficiency say nothing directly about how much, in total, of any productive factor should be devoted to one use rather than another, nor do they hinge on which use generates the most value in total. As in so much of economic analysis, the emphasis is on adjustments *at the margin,* as demonstrated in the *equi-marginal principle.* The issue is seldom whether to produce one product rather than another; it is whether to have a little more of one and a little less of the other. Similarly with inputs – the problem is not to find the use to which land or labour or any other input can be most advantageously put in total, but *how much* of it should be allocated to each of its productive uses. Thus, when a variety of products and services can be produced from forests, or inputs like land, labour, and capital can be used in a variety of forestry or other productive activities, attention must be focused on the costs and benefits *at the margin* in each form of production.

REVIEW QUESTIONS

1 What is the "opportunity cost" of land and labour used in producing timber? How does competition for these inputs help ensure that their prices reflect their opportunity costs?

2 Give examples of how land, labour, and capital are substitutable in timber production. Are these inputs subject to the law of diminishing returns?

3 Describe how the substitutability of one input for another must be compared with their relative prices in order to determine the least-cost way of achieving a particular level of production.

4 The application of fertilizer to a tract of forestland can increase the yield of timber depending upon how many kilograms of fertilizer are applied per hectare, as follows:

Amount of fertilizer (kg/ha)	Increase in timber yield (m³/ha)
50	7.5
100	12.5
150	15.0
200	16.0

The cost of fertilizing consists of a fixed cost of $25 per hectare plus $0.50 per kilogram of fertilizer used. The timber is valued at $10 per cubic metre (which we will assume is realized in the same year that the fertilizer is applied).

Calculate and plot on a simple graph the total revenue and cost of applying these different amounts of fertilizer, and on another graph the corresponding marginal revenue and cost. What is the most efficient level of fertilization? Does this example indicate diminishing returns to fertilization?

5 If fertilizing a forest would cause pollution of waterways, how would the decisions of profit-maximizing forest owners fail to maximize the benefits to society as a whole? What term is used to describe this kind of market failure?

6 Define and provide examples of market failure, externality, and policy failure.

FURTHER READING

Baumol, William J., and Wallace E. Oates. 1988. *The Theory of Environmental Policy.* Cambridge: Cambridge University Press. Chapter 3.

Boyd, Roy G., and William F. Hyde. 1989. *Forestry Sector Intervention: The Impacts of Public Regulation on Social Welfare.* Ames: Iowa State University Press. Chapter 1.

Browing, Edgar, and Mark A. Zupan. 2004. *Microeconomics: Theory and Applications.* 8th ed. New York: John Wiley and Sons.

Gregory, G. Robinson. 1987. *Resource Economics for Foresters.* New York: John Wiley and Sons. Chapters 5 and 6.

Nautiyal, J.C. 1988. *Forest Economics: Principles and Applications.* Toronto: Canadian Scholars' Press. Chapter 3, 5, 6, and 18.

Randall, Alan. 1987. *Resource Economics: An Economic Approach to Natural Resource and Environmental Policy.* 2nd ed. New York: John Wiley and Sons. Part 2.

van Kooten, G. Cornelis, and Henk Folmer. 2004. *Land and Forest Economics.* Northampton, MA: Edward Elgar. Chapter 5.

Chapter 3   Forest Investment Analysis

As noted in Chapter 2, forest management involves important decisions about timing. At what age should the forest stand be thinned or fertilized? How fast should the forest inventory be harvested? How long should the crop be grown? Forests present almost infinite opportunities in silviculture, protection, access development, and other activities spread over time. To weigh the costs and benefits of these activities and to determine the management program that will generate the most net value, we need a consistent method of assessing values that accrue at different times. This chapter explains the assessment techniques and outlines the process of identifying the most economically advantageous management regime.

TIME AND THE ROLE OF INTEREST

A fundamental observation about economic behaviour is that people value a dollar today more than a dollar sometime in the future. The critical link between the value of a dollar today and a dollar 1 year, 2 years, or 10 years hence is the *rate of interest*. The rate of interest is therefore the key to comparing values that accrue at different times.

There are two explanations for putting more weight on present values than on future values. First, capital has an opportunity cost. Like most other productive resources, it can generate returns in alternative uses. When capital is tied up in a particular use, the sacrifice in other production must be taken into account. This *opportunity cost of capital* over time is measured by interest. For example, if a million dollars' worth of timber is held unharvested from one year to another, the opportunity cost of doing so is the interest that the owner could have earned if she had liquidated the timber and invested the million dollars wherever it would yield the highest return.

The second explanation is *time preference*, which refers to the degree to which individuals prefer present over future consumption. People generally prefer something today rather than tomorrow or next year. Saving behaviour demonstrates that savers demand some compensation for postponing consumption. Interest, the return on saving, is the reward for deferring consumption.

In addition to measuring the effect of time on values, there are two other functions of interest. One is to account for risk and the chance of failure. Investors must be compensated for taking risks, and the differences in market rates of interest reflect varying degrees of riskiness in economic ventures. The riskier the venture, the greater the likelihood of failure, hence the higher the interest rate investors demand, resulting in a higher reward that they can earn if the venture succeeds.

The other use of interest is to correct for changes in the value of money that result from inflation and deflation. For example, if inflation erodes the value of money by 4% per year, an investor who demands a *real* rate of return of 8% will have to find investment opportunities that yield a *nominal* rate of return of at least 12%. Therefore, the real rate of return is calculated as the nominal rate less the inflation rate.

Although the determination of prices in markets is formally introduced in Chapter 4, we point out here that the market rate of interest is the price that must be paid to borrow investment capital. The supply of capital funds is provided by savers whose time preference is less than the interest they can earn on their savings. The demand for capital arises from investors who want funds to invest in projects that will yield rates of return in excess of this cost of capital. Thus, the rate of interest simultaneously reflects the return on investment and the reward for saving. The supply and demand for capital respond in the usual manner: suppliers offer more when the price is high and demanders want more when the price is low. The equilibrium price is the interest rate at which supply and demand are equal. The interest rate thus rations the available supply of capital among those who demand it.

If the market for capital works effectively, all projects capable of yielding a return in excess of the market rate of interest will be undertaken, whereas projects with lower rates of return will not. Thus, interest ensures that capital is allocated to its most productive uses. Moreover, as long as the supply is balanced with the demand for investment capital, interest serves to allocate resources efficiently over time because it reflects the trade-off that savers and investors are willing to make between present and future returns.

For example, someone who demands a rate of interest of 8% is indicating that he or she is willing to give up $1.00 today in return for $1.08 one year from now. In other words, $1.08 next year has a *present value* of $1.00. At a rate of interest, or *discount rate,* of 8%, $1.00 today and $1.08 one year hence have equal value. In this manner, interest enables us to bridge time differences when comparing values.

The appropriate rate of interest to use is discussed later in this chapter. First, however, we turn to the linkage between present and future values in more detail.

COMPOUNDING, DISCOUNTING, AND PRESENT VALUES

If interest rates enable us to compare a value in the present with a value expected to accrue in the future, then the present value can be *compounded,* at the appropriate rate of interest, to indicate its equivalent value at a time in the future. Conversely, the future value can be *discounted,* using the same rate of interest, to obtain its equivalent present value. Either way, the two values can be compared at the same point in time. Compounding involves increasing a present value to its equivalent worth at a future time, whereas discounting is the reverse.

Compounding

We often want to know how much an amount invested today will yield at some future date. Take the simple case of investing $1.00. At 8% annual interest, it will amount to $1.08 after one year. If it is invested for two years, the $1.08 accruing at the end of the first year will grow by another 8 percent at the end of the second year, that is, $1.08(1 + 0.08) = $1(1.08)^2 = 1.17. A three-year investment would add another 8% to this total: $1.17(1 + 0.08) = $1(1.08)^3 = 1.26. This process is called *compounding* because after the first year, interest is earned (compounded) on the initial amount invested (the *principal*), and on the interest earned in previous years.

The general formula for these calculations is

$$V_n = V_0(1 + i)^n \tag{3.1}$$

where V_n is the value to which an initial amount V_0 will grow when invested for n years at an interest rate of i. For example, $100 invested at 8% for 10 years has a value, 10 years hence, of $V_n = \$100(1.08)^{10} =$

$215.90. Compounding is a straightforward mathematical calculation that can be carried out with the help of a hand calculator or a spreadsheet program such as *Microsoft Excel.*

Compounding can be used to compare values accruing at different times. We have already compared $1.00 today with $1.08 a year hence. For another illustration, suppose someone offered you either $100 today or $200 in 10 years. Which would you choose? If you knew you could invest money at 8%, you would choose the $100 today because, as we have already seen, at an interest rate of 8% it is capable of growing to $215.90 over a period of 10 years.

Discounting

The reverse problem is to determine the present equivalent value of a future payment. For this purpose, we can simply transpose equation 3.1 to

$$V_0 = \frac{V_n}{(1 + i)^n} \tag{3.2}$$

which converts the amount V_n to be received n years hence to its present worth, V_0. Often the term $1/(1 + i)^n$, which converts, at discount rate i, the amount V_n to be received in n years to its present worth, V_0, is called the *discount factor.* Using this formula, we can calculate the present worth of $1.00 received one year hence. At a discount rate of 8%, $V_0 = \$1.00/(1.08)^1 = \$0.93$. If the $1.00 were received two years hence, its present worth would be $0.86; if it were received 10 years hence, its present worth would be only $0.46.

This process of *discounting* future values is another means of comparing present with future values. In an earlier example, we compounded a present value of $100 to compare it with $200 in 10 years. Alternatively, we can compare these values by discounting the one receivable in the future to obtain its present worth. The present value of $200 receivable 10 years in the future is $\$200/(1.08)^{10} = \$92.64$, which is less than the alternative of $100 receivable today.

Compounding and discounting enable us to measure and compare, in consistent terms, values that accrue at different times. They enable us to represent the values not only in the same terms – dollars – but also at the same point in time. So, for example, to evaluate a particular plan for a forest stand we may have to immediately account for costs of establishing the crop, while accounting for management costs as they are incurred annually or periodically, revenues from thinning sometime during the growing cycle, and revenues from the final harvests at the end of the project. In order to compare the revenues with the costs to see whether the plan for the stand is advantageous, or better than an alternative plan, we must reduce all the values to the common dollar denominator at the same point in time.

Compounding and discounting enable us to choose any point in time to evaluate such a plan. For some purposes, it may be convenient to evaluate all revenues and costs at the time the forest is to be harvested. Alternatively, all values can be reduced to their equivalent value at the date the forest is to be established. The point in time that is most commonly chosen for evaluating forest management plans is simply the present, however.

The calculation of *present value* is usually divided into a calculation of the present value of the revenues or benefits of the project and the present value of the costs of the project. The excess of the benefits, B, over the costs, C, is the *net present value*, V_0, of the project:

$$V_0 = B - C \tag{3.3}$$

In the following subsections, we review the most common forms of present value problems in forestry and how they are calculated.

Present Value of Future Revenues and Costs
A common problem is whether it is advantageous to grow a crop of timber on a tract of vacant land. If, at the outset, it would cost an amount C_p to plant the crop, and if, after the final harvest, it is expected to generate a net return of V_n in n years' time, then the formula for calculating the net present value of the investment is obtained by combining equations 3.2 and 3.3:

$$V_0 = \frac{V_n}{(1 + i)^n} - C_p \tag{3.4}$$

So if the planting cost (C_p) is \$1,000, the growth period (n) 40 years, the value of the harvest (V_n) \$100,000, and the discount rate (i) 6%, then the net present value of the venture is:

$$V_0 = \frac{100,000}{1.06^{40}} - 1,000 = \$8,718$$

This indicates a net gain of $8,718 generated at an initial cost of $1,000, or nearly $9 of benefits per $1 of cost when all these values are measured in their present value equivalents.

Present Value of a Perpetual Annuity

Sometimes an asset like a farm or a forest is expected to yield a regular annual return – an annuity – in perpetuity, while its management costs recur annually as well. In this case, we need to calculate the present lump sum equivalent of the amount that occurs annually in perpetuity. For example, if a tract of timberland is expected to yield an annual harvest worth $1,000, the present value of this recurring revenue, at an interest rate of 8%, is $1,000/0.08 = $12,500.

The formula for calculating the annual return, a, on an amount, V_0, invested at i percent is

$$a = i(V_0) \tag{3.5}$$

Transposed and solved for V_0, this formula becomes

$$V_0 = \frac{a}{i} \tag{3.6}$$

which gives the present worth of an amount receivable each year in perpetuity at a given discount rate. This formula can also be obtained from equation 3.2 above by adding together the present worth of a geometric series of equal amounts receivable each year in perpetuity:

$$V_0 = \frac{a}{1+i} + \frac{a}{(1+i)^2} + \ldots + \frac{a}{(1+i)^\infty}$$

which can be reduced to equation 3.6.[1]

Present Value of a Finite Annuity

If the future series of annual payments is limited to a finite number of years – say, n – we need a formula for a geometric series with n terms:

$$V_0 = \frac{a}{1+i} + \frac{a}{(1+i)^2} + \ldots + \frac{a}{(1+i)^n}$$

This can be simplified to:[2]

$$V_0 = \frac{a[(1+i)^n - 1]}{i(1+i)^n}$$ (3.7)

Equation 3.7 is useful for calculating the present worth of a series of future costs or revenues that will occur each year for a certain number of years. For example, the present worth of a forest that will yield a harvest worth $1,000 each year for six years, at 8% interest, is

$$V_0 = \frac{1,000[(1.08)^6 - 1]}{0.08(1.08)^6} = \$4,622$$

Equation 3.7 can also be used to calculate the equal installments for a loan with fixed payment. For example, on a 10-year term (n), a $50,000 loan ($V_0$) at 12% interest ($i$) will have annual payments (a) of:

$$a = \frac{iV_0(1+i)^n}{(1+i)^n - 1} = \frac{0.12 \times 50,000(1.12)^{10}}{(1.12)^{10} - 1} = \$8,849$$

If the loan is to be paid on a monthly basis, equation 3.7 becomes a typical mortgage payment formula. As long as one has the monthly interest rate (i_m, see below), the number of months (n), and the original loan amount (V_0), the monthly loan payment or mortgage can be calculated using equation 3.7. Because banks often give loans at an annual rate, however, we need to figure out the monthly rate to use this formula:

$$i_m = \sqrt[12]{(1+i)} - 1$$ (3.8)

where i_m is the monthly interest rate and i is the annual interest rate. Equation 3.8 comes from the identity

$$(1 + i_m)^{12} = 1 + i$$

where 12 is the number of months in a year.

Present Value of a Periodic Series
Of the considerable variety of other formulas for special problems, one frequently used in forestry is the formula for determining the present worth of a future series of revenues or costs that will accrue only periodically at regular intervals of several years, for example, at the end of every timber rotation. If the interval is t years, and the net revenue received every t years is V_t, then

$$V_0 = \frac{V_t}{(1+i)^t} + \frac{V_t}{(1+i)^{2t}} + \ ... \ + \frac{V_t}{(1+i)^\infty}$$

This simplifies to:[3]

$$V_0 = \frac{V_t}{(1+i)^t - 1} \tag{3.9}$$

This is the formula required to calculate, for example, the present worth, V_0, of a future infinite series of forest crops having a harvest value of V_t, which will accrue after each crop rotation period t. This is the basis for the site value or soil expectation value of bare land in even-age forest management and the derivation of the Faustmann optimal rotation age that will be introduced in Chapter 7.

A notable difference between equation 3.9 and equation 3.2 is the extra -1 in the numerator of equation 3.9. Further, although both n in equation 3.2 and t in equation 3.9 mean a number of years from the present, the former is a one-time payment at year n, whereas the latter is a series of payments, accruing periodically at a regular interval of t years. Thus, equation 3.2 is used for calculating the present value of a future payment, and equation 3.9 is used for calculating the present value of a periodic series of future payments.

Thus, if a tract of forestland can produce a crop worth $100,000 in 40 years, and every 40 years thereafter, and the interest rate is 8%, then the present worth is

$$V_0 = \frac{100,000}{(1.08)^{40} - 1} = \$4,826$$

Incidentally, the low present worth of these crops results from the effect of discounting at 8% over 40 years or more. Experimentation with this result and equation 3.2 will reveal that the second and subsequent crops contribute little to the present value when the interest rate is as high as 8%.

Now, to incorporate preceding formulas (equations 3.3, 3.6, and 3.9), let us elaborate on this example by introducing two realistic complications. First, it will cost an amount m every year to manage and administer the forest. Second, the forest must be planted at a cost C_p at the outset and at the time of each harvest. The solution then takes the form

$$V_0 = \frac{V_t - C_p}{(1 + i)^t - 1} - \frac{m}{i} - C_p \tag{3.10}$$

The first term on the right-hand side of the equation is the present worth of the series of future crops net of the cost of reforestation. The second term allows for the recurring annual costs of management. The third accounts for the initial cost of planting the first crop. If, for example, harvests are expected to yield \$100,000 (V_t), planting costs (C_p) are \$1,000, annual management costs (a) are \$100, the discount rate (i) is 8%, and the rotation period (t) is 40 years, then the net present value of this program is

$$V_0 = \frac{100,000 - 1,000}{(1.08)^{40} - 1} - \frac{100}{0.08} - 1,000 = \$2,528$$

A variety of other intermediate costs and returns can be incorporated into this equation.

These formulas for compounding and discounting, and some others commonly used in forestry, are summarized in the appendix to this chapter.

For simplicity throughout this book, all formulas and calculations of values over time are based on annual compounding and discounting. It should be recognized, however, that values can be compounded or discounted with any periodicity, which in practice may be as often as semi-annually or even more frequently. The interest on your own bank account, for example, is probably calculated monthly, and added to the principal at that time. Mortgage is typically paid on a monthly basis. All interest rates that are paid periodically (that is, daily, monthly, or quarterly) can be converted to an equivalent annual rate of interest using the following equality:

$$(1 + i_p)^p = 1 + i \tag{3.11}$$

where i_p is the periodic interest rate, p is the number of periods in a year, and i is the annual interest rate. The monthly interest rate formula (equation 3.8) is a special case of this equality.

Continuous Compounding
At any given interest rate, the more frequent the compounding, the greater the impact on the accumulating amount. At the extreme, compounding

TABLE 3.1   Derivation of continuous compounding

Periods per year	End value	Equation
1	$(1 + i)$	3.12a
2	$(1 + i/2)(1 + i/2) = 1 + i + i^2/4$	3.12b
3	$(1 + i/3)^3 = 1 + i + i^2/3 + i^3/27$	3.12c
.	.	
.	.	
n	$(1 + i/n)^n = 1 + i + i^2/2! + i^3/3! + ... + i^n/n!$	3.12d

can be continuous. Table 3.1 shows the value of $1 compounded at an annual rate of i at varying numbers of times over one year.

This last step is not obvious, but is proved in most elementary calculus textbooks. As n grows large, $(1 + i/n)^n$ approaches e^i, where $e = 2.7183...$, the base of natural logarithms. This too is not obvious, but is, in fact, one way of defining e.

For reasons that become apparent later (especially in Chapters 7 and 9), continuous compounding is analytically simpler than annual compounding. Most economics literature now uses continuous compounding and discounting, but discrete rates are usually employed in accounting and finance. Students of forest economics should be familiar with both techniques.

Using the same rate of interest, continuous compounding yields a higher end value than annual compounding, but the former can be converted to the latter by applying an equivalent annual rate (EAR, or i). This is the rate that, if compounded annually, would have the same effect as compounding the initial rate continuously. To find the EAR for a continuous rate r, we equate the end values of annual and continuous compounding in t years (when $t = 1$, it simply disappears in equations 3.13a and 3.13b below):

$$(1 + EAR)^t = e^{rt} \tag{3.13a}$$

$$t \ln(1 + EAR) = rt \text{ because } \ln(e) = 1 \tag{3.13b}$$

$$\ln(1 + EAR) = r \tag{3.13c}$$

$$EAR = e^r - 1 \tag{3.13d}$$

This means that any discounting, compounding, or growth analysis can be done in either annual (discrete) or continuous formats. If one is given an annual rate EAR (that is, i), the equivalent continuous rate r can be found using equation 3.13c; if one is given a continuous rate r, the equivalent annual rate EAR (or i) can be found using equation 3.13d.

For example, the equivalent annual rate of a continuous interest rate of 12% is 12.75%, using equation 3.13d. Thus, continuous compounding at 12% would cause $100 to grow to $112.75 in one year, compared with only $112.00 under annual (discrete) compounding of 12%. We need to be careful not to mix up discrete and continuous interest rates.

In the remainder of this book, we use the symbol i to represent the discrete annual interest rate, and r to represent the equivalent continuous interest rate.

CRITERIA FOR INVESTMENT DECISIONS

To evaluate opportunities for forestry investments, we again concentrate on the criterion of efficiency – the benefits generated relative to the costs. Benefits and costs occur at different times. Investment opportunities typically involve an initial cost, and perhaps additional costs later, which give rise to future benefits that may also be spread over time. Evaluating investments involves weighing the benefits against the costs, taking into account the timing of their occurrences, in order to establish their relative efficiency. The most efficient investment generates the greatest possible benefit for the cost incurred, when both benefits and costs are measured at the same point in time.

The technique for weighing benefits and costs is called *benefit/cost analysis*. It can take several forms, each appropriate for a different purpose.

Identifying Advantageous Investments

In considering an array of possible forestry investments, the first task is to distinguish those that are advantageous from those that are not. Advantageous investments are those that yield benefits in excess of the costs, or a positive *net benefit*. Whenever the relevant values occur at different times, the net benefit is the difference between the present value of benefits and the present value of costs. A project that has positive net benefits is economically *feasible* in the sense that it generates returns in excess of all costs, including the opportunity cost of the capital involved. Where the expected benefits fall short of the costs, the project is economically unfeasible because the net benefit of the undertaking would be negative, implying that the inputs could generate greater value in producing other things.

Thus, the basic rule for identifying investment opportunities is that the benefits must exceed the costs when both are expressed in terms of their present values. In terms of equation 3.3, this condition is

$$V_0 = B - C > 0$$

A positive net benefit is a necessary condition for worthwhile investments, but it is not a sufficient justification for undertaking them when alternative opportunities that also meet this condition exist. We will learn, however, that the manager's budget may constrain choices among those investments with positive net benefits.

Identifying Priorities

Which of all the economically feasible forestry projects should be undertaken? The answer depends on the circumstances. If all the projects were independent of each other, and there were no limits to the capital and other resources available, then all of them might be undertaken because all will generate a net gain. Available resources are usually limited, however, and one project may be a substitute for another, or two projects may be alternative ways of using the same site, so we need criteria for selecting among projects that are all economically feasible.

Net Present Value (Net Benefit)

Net benefit is often called *net present value* because future benefits and costs are discounted to present time. The greater the surplus of benefits over costs, the greater the gain from a project. Where there are alternative ways of using some fixed resource such as a tract of land, the task is to find the alternative that will generate the highest returns. For example, a forest owner might have to decide which of three mutually exclusive uses of a forest tract is most advantageous: timber production, agriculture, or wilderness preservation. The net benefit of each can be compared and the one promising the greatest net benefit would be selected.

For another example, consider a typical forestry calculation, where a tract of 40 hectares has been harvested and the choice is between leaving it to regenerate naturally or planting it with young seedlings. If left to reforest naturally at no cost, in 55 years the tract would yield 375 cubic metres per hectare of an inferior species worth only $15 per cubic metre. If planted immediately at a cost of $400 per hectare, in 55 years the tract would yield 475 cubic metres per hectare worth $25 per cubic metre. The problem is to identify the advantage of planting. Would it be an efficient choice?

The net benefit of the investment in planting is the difference between the net gain due to planting, V_p, and the net gain from natural regeneration, V_g, or $V_0 = V_p - V_g$.

If the discount rate is 3%, then the net benefit can be calculated as:

$$V_p = \frac{40 \times 475 \times 25}{(1.03)^{55}} - (40 \times 400) = 93{,}464 - 16{,}000 = 77{,}464$$

$$V_g = \frac{40 \times 375 \times 15}{(1.03)^{55}} = 44{,}272$$

$$V_0 = 77{,}464 - 44{,}272 = \$33{,}192$$

Given the assumed yields, prices, costs, and discount rate in this example, planting is clearly the more advantageous choice, showing a greater net benefit of $33,192 greater than the natural regeneration alternative. Calculations of this kind can identify the best choice among a wide range of silvicultural regimes to determine how land should be managed to yield the greatest return. In practice, if the optimal rotation age for natural forests is 55 years, the optimal rotation age for the planted forests is likely to be shorter, making planting more advantageous. In Chapter 7, we extend this analysis to determine the optimal forest rotation age.

Benefit/Cost Ratio

Forest managers often find that the main constraint on their activities is a fixed budget, making financial capital their limiting factor. In this case, the challenge is how to generate the maximum possible benefits from the funds available, or how to generate the maximum possible benefits for every dollar invested. This leads to a second criterion for setting priorities among alternative projects – the *benefit/cost ratio*. This ratio measures the benefits per dollar of costs, or the efficiency with which funds are expended. Corresponding to the criterion of efficiency described in Chapter 1, the benefit/cost ratio measures the economic efficiency with which resources are used.

Our earlier example of a new forest plantation cost $16,000 and generated $49,192 (93,464 – 44,272) in expected benefits in excess of the benefits that natural regeneration can bring, so its benefit/cost ratio is 49,192/16,000, or 3.07.

Suppose the forest manager could alternatively spend his funds elsewhere on the precommercial thinning of 60 hectares of juvenile stands, which would cost $550 per hectare, increase yields in 50 years from 425 to 550 cubic metres per hectare, and raise the value of the harvest from $15 to $25 per cubic metre. At a discount rate of 3%, the present net benefit arising from the precommercial thinning is

$$V_0 = \frac{60 \times 550 \times 25}{(1.03)^{50}} - \frac{60 \times 425 \times 15}{(1.03)^{50}} - (60 \times 550)$$

$$= 100{,}937 - 33{,}000$$

$$= 67{,}937$$

This project is also economically feasible, and it yields a larger net benefit than the planting project. The thinning project, however, produces a benefit/cost ratio of 100,937/33,000, or 3.06, which is slightly less than the benefit/cost ratio for the planting project. Thinning generates a lower return per dollar invested than the planting project.

The benefit/cost ratio is one criterion for selecting among a variety of independent projects. If all projects were designed to their optimal scale, then giving priority to higher benefit/cost ratios will generate maximum net benefits from limited investment funds. Ranking according to benefit/cost ratios seldom corresponds to ranking based on net present values, however.

Internal Rate of Return on Investment

Investors are often interested only in the return on the capital they invest, and therefore prefer to assess projects and determine their priority according to their *internal rate of return.* Instead of comparing the present value of benefits and costs discounted at some predetermined interest rate, this approach involves finding the interest rate, or discount rate, that equates the present value of benefits with the present value of costs. The higher the indicated rate of return on the investment, the more attractive the project.

The internal rate of return of a project is the percentage rate at which the initial investment grows over the investment period to equal the value of the eventual expected benefits. Unlike the preceding criteria, which relate benefits to the cost of all inputs, including capital, this approach treats the return on capital as the residual after all other costs are accounted for.

Consider again our first example above, the opportunity to plant trees to increase the value of the harvest. The net present value of the project can be restated as

$$V_0 = \frac{(40 \times 475 \times 25) - (40 \times 375 \times 15)}{(1 + i)^{55}} - (40 \times 400)$$

To use the internal rate of return criterion, we must solve for the discount rate, i, at which benefits equal costs (or $V_0 = 0$):

$$(1 + i)^{55} = \frac{250,000}{16,000}$$

$$i = 5.12\%$$

This internal rate of return (5.12%) is the rate at which $16,000 grows to $250,000 in 55 years.

In the thinning example above, the initial cost of $33,000 produced an expected gain of $442,500 [(60 × 550 × 25) – (60 × 425 × 15)] in 50 years, yielding an internal rate of return of

$$(1 + i)^{50} = \frac{442,500}{33,000}$$

$$i = 5.36\%$$

The internal rate of return criterion is used mainly by investors seeking to maximize the return on their financial capital. It eliminates the need to select a rate of interest in advance, although sometimes investors consider only investment opportunities that achieve a rate of return in excess of some minimum acceptable rate. The internal rate of return has the additional appeal of simplicity; most people understand that a higher rate of return on an investment is preferable to a lower one.

Nevertheless, this criterion presents serious problems for many applications, and it can give misleading results. It is suitable only for projects that involve an initial investment and later returns. Some projects generate some early returns and impose later costs, and this criterion makes projects with early benefits and late costs appear more favourable. Moreover, calculation of the internal rate of return becomes exceedingly complicated, and sometimes indeterminate, when a project involves positive and negative returns scattered through time.

Comparing Criteria

A third numerical example will help illustrate the relationships among these investment criteria. Consider a 100-hectare tract of forest scheduled for harvest in 15 years, with an expected yield of 350 cubic metres worth $15 per cubic metre, and an insect infestation threat that can destroy 25% of the timber unless control measures are undertaken immediately. The cost of spraying is $350 per hectare.

The benefit of insect control is the present worth of the 25% of the stand that would be saved and harvested in 15 years. The net benefit of the control activity, at a discount rate of 3%, is

$$V_0 = \frac{100 \times 350 \times 15 \times 0.25}{(1 + 0.03)^{15}} - (100 \times 350)$$

$$= 131,250/1.03^{15} - 35,000$$

$$= 84,244 - 35,000$$

$$= \$49,224$$

The benefit/cost ratio for this project is 84,244/35,000, or 2.41, and the internal rate of return is computed as $[(1 + i)^{15} = 131,250/35,000]$, yielding an i equal to 9.21%.

Table 3.2 summarizes the results of applying each of the three evaluation criteria to each of our three examples above.

All three projects are economically feasible insofar as they generate returns exceeding their costs, but the three different criteria each rank the projects in a different order. Net benefits are greatest for the spacing project. The benefit/cost ratio is largest for the planting project, but this project has the smallest net benefit and the lowest internal rate of return. The internal rate of return is highest for the pest control project, yet this project has the lowest benefit/cost ratio.

TABLE 3.2   Comparison of investment projects using alternative evaluation criteria

	Project		
Criterion	Planting	Spacing	Pest control
Net benefit (B – C)	$33,194.00	$67,937.00	$49,244.00
Benefit/cost ratio (B/C)	3.07	3.06	2.41
Internal rate of return (i)	5.12%	5.36%	9.21%

Obviously, these results are dependent on the economic and physical conditions specific to our examples. Different prices and costs, discount rates, and technical production relationships would give different results, and therefore different rankings of priorities among planting, spacing, and pest control.

The different criteria will always agree on the thumbs-up/thumbs-down question of whether a project is advantageous at all, but they often produce different rankings. The main point in our examples and Table 3.2 is that, for any set of prices and costs, discount rates, and technological relationships, the *relative* attractiveness of any project may be different under different evaluation criteria. Because the criteria often suggest different investment priorities, it is important to understand why they differ and the appropriate use of each criterion.

The problem of choosing between investment criteria is aggravated by *indivisibilities* in project scale. Indivisibilities refer to the technical characteristics that restrict projects to certain discrete sizes. Water projects provide excellent examples, although there are also good forestry examples. For example, local geographic and topographic configurations may make it possible to build a dam 20 or 50 metres high, but intermediate heights may be unfeasible.

Or consider our three forestry examples. Planting or spacing can be carried out to almost any density in terms of stems per hectare, so the divisibility and scale of such projects are not problematic. It may, however, be impractical to protect only half a hillside from pests.

The net benefit criterion often used to find the maximum returns to some fixed factor, such as a tract of land, favours large projects. If our planting example were three times larger, its net benefits would be three times larger, and its ranking under the net benefit criterion would shift from third to first – without changing its internal rate of return or its benefit/cost ratio.

When managers have budget limitations, they may not be able to afford to expand all investments until their marginal benefits equal their marginal costs, or until the marginal benefit of the last dollar invested in one project equals the marginal benefit of the last dollar invested in all other acceptable projects. They must apply alternative criteria, such as the benefit/cost ratio.

Furthermore, both present net benefits and benefit/cost ratios are sensitive to the interest rates used. Higher interest rates favour projects that yield earlier returns and incur later costs. Changing the interest rate changes the relative attractiveness of a project as judged by the present net worth or benefit/cost criteria. Investors seeking to maximize their return from a fixed budget may therefore choose to rank projects by their internal rates of return.

We also need to consider the underlying assumptions in these criteria about different time horizons and time scales. It should be pointed out that when they have different time horizons, net benefit and the benefit/cost ratio assume that returns are reinvested at the discount rate, whereas internal rate of return assumes that returns are reinvested at the project rate (or internal rate of return). This is important in forestry, as the time horizon of a forestry project may be infinite (that is, repeating the rotation forever).

When there are different project scales, net benefit assumes that funds not used for the smaller project are invested in projects earning the discount rate, whereas benefit/cost ratio and internal rate of return (which are "scale-free") assume that the project can be duplicated so that funds not used for the smaller project are invested in similar projects earning the project rate.

Finally, one may face two constraints: mutually exclusive projects for the same sites and a limited total budget. In these cases, a reasonable rule of thumb is to select the project with the highest net benefit for each site and then rank projects in order of decreasing benefit/cost ratio until the budget is exhausted.

INTEREST RATE, INFLATION, RISK, AND UNCERTAINTY

We have seen that the rate of interest is critically important in evaluating investments. The appropriate rate to use is often problematic, however. In principle, the appropriate rate of interest is the investor's opportunity cost of capital – that is, the rate that he or she must pay to borrow or the rate that he or she can earn on capital invested elsewhere at the margin. This rule applies to both public and private investors.

Financial markets reveal a spectrum of interest rates, however, reflecting varying allowances for risk, expected inflation, distortions of the tax system, and other market imperfections, as well as real returns to capital. Market rates also fluctuate continuously, so it is difficult to identify the opportunity cost of capital to apply to a specific investment, and even more difficult to determine how much society as a whole should discount future over present values.

Market Interest Rates and the Rate of Time Preference

A first approximation of the opportunity cost of capital free of the distortions due to risk, short-term disturbances, and inflation is often made from the historical average yield on long-term government bonds. These securities, bought and sold in large quantities in competition with private securities, are considered virtually risk-free because they represent loans to governments, and governments in large and stable countries are unlikely to default on their debts because of their control over tax revenues and the money supply. Moreover, because these securities are widely available, the return they offer represents a risk-free opportunity cost of capital.

In Canada and the United States, this rate, after adjusting for inflation (see below), has historically been in the order of 1% or 2%. Real rates of return on private equity capital have been considerably higher, in the order of 6%. Private rates must be higher than the rates for long-term government securities because, unlike government projects, private incomes are taxed, and because private securities are riskier than government securities.

Our general rules for economic efficiency, discussed in Chapter 2, suggest that efficient allocation of resources calls for the same return, at the margin, to public and private investments, after appropriate allowances for differences in risk and taxation. There remains a fundamental question, however, about the rate of discounting of future values that best serves the interests of a society at large, or a nation, or a government, and whether the *social rate of time preference* bears any relationship at all to the market rates of interest generated by independent savers and investors. Despite extensive economic and philosophical debate about this issue, the social rate of time preference remains elusive. In any event, to the extent that it differs from the market rate of interest, it must be decided collectively through the political process, but this is never done explicitly. There is, however, a general agreement that the social rate of time preference is lower than the market rates of interest and is what should be used when intergenerational investments are made, as is the case with some forestry investments. More controversial proposals call for using zero interest rate or time-declining interest rates in assessing long-term public investment projects.

Thus, the search for an appropriate social rate of interest can take only limited guidance from observed market rates. In practice, however, the task facing most government resource management agencies and private forest owners is simpler; they are guided by their opportunity cost of capital reflected in the rate at which they can borrow or the rate they can earn on other investments. On this basis, governments and private decision makers often specify a required rate of return for evaluating their investment opportunities.

Because the investment periods in forestry are long, they are highly sensitive to the chosen interest rate. The US government uses a real rate of 4% for evaluating forestry investments. Such a modest rate is sometimes justified on grounds of low risk, as well as the social benefits that these projects generate. Private investors and forest owners typically employ higher rates, reflecting in part their greater allowance for risk. The riskiness of forest investments varies widely, however, and should be considered in the circumstances of each case.

Accounting for Inflation

Inflation is a rising general price level or, put another way, a declining value of money. It distorts dollar values over time. Thus, benefits or costs accruing at different points in time must be corrected for inflation before they can be compared on a consistent basis.

Over the long periods of time commonly considered in forest management, even a modest rate of inflation can have a large impact on prices and costs. For example, an annual inflation of 3%, compounding over time like interest on capital, will double prices in less than 25 years. In this case, if an amount receivable 25 years hence, expressed in *current dollars* of that time, were to be compared with today's values, it must be reduced by more than half for the values to be comparable in *constant dollars.*

"Constant dollars" means expressing all benefits and costs that accrue at different points in time (and in the current dollars of the time) in the dollar value of one fixed particular time. For example, if the inflation rate is 3% per year, to evaluate a project in terms of the value of money in year 1, an amount incurred in year 25 (in the current dollars of that time) must be reduced, or *deflated,* by more than half to express it in dollars of the same value as dollars in year 1.

Values accruing in different years, expressed in their current dollar values, can be corrected for inflation by deflating them, using a *price index* that measures the change in the price level over the years. For example, with an inflation rate of 3%, a price index based on a value of 100 in the current year will be 103 next year and 209 in 25 years. Future values can all be converted to their equivalent values in dollars of the current year by dividing each future value by the ratio of the price index for the year in which it accrues to the price index of the current year.

The best-known index of inflation is the consumer price index, which is based on the change in price of a representative sample of consumer goods and services. The index is used to calculate the rate of inflation in Table 3.3. The consumer price index is an appropriate index for correcting for inflation in the prices of final consumer products, such as

TABLE 3.3   Annual rates of return for the NCREIF timberland index,
S&P 500 index, and US government bonds, as well as annual rates of
inflation, 1987-2007

Year	NCREIF timberland index (%)	S&P 500 (%)	US government bonds (%)	Rate of inflation (%)
1987	24.58	5.67	6.77	3.65
1988	27.44	16.61	7.65	4.14
1989	33.43	31.69	8.53	4.82
1990	10.64	−3.10	7.89	5.40
1991	19.06	30.47	5.86	4.21
1992	34.25	7.62	3.89	3.01
1993	21.57	10.08	3.43	2.99
1994	14.74	1.32	5.32	2.56
1995	13.23	37.58	5.94	2.83
1996	10.42	22.96	5.52	2.95
1997	17.86	33.36	5.63	2.29
1998	5.76	28.58	5.05	1.56
1999	10.62	21.04	5.08	2.21
2000	4.36	−9.10	6.11	3.36
2001	−5.16	−11.89	3.49	2.85
2002	1.87	−22.10	2.00	1.58
2003	7.48	28.69	1.24	2.28
2004	10.83	10.88	1.89	2.66
2005	18.44	4.91	3.62	3.39
2006	13.11	15.79	4.94	3.23
2007	17.45	5.49	4.53	2.85
Average	14.86	12.69	4.97	3.09
Standard deviation	9.90	16.28	1.94	0.96
Coefficient of variation	0.67	1.28	0.39	

Sources: National Council of Real Estate Investment Fiduciaries (NCREIF), http://secure.ncreif.org/
ncreif.org/timberland-returns.aspx; Standard & Poor's (S&P) Index Services, http://www2.
standardandpoors.com/spf/xls/index/MONTHLY.xls; market yield on US Treasury securities at
one-year constant maturity, quoted on investment basis, Federal Reserve Board, http://www.
federalreserve.gov/releases/h15/data/Annual/H15_TCMNOM_Y1.txt; rate of inflation: Bureau of
Labor Statistics (BLS), US Department of Labor, ftp://ftp.bls.gov/pub/special.requests/cpi/cpiai.txt.

food, clothing, and recreation, and for the general decline in the value of
money over time. Other indices, however, such as the wholesale price
index or producer price index, are more suitable for estimating changes
in particular products, such as timber, that are used in further industrial
production.

A second way to account for inflation is to leave the prices stated in current values while adjusting the interest rate used in project evaluation upward to allow for inflation. The *nominal* rates of interest that we observe in money markets are the sum of two components: the *inflation rate* and the *real rate.* The inflation rate, an allowance for inflation, is the rate at which a value must grow in current dollars to maintain its value in constant dollars. The real rate of interest, the return on capital after allowing for inflation, is the difference between the nominal rate of interest and the inflation rate.

If the real rate of return is i_r and the rate of inflation is f, then the nominal rate of return (i) can be shown as follows:

$$i = (1 + i_r)(1 + f) - 1 = i_r + f + i_r f \qquad (3.14)$$

Or, if we know the nominal rate of return and the rate of inflation, then the real rate of return can be shown as follows:

$$i_r = \frac{(1 + i)}{(1 + f)} - 1 \qquad (3.15)$$

Projects can be evaluated in terms of expected future nominal costs and benefits that have been adjusted for inflation. Adjusting expected costs and benefits for inflation enables one to use the nominal rate of interest to come up with the present values of the projects.

The future rate of inflation may be unknown, however, and, in any event, the rate of price and cost change for different outputs and inputs over the investment period may differ from the rate of inflation in the general economy. Thus, projects are more often evaluated in constant dollars. In this case, all costs and benefits are estimated in dollars of a given base year, usually the current year or the year the project would be undertaken. Then, using an inflation-free *real* rate of interest, the stream of project benefits and costs over time can be reduced to their present values.

These two approaches yield the same result. To be consistent, the analyst must express benefits and costs in constant dollars and use a real (inflation-free) discount rate; alternatively, current (or nominal) values can be used for both. We shall return to this issue in the next section.

Uncertainty and Risk

So far we have discussed the revenues and costs associated with an investment project as though they were known, or could be predicted accurately. Future revenues and costs are always more or less uncertain, however, and the further into the future they are expected to occur, the more uncertain they are.

Future costs and revenues associated with long-term forestry projects are often highly uncertain, especially when they are based on predictions spanning several decades. Knowledge about how stands grow and respond to treatments is always limited. Expectations about future harvests can be upset by unpredictable events such as fire and other natural catastrophes. And the technology, product prices, and production costs assumed in making predictions are likely to change in unforeseeable ways.

Sometimes a distinction is made between risk and uncertainty on the basis that the former lends itself to prediction while the latter does not. For example, long-term future prices of timber may be uncertain when there is no statistical basis for predicting them. In contrast, the risk of forest fires can be statistically estimated. Moreover, risk can be spread, and thereby eliminated.

To illustrate, if a tract of forest is the only asset of a small landowner, the risk that it may be destroyed by fire, though small, is likely to be of major consequence to the owner. In contrast, if it is only one of thousands of such tracts held by a large landowner, the owner's average fire losses are likely to be statistically predictable and can be allowed for as a predictable and regular cost of doing business without much risk.

This is analogous to a home insurance company that assumes the risk of fire losses of thousands of individual homeowners. By spreading the risk over a large number of homes, the insurance company makes its losses predictable and thereby eliminates the element of risk in its business. This means that although the inherent risks associated with a project must be accounted for by any investor, individual investors' responses will be influenced by their circumstances. In general, large, diverse investors, especially governments, are likely to be less averse to risk to the extent that they can spread its effects.

Investors are generally *risk-averse* – that is, faced with two investment opportunities having equal expected net returns but with one being riskier than the other, most investors prefer the less risky one. Accordingly, investors demand higher returns from riskier ventures. Risk thus adds another element of cost to a project, a *risk premium*. The more uncertain the project's outcome, the greater the risk premium investors will demand on their investments.

There is no single accepted technique for allowing for uncertainty and risk in analyzing investments, but several criteria have been developed to assist managers and investors in making consistent decisions in the face of them. The appropriate choice among them depends on investors' objectives and their attitude towards risk taking.

Risk Premiums

One way to take uncertainty into account is to increase the interest rate used in the evaluation by a premium sufficient to compensate the investor for the riskiness of the project. The size of the risk premium that should be added to the risk-free interest rate depends on the inherent riskiness of the project and the investor's aversion to risk taking.

Adding a risk premium to interest rates is a simple procedure. The so-called risk-adjusted discount rate (RADR) is simply equal to a risk-free rate (i_f) plus a risk premium (k). The risk premium (k) is positive for risk-averse individuals, equals zero for risk-neutral individuals, and may be negative for risk-loving individuals or risk seekers.

$$\text{RADR} = i_f + k \tag{3.16}$$

This procedure presumes that the uncertainty surrounding future revenues and costs is perfectly correlated with their time of expected occurrence. This is not usually the case. Some future costs and revenues, such as insurance premiums and property taxes, can be estimated quite closely for years into the future, while others, such as timber prices and fire-fighting costs, are often difficult to predict even over a relatively short time span. Applying a risk premium to the interest rate means discounting all events in all time periods by a factor that depends only on how far into the future they will occur, thus blurring their differences in uncertainty.

To determine the appropriate risk premium, k (and thus the RADR), for a risky investment project, one can begin with the investor's expected revenue, $E(R)$, from the project. This can be compared with the amount of certain income – the *certainty-equivalent* (CE) – that would give the investor the same satisfaction. This certainty-equivalent refers to the amount of income from an investment, such as government bonds, that pays the investor a risk-free rate of interest (i_f). By equalizing the present values of these two investment alternatives (one risky and one risk-free) that have the same investment period (n years), one can find the RADR (and then k):

$$\frac{CE}{(1 + i_f)^n} = \frac{E(R)}{(1 + \text{RADR})^n} \tag{3.17}$$

Equation 3.17 implies that, given certainty-equivalent and a risk-free rate of discount, RADR (and thus the risk premium, k) depends on the investment period, n, and the amount of risk in the revenue, $E(R)$.

The foregoing discussion demonstrates that there is a trade-off between risk and expected return (along with taxes) among various investment alternatives. Stocks, for example, have a higher level of risk and higher rate of returns than government or corporate bonds. In finance, this relationship is often described in various financial asset pricing models, including the widely used *capital asset pricing model* and more sophisticated multi-factor models. Interested readers may find detailed descriptions of these models in modern finance textbooks.

Recent empirical studies have shown that, in the past two decades, US timberland had a higher rate of return and lower level of risk than common stocks and an efficient portfolio consisting of common stocks, small company stock, corporate bonds, US government bonds (immediate-term and long-term), and US treasury bills. With the usual qualification that past performance is not a guarantee of future return, it appears that timberland is a relatively secure investment vehicle, implying that the appropriate discount rate for such investments is relatively low, and perhaps lower than previously believed.

Table 3.3 shows annualized rates of return for timberland held by some institutional investors and the S&P (Standard & Poor) 500 index of common stocks, which represents the broader performance of the stock market, as well as the rate of inflation and return to long-term US government securities. All are in nominal terms. The rates of return for timberland are generated from a timberland index maintained by the National Council of Real Estate Investment Fiduciaries (NCREIF). One can use equation 3.15 to calculate the real rates of return for the three investment vehicles listed in Table 3.3, which reveals that in some cases the real rates of return are negative.

Note that standard deviation (or variance) is a measure of risk. For any given expected value or rate of return, the higher the standard deviation, the greater the risk. If the expected values of investment alternatives vary, one needs to calculate and compare their relative risk. The *coefficient of variation*, which is the standard deviation divided by the expected value, can be used for this purpose: the higher the coefficient of variation, the greater the risk.

Using the coefficient of variation as a measuring stick, we find, based on 21-year data, the riskiness of the three investment alternatives listed in Table 3.3. Ranked in increasing order, they are US government bonds, timberland investments, and stocks.

Payback Period

A simple rule used by some investors is that an acceptable project must generate returns sufficient to cover the cost of the investment within a certain period. Fixing a maximum payback period does not explicitly recognize the uncertainty of future costs and returns; it is simply a decision rule that recognizes the most likely outcome within the period, and nothing beyond it. The payback period used by private investors is often short, frequently five years or less. This provides assurance that the investment will be profitable but it rules out superior (and even more certain) projects that would yield returns after the payback period.

Investors in some developing countries often rely on short payback rules, not because of uncertainties about the projects themselves but because of the risk associated with the political and regulatory environment.

Analyzing the Range of Possible Outcomes (Sensitivity Analysis)

Another way to assist decision makers considering uncertain projects is to evaluate them not only in terms of their most likely outcomes but also for the full range of all possible outcomes, from the most pessimistic, or "worst-case," scenario to the most optimistic. The results give investors a feel for the range of possible acceptable and unacceptable outcomes, and for their dispersion around the outcome considered most likely.

To provide additional guidance, the analyst may, for each possible outcome, provide an estimate of the probability that it will occur. Then each possible outcome can be weighted by its probability in order to portray the likelihood that the project will yield acceptable results. The empirical information suitable for assigning probabilities is usually scanty, however, and estimates have to be made subjectively. Nevertheless, the process forces analysts and decision makers to be explicit, and hence more consistent, in recognizing the relative probabilities of different possible outcomes.

Once probabilities have been attached to the alternative possible outcomes, decision makers can use a variety of criteria to select their response to risk. One way of organizing a decision involving uncertainty is the so-called decision tree, which is a diagrammatic representation of the choices facing the decision maker, the order in which events take place, the probabilities associated with them, and the possible outcomes.

Figure 3.1 illustrates a decision tree for the pest control problem described earlier in this chapter, with introduction of uncertainty into the problem. The branches from the square indicate the decision maker's choices; branches from circles represent uncertain events. The branches

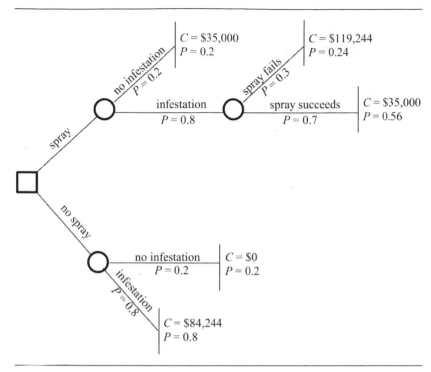

Figure 3.1 Decision tree for a pest control project.

from the square on the left indicate that the decision maker can choose either to spray or to do nothing. The upper branches indicate the possible outcomes if the spraying is undertaken: the expected infestation may not actually occur, in which case the control measures would be wasted, or it will occur and then the spraying may either fail or succeed. Correspondingly, the lower branches indicate that if the decision maker chooses not to spray, the expected infestation still may or may not occur.

The end points of the decision tree show the total cost or loss, C, of each possible outcome, j, which is denoted as C_j below. If the forest is sprayed and no infestation occurs, the loss is the cost of the spraying, $35,000. This same cost is incurred if the infestation occurs and the spraying is successful. If the spraying is unsuccessful, however, the total cost will be the sum of the cost of spraying and the loss of timber value – that is, $35,000 + $84,244 = $119,244. If no spraying is done and infestation occurs, the cost will be the loss in timber value alone, $84,244.

The probability, P, of each uncertain event, j, occurring is indicated along each branch in Figure 3.1 and is denoted as P_j below. The probabilities for the possible events at each stage add up to 1.0. The probability of each final outcome, also indicated at the endpoints of the tree, is the product of the probabilities of the events leading to it. The expected outcome (the expected cost in this case) is simply a sum of all possible outcomes (C_j) multiplied by their probabilities (P_j) of occurrence. In general terms:

$$\text{Expected cost} = E(C) = \sum_{j=1}^{n} P_j C_j \tag{3.18}$$

In our case, the expected cost of spraying is

0.2(35,000) + 0.24 (119,224) + 0.56(35,000) = $55,219

and the expected cost of not spraying is

0.2(0) + 0.8 (84,244) = $67,395

Potential investors are sometimes interested in seeing the expected outcome and the variance of the outcome in bar graphs or probability histograms where probabilities of outcomes are plotted on the vertical axis and possible outcomes along the horizontal axis. Interested students can draw such a histogram for this spraying-or-not-spraying example.

The expected cost of not spraying exceeds the expected cost of spraying, so the criterion of minimizing expected losses (or maximizing expected values) suggests that the preferred choice is to spray.

Other criteria or decision rules can be adopted in recognition of decision makers' differing attitudes towards risk. One is the so-called "minimax" criterion, which is to choose the alternative for which the worst possible outcome is least unfavourable. In our example, spraying could result in a loss of $119,224 while the maximum loss without spraying is only $84,244, so the latter is preferred by the minimax rule. This rule is suitable only for investors who are highly averse to risk, whereas the previous criterion (maximizing expected values) is best used by investors who are risk-neutral. By focusing only on minimum possible returns and ignoring probabilities, the minimax criterion biases decisions against attractive projects even if they have only a small probability of poor returns.

A variety of other rules of thumb can be used to assist decision making in the face of uncertainty. None can be said to be the correct rule in all cases because the best choice is that which best reflects the decision maker's own attitude towards risk. It is important, however, for decision makers to select a specific decision rule so that all the relevant information can be brought to bear appropriately and decisions can be made consistently. Because uncertainty is inherent in forestry investment decisions, it is particularly important to account for it explicitly and consistently in evaluation procedures.

INTEGRATING ECONOMICS AND SILVICULTURE

Forest economics involves applying economic analysis to forest management operations and opportunities to determine the most advantageous courses of action. A forest economist must therefore understand both the techniques of economic analysis and the dynamics of forest growth and how it can be managed. Economic analysis of forest management does not always require sophisticated computer-based technology, as illustrated by the examples presented earlier in this chapter, but they all require a clear understanding of the basic concepts of economic analysis.

Modern forest managers depend heavily on forest growth-and-yield models to guide their management plans. These models are integrated sets of relationships, usually expressed as equations and embodied in computer programs that provide estimates of a forest's future condition, its growth, how it will respond to various silvicultural treatments, and how these will affect future harvests. By applying appropriate costs, prices, interest rates, and other economic data and analysis to these models, forest managers can evaluate alternative management regimes to guide their decision making. Commercial software such as *Padita3* and *Forman* provide models that integrate economics, forest growth-and-yield relationships, and silviculture, and are increasingly used in forest management. Even if forest managers are relatively certain about prices and quantity as well as interest rates, they need to avoid the common error in forest investment analysis of mismatching interest rate and prices.

For example, some forest managers conduct forest investment analysis using current timber prices and the historical rate of return in financial markets as the discount rate. A farmer, when considering converting his marginal farmland into forest plantations in South Alabama in the late 1990s, told one of the authors of this book that his required rate of return was 9%, as this was the historical rate of return in US stock markets that represented his opportunity cost of capital.

This historical rate of return of 9% includes, however, an average annual rate of inflation of around 3%, thus his real rate of return should be around 6%. Yet if one uses current timber prices as indicative of future

	Real	Nominal
	Price	
Real	✓	✗ (inflate the returns of forest projects)
Nominal	✗ (biased against forest projects)	✓

(left axis label: *Interest rate*)

Figure 3.2   Correct and incorrect match of interest rate and timber price in forest investment analysis.

timber prices in forest investment analysis, the current timber prices are real prices, because they do not have an inflation component. If one uses 9% as the interest rate and current prices to conduct forest investment analysis, the result would be biased against forest investment projects. As forest investment often involves many years, this kind of mismatch in interest rates and prices can have serious consequences.

Figure 3.2 presents the right match and wrong match in interest rate and revenue (price). In essence, one has to be consistent with either real interest rate and real price or nominal interest rate and nominal price.

If we discount a future nominal value, V_n, with a nominal interest rate, i, we will get the same present value as we would if we discount the real value (V_n^r) with a real interest rate (i_r). Equation 3.19 shows this identity:

$$V_0 = \frac{V_n}{(1+i)^n} = \frac{V_n^r(1+f)^n}{(1+i_r+f+i_rf)^n}$$

$$= \frac{V_n^r(1+f)^n}{(1+i_r)^n(1+f)^n} = \frac{V_n^r}{(1+i_r)^n} \tag{3.19}$$

where V_0 is present value and f is rate of inflation.

In the second segment of equation 3.19, both numerator and denominator are in nominal terms. In the third segment, the numerator in current value consists of a term of real value and an inflation component, and the nominal interest rate now has three components: a real interest rate, a rate of inflation, and a joint term of real interest rate and inflation (from equation 3.14). In the fourth segment, the term $(1 + f)^n$ in the numerator and denominator cancels out, expressing present value in real terms.

Equation 3.19 states that if one uses the same inflation rate in the revenue (cash flows) and the interest rate, she will get the same result as using real revenue and real interest rate. Note that, as was shown in equation 3.14, $i_r + f + i_r f$ is the nominal interest rate. Further, the term $i_r f$ is a small but important component, especially when the rate of inflation (f) is high.

As timber volume is always in real terms, the timber price and cost components determine whether the revenue term is in real or nominal terms. Should prices and costs in a particular project be in real or nominal terms, the interest rates should be adjusted accordingly.

Integrating economics, growth and yield, and silviculture can help us understand where the economic returns of forest investments are coming from. The returns to timberland, for example, come from timber production, bare land value appreciation, and non-timber values. For timber production, the return is determined by stumpage prices, timber growth (which includes natural growth and in-growth), and silvicultural expenditures, and by changes in prices and costs over time. When the forestland continues to be used in timber production, its value could change due to changes in stumpage price, silvicultural cost, and timber growth rate over time. Bare forestland value may increase if additional (often non-timber) uses, such as minerals or hunting leases, are found. The timberland returns reported in Table 3.3 include returns from these sources for a group of institutional timberland owners in the US. The returns for small private forest landowners in various parts of the country might deviate significantly from these results.

The next section presents a good example of the integration of silviculture with economics: the valuation of pre-merchantable timber.

VALUING PRE-MERCHANTABLE TIMBER

Pre-merchantable timber stands are stands composed of trees too immature to yield timber that can be profitably harvested and sold for manufacturing forest products. This section examines how the present value of such stands can be determined, either with or separately from the bare land value.

In Chapters 4 and 11, we discuss how mature timber is valued and priced. Conceptually, this is a straightforward calculation of the value of the timber that can be harvested and delivered to a market minus the logging and hauling costs. In other words, the value of mature standing timber (stumpage prices) is *residual* from the perspective of buyers. This computation of stumpage prices is the *residual valuation method*. Of course, actual stumpage prices are set by bargaining between buyers and sellers, and the sellers may set the price higher than those seen in comparable sales for non-timber values.

Evaluating pre-merchantable timber is more complicated, partly because competitive market transactions in immature timber stands are relatively rare, in many areas of North America at least. As a result, there is often little ready market information to guide buyers and sellers, so one needs to calculate or appraise the value of such stands. The term *valuation* describes the procedure for finding an investor's value of an asset, whereas *appraisal* is the procedure for finding its market value – the price it could be expected to fetch if offered for sale. These terms are sometimes used interchangeably, however.

Valuations or appraisals are not only useful in setting sales prices or determining the value of an investor's asset but are also needed for mortgages, taxes, bequests, and other purposes. Broadly speaking, there are four methods of valuing or appraising pre-merchantable timber stands.

Comparable Sales (Hedonic) Approach

The market value of an asset is the price for which the asset would sell in a competitive market. Appraisers often use comparable sales to appraise residential and commercial properties. Typically, when the market of an asset (such as residential houses) is competitive, the appraisers are able to find many comparable sales and can assign a value to every aspect (age, number of bedrooms, square footage, size of yard, type of neighbourhood, and location) of the asset (house). Most appraisers prefer to use comparable sales information to estimate the value of an asset when such information is available. The hedonic approach, also known as "hedonic regression" or "hedonic pricing model" in economics, is a method of estimating demand or value. It decomposes the good being researched into its constituent characteristics, and obtains estimates of the contributory value of each characteristic.

Pre-merchantable timber stands can be valued or appraised in a similar fashion if comparable sales exist. When using this valuation approach, one needs to control for characteristics of the timber stand such as age, species, planting density, seedling quality, growth conditions, and the characteristics of the land, such as location, size, site productivity (site index), and potential for higher and better uses. A major difficulty is that information about sales of pre-merchantable timber with comparable site and stand characteristics is often not available in the relevant market area, even though we have seen that some appraisers use as few as four to five comparable sales to appraise another pre-merchantable stand.

When prices and information about timber stand and site characteristics are available, one can use regression analysis to estimate the value of pre-merchantable timber stands. Here we present a preliminary study from southwestern Alabama, where some of the best southern pine timber stands are found and stumpage markets are competitive. Data used in the study, 77 sales of pre-merchantable southern pine stands in seven counties in southwestern Alabama from 2001 to 2007, were provided by a consulting forester.

Table 3.4 shows the descriptive statistics of the variables used. The average sale price per acre is $1,487 (in 2001 constant dollars), the average tract size of the pre-merchantable stands is about 260 acres, the site index is about 88 feet (measured in 25 years), the average stand age is about five years old, and some 60% of the stands have direct access to public or private roads. The consulting forester also provided his opinion regarding the potential for higher and better use of the stand (hunting and other recreational uses, home sites). Other information, such as species, stand density, quality of seedlings, and growing conditions, are not available.

Since the number of observations (77) is limited and some variables are missing, the regression model was not expected to have a very good

TABLE 3.4   Descriptive statistics for variables used in a hedonic study of pre-merchantable timber stands in southwestern Alabama

Variables	Mean	Standard deviation	Minimum	Maximum	Expected sign
Dependent variable: per-acre price of pre-merchantable stand (in 2001 $)	1,487.38	556.38	619.61	3,256.5	
Size (acres)	259.63	285.46	20.00	1,500.00	?
Site index	88.39	4.16	80.00	95.00	+
Age of stand (years)	5.05	5.16	0	15	+
Time of sale (2001 = 1; 2007 = 7)	4.56	1.67	1	7	+
Road frontage (1, 0)	0.60		0	1	+
Higher and better use (1, 0)	0.09		0	1	+

fit. Nonetheless, tests show that the model is not mis-specified. In the jargon of econometrics, there are no multicollinearity, auto-regression, or heteroskedasticity problems associated with the model that was run in the semi-log form.

Table 3.5 presents the regression results. Although the F-value shows that the overall model is significant, the goodness-of-fit for the model is only 29%. This means that the model explains only 29% of the variations in sale prices of pre-merchantable timber stands in these counties of southwestern Alabama. As for individual variables, the tract size, potential for higher and better use, stand age, and time of sale are statistically significant at the 10% level or better. Site index and access to road have the correct signs, but are not statistically significant. Lack of variation in the site index variable perhaps explains why it is insignificant. More data are needed in order to improve the level of fitness in this study.

In spite of its weakness, this preliminary study shows how a hedonic study is done. It also shows the difficulty in finding enough observations, and all the variables needed to run a regression.

Unlike the next three methods, which provide an estimate of the value of pre-merchantable timber only, this method often generates a value for both pre-merchantable timber and land because actual sale price data often include both. Thus, one needs to consider the characteristics of the land, such as size and potential for higher and better use, in addition to the timber stand characteristics listed above.

TABLE 3.5   Regression results using ln(price/acre) as the dependent variable

Variable	Coefficient (β_i)	Standard error	Predicted effect[a]
Size	0.0002*	0.0001	0.0002
Site index	0.0085	0.0086	0.0887
Age	0.0155**	0.0074	0.0156
Time of sale	0.0751**	0.0222	0.0780
Road frontage	0.0502	0.0772	0.0515
Higher and better use	0.2974**	0.1260	0.3466
Constant	5.9465**	0.7678	381.41
Adjusted R^2	0.2908		
F-value	4.78**		
Number of observations	77		

* $p > .1$, ** $p > .01$

a   Predicted effect = $\exp(\beta_i) - 1$.

Replacement Cost Approach

This approach generates an estimate of the cost of replacing an asset. For pre-merchantable timber stands, this is the cost of replacing the stands. Under this approach, all of the pre-merchantable timber stand's production costs are compounded to the stand's current age by applying the compound interest formula to site preparation, planting, and other costs incurred in establishing and maintaining the stand.

For example, assume that a southern yellow pine plantation was established for $250 per acre seven years ago, and that the owner's guiding rate of return was 8% in nominal terms. Using the compounded cost approach, the investment value of the seven-year-old stand is

$$\$250/\text{acre} \times 1.08^7 = \$428.46/\text{acre}$$

The compound interest formula used to derive the $428.46 is the same formula used by banks to determine account value. That is, if the $250 had earned 8% each year in an interest-bearing account, the account would have a balance of $428.46 after seven years. This $428.46/acre estimated stand value does not include the current value of the land, which can be estimated by comparable sales or by applying the compound interest formula to the cost of the land purchased seven years ago.

In general, the younger the stand, the more accurate the investment value estimate will be using compounded costs. This method, however, typically yields relatively low estimates of stand value because it underestimates the potential of the stand in producing a return to the landowner. In addition, the objective of most owners of forest plantations is not to grow and sell pre-merchantable timber but to grow and sell more valuable mature timber products.

One should not confuse this approach with the message contained in the old adage "Sunk costs do not matter." This approach (and other approaches) is used by investors to indicate the value of assets that they may want to buy, or by owners to ascertain the likely price they could get if they were to sell. In either case, sellers or buyers can hire an experienced appraiser (in our case, a forester) to provide an estimate of the market value of the asset. The appraised value is a piece of information that helps sellers or buyers decide whether to sell or buy the asset, and at what price.

On the other hand, *sunk costs* are previously incurred, irretrievable costs of a currently owned resource, and are considered by many economists to be irrelevant to current decisions. We often hear the admonition not to "cry over spilled milk." This is because spilled milk is a sunk cost, and consequences of earlier decisions (a mistake, in the case of milk spill) cannot be reversed.

An example will further illustrate these points. Suppose that a landowner in Texas spent $250 per acre to establish a 100-acre southern pine stand two years ago, expecting to harvest the stand in 30 years. Because the trees have not grown well, he has realized that it is unlikely to be profitable for him to continue with this plantation for another 28 years. He considers three options: do nothing (which means that he would not be able to recover his investment), sell the land and his two-year-old pine stand, or convert the property to pasture, which is likely to be profitable.

If he decides to sell, should his asking price for the pre-merchantable timber be dictated by the $25,000 that he spent on establishing the stand? Or if his interest rate were 8%, should his minimum acceptable price for the stand be $25,000(1 + 0.08)^2, or $29,160, which is the assessed value based on the replacement cost approach? Or should he continue with the tree-growing business because he has spent some money, even though he now knows that it is unlikely to be profitable, whereas to the best of his knowledge converting the land use to pasture will be, or is more likely to be, profitable?

The answer to all three questions is no. At least a portion of the $25,000 or $29,160 in costs that he has already incurred is now sunk cost and cannot be recaptured. Some of his previous work (such as tree planting) may have no value if the property is put to a use other than tree growing; in fact, some changes may actually reduce the property's value for other use. He must realize that some of the money he invested is gone forever. To decide what to do with the property and to determine the economically most advantageous use of it *now*, he must consider only the expected future benefits of each alternative and weigh them against the expected future costs of each. If the expected benefits of converting it to pasture outweigh the expected costs, he should convert it to pasture. On the other hand, if the trees grow well and he plans to stay in the tree-growing business for another 28 years, $29,160 is likely to be his minimum acceptable price if somebody offers to buy his two-year-old timber stand.

Income Approach

This approach produces an estimate of the value today of the asset's projected income. Under this approach, the value of pre-merchantable timber is equal to the timber's projected (future) revenues discounted to the stand's current age.

If the southern pine plantation in the previous example was expected to be harvested for sawtimber at age 30, what is the value of the stand at age 7 to an investor who requires a return on his investment of 8%? If we assume that the stand will yield 10,000 board feet of timber per acre (10 MBF/acre; pulpwood and chip-n-saw are ignored here for simplicity) at age 30, and if we assume a current price of $400/MBF and the price is expected to appreciate at 2% per year in real terms, the expected stumpage value is

$$(10 \text{ MBF/acre}) \times (\$400/\text{MBF})(1.02)^{30-7} = \$6,307.60/\text{acre}$$

Using the discounted revenue approach, the investment value of the seven-year-old stand is

$$\$6,307.60/1.08^{30-7} = \$6,307.60/1.08^{23} = \$1,074.30/\text{acre}$$

The formula above simply discounts the projected revenue of $6,307.60/acre for 23 years at 8%. Again, this method is similar to the way banks calculate the growth in savings accounts – if one places $1,074.30 in an 8% interest-bearing account, the account would have a balance of $6,307.60 after 23 years.

The mathematical formula for this approach is

$$\frac{P(t)Q(t)}{(1+i)^{t-x}} \tag{3.20}$$

where $P(t)$ is the stumpage price at mature age t, $Q(t)$ is the timber volume at mature age t, i is the discount rate, and x is the current age of the pre-merchantable timber.

In general, the older the stand, the more accurate the value estimate will be using discounted revenues. The discounted revenue method often yields higher estimates of the investment value of pre-merchantable timber than the replacement cost approach. As in the replacement cost approach, the cost of the money tied up in land is not accounted for in the calculation of discounted revenues in equation 3.20. Therefore, if one is valuing the timber and land (not just timber), one needs to add a "bare land value," which can be obtained from current sales of similar land in the area or by discounting future land value to the present. The formula for the latter is

$$\cdot \quad \frac{L(t)}{(1+i)^{t-x}} \qquad\qquad (3.21)$$

where $L(t)$ is the amount (future value) that the bare land can be sold for at year t, when the timber is sold and harvested. This formula represents the present value of bare land.

If other harvest options (such as other rotation ages, t) are considered, one could have several discounted values for a pre-merchantable stand. The value of the pre-merchantable timber to a particular investor would be the highest value obtained using the investor's expected rate of return as the interest rate.

For example, one can use the value of 15-year-old stands – just as the stands become commercially merchantable in producing pulpwood – to appraise 5-year-old loblolly pine stands. One may also use the value of 30-year-old stands that are ready for final harvest to appraise 5-year-old stands.

The difference in value between these two appraisals can be substantial. This is because after becoming merchantable at age 15, timber stands usually grow quickly in volume and appreciate even more quickly in unit value (dollars per thousand board feet or per ton, for example) as trees grow from pulpwood to chip-n-saw to predominantly sawtimber until they approach final harvest age. The combination of accelerated volume growth and unit-value appreciation can triple or even quadruple the value of a southern pine stand in the 15 years between ages 15 and 30. Thus, using 15- and 30-year-old stand values can give two quite different values for the same pre-merchantable stands.

Figure 3.3 shows this difference. Point A on the graph is the value of a 5-year-old pre-merchantable timber stand discounted from its value at age 15, when the stand is worth $1,500 per acre. Point B is the value of the same 5-year-old stand discounted from its value at age 30, when the stand is ready for final harvest and is worth $5,100. Using a 7% discount rate, the stand is worth $762 and $940 per acre, respectively.

Rational, profit-maximizing landowners would not sell (or appraise) their pre-merchantable timber stands for anything less than the net present value of future earnings from the best management plan they can come up with given their minimum acceptable rate of return. In our example, they will not sell their pre-merchantable timber stands at an appraised price based on age 15 when their stands have just become merchantable for pulpwood. Rather, it would be more appropriate to sell the stands at or more than the higher appraised price based on age 30, when the stands are ready for final harvest.

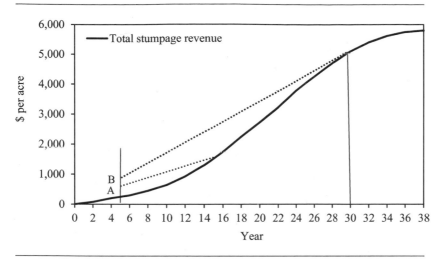

Figure 3.3   Value of a pre-merchantable loblolly pine timber stand: difference between discounting at age 30, when the stand is ready for a final harvest (valued at B), and at age 15, when the stand just becomes merchantable (valued at A).

The Internal Rate of Return Approach

In the previous subsections, we have noticed that the appraised values of the seven-year southern pine stand are quite different: $428.46 per acre under the replacement cost approach and $1,074.30 per acre under the income approach. The difference between these two appraised values reveals the inconsistency of these approaches in evaluating pre-merchantable stands, even using the same cost, revenue, and timing information. If either of these figures were accepted in a sale of the growing stock, it would bias the transaction in favour of one party over the other.

If the growing stock adds $428.46 per acre to the value of the land, the seller will receive 8% annually on a $250 planting investment, whereas the buyer, who invests the $428.46, will receive $6,307.60 twenty-three years later, for a significantly larger return of 12.4% per year.

On the other hand, if the growing stock adds $1,074.30 to the value of the land, the buyer will receive 8% on his growing stock investment, whereas the seller will receive a whopping 23.16% on the $250 planting cost ($1,074.30 after seven years).

In the context of real estate transactions, negotiations between the buyer and seller, perhaps assisted by a mediator, might lead to a settlement somewhere between $428.46 and $1,074.30 per acre, but such compromises will probably only reduce rather than eliminate the bias. This problem can be solved by evaluating the forest stand using the *internal rate of return approach.* If the growing stock in this case were evaluated at the investment's own internal rate of return of 11.36%, calculated as

$$\sqrt[30]{\frac{6307.6}{250}} - 1 = 11.36\%$$

then the compounded cost value (the replacement cost approach) and the discounted future income (income approach) will be identical ($530.95), and the bias will be completely eliminated. At this value, both buyer and seller will receive the 11.36% (the internal rate of return).

If one connects the value after the stand was just established ($250 per acre at age 0) and the value when the stand is mature ($6,307.60 per acre at age 30) using the inherent internal rate of return, the value of the pre-merchantable stand will follow a smooth exponential curve, increasing with stand age.

TAXES AND SOCIAL CONSIDERATIONS

Much of the methodology for evaluating investment opportunities has been developed to assist private investors whose concerns focus relatively narrowly on potential returns and associated risks. These returns are calculated without taxes or under the assumption that taxes have been subtracted. Investors who want to explicitly recognize the taxes associated with various investment opportunities would simply treat taxes as a cost item and calculate their before-tax and after-tax rates of return. Forest taxes and other charges are dealt with in Chapter 11.

Analyses of government investments must also deal with these concerns, but they often raise additional issues.

First, public investors are usually more broadly concerned with efficient use of resources generally, in contrast to private investors, who may be interested in maximizing returns to their land, capital, or other particular assets. Because of this broader interest of governments, the benefit/cost ratios are often considered more appropriate for establishing priorities for public investments than other criteria.

Second, the benefits to be taken into account must correspond to the decision maker's objectives, which may be more diverse for public investments. Effects on employment, impacts on long-term regional economic stability, consequences for other resource users, and changes in taxes and transfer payments, none of which bear on most private decisions, may figure importantly in public decision making.

Third, public investment projects often raise more complicated questions about whose benefits and costs are to be considered. Activities in one forest can generate benefits and costs for owners of adjacent land, or for people whose interest in the forest is indirect, such as those who depend on the watershed downstream. These are externalities that are often ignored by private decision makers but that must be included in any analysis that assumes the perspective of society as a whole.

The *reference group* problem draws attention to the importance of clearly defining the scope of the agent or institution whose costs and benefits are to be considered. It is especially important where governments of differing levels are concerned with the same project. For example, the benefits of a public forest project may accrue to local residents, while its costs are borne largely by taxpayers outside the area. An evaluation of the benefits and costs to the local community might reveal substantial net benefits, whereas an evaluation from the national viewpoint might indicate that costs exceed benefits. The appropriate reference group must therefore be defined for each governmental decision maker.

The principles outlined in this chapter for analyzing the relationships between costs and benefits over time are fundamental tools of forest economics, and they are employed throughout the rest of the book.

NOTES

1 Equation 3.6 can be obtained from this equation by multiplying all terms by $1/(1 + i)$, subtracting the new equation from the original one, and simplifying:

$$V_0 = \frac{a\left[\frac{1}{1+i}\right]}{1 - \frac{1}{1+i}} = \frac{a}{i}$$

2 Equation 3.7 is obtained by multiplying all terms in the equation by $1/(1 + i)$, subtracting the new equation from the original one, and simplifying:

$$V_0 = \frac{a\left[\frac{1}{1+i} - \frac{1}{(1+i)^{n+1}}\right]}{1 - \frac{1}{1+i}}$$

$$= \frac{a\left[1 - \frac{1}{(1+i)^n}\right]}{i}$$

$$= \frac{a[(1+i)^n - 1]}{i(1+i)^n}$$

3 The simplification involves multiplying all terms by $1/(1+i)^t$ and then subtracting the new formula from the original one to obtain

$$V_0 = \frac{V_t\left[\frac{1}{(1+i)^t}\right]}{1 - \frac{1}{(1+i)^t}}$$

APPENDIX: COMPOUNDING AND DISCOUNTING FORMULAS COMMONLY USED IN FORESTRY

Compounding and Discounting Single Values
- The amount to which a value will grow over a period

$$V_n = V_0(1 + i)^n \tag{3.1}$$

- The present value of a future amount

$$V_0 = \frac{V_n}{(1+i)^n} \tag{3.2}$$

Discounting an Annual Series of Values
- The present value of a perpetual annual series beginning in one year

$$V_0 = \frac{a}{i} \tag{3.6}$$

- The present value of a terminating annual series beginning in one year

$$V_0 = \frac{a[(1+i)^n - 1]}{i(1+i)^n} \tag{3.7}$$

Discounting a Periodic Series of Values

- The present value of a perpetual periodic series beginning in one period

$$V_0 = \frac{V_t}{(1+i)^t - 1} \qquad (3.9)$$

- The present value of a terminating periodic series beginning in one period

$$V_0 = \frac{V_t[(1+i)^n - 1]}{(1+i)^n[(1+i)^t - 1]}$$

Compounding a Series of Values

- The final value of an annual series beginning in one year

$$V_n = \frac{a[(1+i)^n - 1]}{i}$$

- The final values of a periodic series in one period

$$V_n = \frac{V_t[(1+i)^n - 1]}{(1+i)^t - 1}$$

Definition of Terms

i	annual rate of interest expressed as a decimal or a fraction
n	number of years of compounding or discounting
a	value recurring annually
t	number of years between periodic recurrences of V_t
V_t	value recurring periodically at intervals of t
V_0	present or initial value (at beginning of year 1)
V_n	future or final value (at end of year n)
annual	each year
periodic	at intervals of two or more years (t)

REVIEW QUESTIONS

1 Why is $100 received today worth more than $100 to be received five years hence? Given an interest rate of 10%, how do these two payments compare in value? By how much would the future receipt have to be increased to make it equivalent in value to the $100 today?

2   What is the present value of a forest park that generates recreational benefits of $250,000 per year, given an interest rate of 7%?

3   A developer has a standing offer to purchase a parcel of forestland for $1,800 per acre. The forest supports a 20-year-old crop that would yield $3,000 per acre if harvested in 10 years and a similar amount, net of all costs, every 30 years thereafter. If the forest owner's interest rate is 8%, is it advantageous to him to sell the land to the developer now, after harvesting the present crop in 10 years, or never?

4   A forestry graduate had an initial salary of $25,000 per year in 1995. In 2005, his salary increased to $40,000. In the 11 years, the consumer price index changed from 150 to 210. Was the forestry graduate worse off or better off in 2005 than in 1995?

5   Recalculate the optimal level of fertilization in Review Question 4 in Chapter 2, assuming that the timber values would be realized five years after application of the fertilizer, given an interest rate of 4%.

6   If you had a limited budget and you wanted to allocate it among a variety of potential silvicultural projects to maximize the economic benefits, what criterion would you use to establish priorities among projects?

7   What is the real rate of interest if the market or nominal rate is 12% and the inflation rate is 4%? Using equation 3.15, calculate the real rate of return for the NCREIF timberland index and the S&P 500 index.

8   Describe the general trade-off between risk and expected rate of return for a capital asset.

9   The actual management cost and estimated revenues (current market prices × future timber volume) of an 8-year-old pre-merchantable loblolly pine stand are presented in the following table. If there is no stumpage price appreciation in the next 22 years, what is the appraised value of this stand using the internal rate of return approach? Using a (real) interest rate of 5%, what is value of this stand using the replacement cost approach and the income approach?

Age of growing stock	Actual cost or estimated revenue ($/acre)
0 (establishment cost)	200
5 (herbaceous weed control)	60
15 (income from first commercial thinning)	160
22 (income from second commercial thinning)	220
30 (income from final harvest)	3,500

FURTHER READING

Bullard, Stephen H., and Thomas J. Straka. 2000. *Basic Concepts in Forest Valuation and Investment Analysis.* 2nd ed. Clemson, SC: Department of Forestry, Clemson University.

Davis, Lawrence S., K. Norman Johnson, Pete Bettinger, and Theodore E. Howard. 2005. *Forest Management: To Sustain Ecological, Economic, and Social Values.* 4th ed. New York: Waveland Press. Chapters 7-9.

Foster, B.B. 1986. Evaluating precommercial timber. *Forest Farmer* 46 (2): 20-21.

Gunter, John E., and Harry L. Haney, Jr. 1984. *Essentials of Forestry Investment Analysis.* Corvallis, OR: Oregon State University Bookstores.

Harou, Patrice A. 1982. Evaluation of forestry programs: The with-without analysis. *Canadian Journal of Forest Research* 14 (4): 506-11.

Mishan, E.J., and Euston Quah. 2007. *Cost-Benefit Analysis.* 5th ed. New York: Routledge. Part 5.

Pearce, David W. 1983. *Cost-Benefit Analysis.* 2nd ed. London: Macmillan. Chapters 5-8.

Zhang, Daowei, and Li Meng. 2009. *Valuing Forest Land and Pre-merchantable Timber Stands: A Hedonic Pricing Approach.* Working Paper. Auburn, AL: School of Forestry and Wildlife Sciences, Auburn University.

PART 2

The Forest Sector:
Land, Timber, and Unpriced Forest Values

Chapter 4

Timber Supply, Demand, and Pricing

Of the many products and services produced from forests, timber is
the dominant commercial product. Accordingly, much forest management effort is directed towards producing timber for industrial raw material. This chapter examines the economic forces that govern the value
of timber.

The simple answer to the question of what determines the value of
timber is the same as for other goods and services: supply and demand.
The market price of timber will adjust to the level at which supply
meets demand. The concepts of supply and demand are not as straightforward for timber as they are for many other products, however, and
so call for some special consideration. Thus we begin this chapter with
the general concepts of supply, demand, price equilibrium, and elasticity.
We then examine the special characteristics of forest products markets,
and methods for estimating the supply and demand for timber, logs, and
manufactured forest products. Finally, we distinguish between short-
term and long-term market adjustments, review the determinants of
timber value in local markets, and note possible price distortions in timber markets.

SUPPLY, DEMAND, AND PRICE EQUILIBRIUM

The Law of Supply
In Chapter 2, we noted that when the price of its product rises, a firm attempting to maximize its profit must expand its output, moving upward
along its marginal cost curve to maintain equality between its marginal
revenue and marginal cost. Thus, the competitive firm's supply curve

slopes upward, following its marginal cost curve. Since the market supply curve is the horizontal summation of the supply curves of all the producing firms, it, too, slopes upward.

This illustrates the *law of supply,* which states that, other things being equal, firms and industries will produce and offer to sell greater quantities of a product or service as the price of that product or service rises. The *quantity supplied* is directly related to its price: as price rises, the quantity supplied increases; as price falls, the quantity supplied decreases. The assumption of other things being equal ensures that the important relationship between price and quantity supplied is precisely measured. This relationship is a supply schedule or *supply curve,* which shows the quantities of any good that firms or industries will supply over a range of prices within a specified period of time. These "other things," including input costs and technology, determine the intercept, and govern the shape, of the supply curve.

Economists often use the term *ceteris paribus* to signify that all relevant variables except those being studied at that moment are held constant. This Latin phrase literally means "other things being equal," referring to a hypothetical situation in which some variables are assumed to be constant. In the real world, however, many things change at the same time. For this reason, when we use the term "supply" (or "demand") and use it to analyze events or polices, it is important to keep in mind what is being held fixed and what is not.

As this subsection focuses on supply, the variables assumed to be constant are all supply-side variables, and any change in price is due to a change in something that affects demand, such as a change in the income of consumers, a new substitute product, or a change in consumer tastes. Likewise, in the next subsection, which focuses on demand, the price changes are due to changes in things that affect supply, such as input prices, technologies, or a discovery of new raw materials. Natural or human-made events (such as labour strikes) that disrupt the production process are supply factors as well.

Note that whereas a change in price will cause a *change in quantity supplied,* indicated by a movement along the supply curve, a change in other supply factors that affect costs will cause a *change in supply,* indicated by a shift in the supply curve to the right or left.

One of these "other" factors in timber supply deserves special attention. In most industries, when the price of the product rises, producers can increase output without worrying about the effect this has on their output in future years. In a forest enterprise, however, timber harvested today will not be available in the near future. Admittedly, a new forest may begin to replace the stands harvested; nevertheless, timber harvested today almost always means less available for harvesting tomorrow *in the short term*. To determine whether it is most efficient to harvest more today and less tomorrow, *or vice versa,* the values that can be realized today must be compared with the value that could be realized if same resources were harvested sometime in the future. This is the problem of determining the optimal pattern of resource use over time.

Chapter 3 examined methods of solving this problem: the value generated from the resource will be increased by harvesting more in the present as long as its net value, at the margin, exceeds the value it would generate if harvested instead in the future period. The future value must be discounted at the owner's interest rate to indicate its equivalent present value, however. For example, if the owner's interest rate is 5%, timber harvested next year yielding $100 per cubic metre has a present value of $95 per cubic metre. This is the *user cost* of harvesting today, the sacrifice in future returns resulting from harvesting in the present. To realize the maximum value from the resource, harvesting should be postponed as long as the owner's net return from harvesting in the present, at the margin, falls short of this user cost. By thus balancing present and future returns at the margin, an owner can maximize the value realized from the resource over time.

The response of producers to a change in price may not always follow the orderly pattern implied by a smooth supply curve because much depends on their expectations about how long the new price will prevail. For example, if an increase in the price of timber is believed to be a temporary market fluctuation, producers will find it advantageous to harvest more today at the expense of future production. If the new price is expected to prevail indefinitely, however, future harvests will be more profitable also, raising the user cost of harvesting today and reducing the incentive to shift production towards the present. A higher price that is expected to be permanent will also make it attractive to invest more on silviculture and forest growth, to harvest stands that were previously beyond the margin of economic recoverability, and to recover more of the trees harvested. Thus, a price increase that is expected to be permanent will generate a variety of responses that will almost always have the net effect of increasing timber production in all periods.

The Law of Demand

We also noted in Chapter 2 that producers will demand more inputs if input prices fall. Thus, the demand curves for inputs in production slope downward to the right, as in Figure 4.1. This relationship is also true for

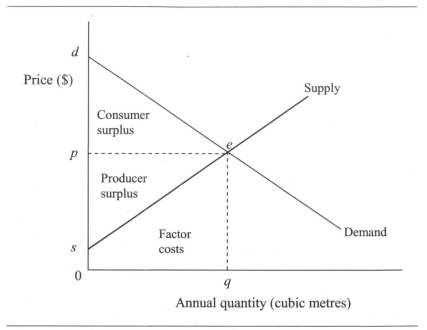

Figure 4.1   Market supply, demand, and net value of a forest product.

other goods and services: the quantity of goods and services that consumers are willing to purchase is greater when their prices are lower.

This illustrates the *law of demand:* other things being equal, the *quantity demanded* of any good or service is greater the lower its price. In
other words, the quantity demanded is inversely related to the price of
the good or service in question, as indicated by its downward-sloping
demand curve.

The other things assumed to be equal or constant in this relationship
include the income of buyers, the price of substitutes, and the preferences or tastes of buyers. Market demand also responds to price expectations. When prices are expected to rise, demanders tend to buy more
now, bidding up the current price. As in the case of supply, a change in
price causes a change in the *quantity demanded,* while a change in other
factors, including price expectations, causes the demand curve to shift.

Price Equilibrium
The interplay of supply and demand produces an *equilibrium price,*
where the quantity demanded is equal to the quantity supplied. The
interaction of the supply and demand for a forest product at a particular
time is summarized in Figure 4.1, where supply and demand determine

an equilibrium price, *p*, and a corresponding equilibrium annual rate of production, *q*. If the demand were to increase, shifting the demand curve upward, the shift to the new equilibrium would increase both the price and the quantity produced. If production costs were to fall, shifting the supply curve downward, the quantity produced would increase but the price would fall. In both cases, the results would be reversed if the change were in the opposite direction.

Net Benefit of Production

The total value of the product produced is the full amount that the demanders would be willing to pay for it. This is reflected in the total area under the demand curve to the left of point *e* in Figure 4.1, the area 0*deq*. As long as all production is sold at the equilibrium price, however, the total payments from buyers to suppliers is less than that; price multiplied by the quantity sold, represented by the rectangle 0*peq*. The supply schedule indicates that the cost of the factors of production that the suppliers need to produce this product is represented by the area 0*seq*. These *factor costs* account for only part of the suppliers' total receipts, however. The remainder, represented by the triangle *spe*, is *producer surplus*. The remaining area (area *dpe*) between the price line and the demand curve to the left of point *e* is *consumer surplus*, or the amount that demanders would be willing to pay in excess of the market price.

In terms of Figure 4.1, the *net* value of the forest product production is the gross value minus the supply costs, or the sum of producer and consumer surpluses, which is often called *net social benefit*. This is the measure of the net benefit, to society as a whole, of producing the product.

It is important to take account of producer and consumer surpluses when the market's entire production, or a significant portion of it, is at stake. Note, however, that changes in consumer surplus, producer surplus, and net social benefit due to shocks (such as weather events, policy change, changes in consumer tastes, technological advancement, and new advertising campaigns) that shift the supply or demand curve have a lot to do with the relative size of supply and demand elasticity, whose definition and importance will be illustrated in the following subsection.

Elasticity

The responsiveness of the quantity demanded (supplied) to a change in price is the *elasticity of demand (supply)*. It is measured at any point along a demand (supply) curve as an elasticity coefficient, E_d (E_s), calculated as the percentage change in quantity demanded (supplied) divided by the corresponding percentage change in price:

$$E_d \text{ (or } E_s) = \frac{dQ}{Q}\Big/\frac{dP}{P} = \frac{dQ}{dP}\cdot\frac{P}{Q} \qquad\qquad (4.1)$$

Thus, if a 5% increase in price results in a 5% decrease in quantity demanded, the (demand) elasticity coefficient has a value of 1 ($E_d = -0.05 \div 0.05 = -1$). Demand elasticity varies along the demand curve, and supply elasticity differs at different points along the supply curve. Thus, what is important and most relevant is the elasticity at the equilibrium price.

When the elasticity coefficient is less than 1, demand is said to be inelastic; when it is greater than 1, demand is said to be elastic. At the extreme, a coefficient of 0 implies no demand response to a price change, while an infinite value implies that a small increase in price would eliminate all demand for the product. The former is represented by a vertical demand curve and the latter by a horizontal demand curve. The elasticity of market demand for a good or service almost always lies between these extremes.

Corresponding measurements can be made of the *elasticity of supply*. Under perfect competition, the supply curve will be horizontal and the supply elasticity is infinite. In another extreme case, supply could be totally irresponsive to price. In most cases, however, the elasticity of supply is between these two extremes.

As elasticity measures the responsiveness of a function to changes in parameters in a relative way, it is a popular tool among empiricists because it is independent of units and thus simplifies data analysis and facilitates comparison of results across different analyses. Besides elasticity of supply and demand, commonly analyzed are income elasticity (the ratio of the percentage change in quantity demanded to the percentage change in income), cross-price elasticity (the ratio of the percentage change in quantity demanded for one good in response to the percentage change in price of another good), and elasticity of substitution (a measure of the ease with which the varying factor can be substituted for others; the ratio of proportionate change in the relative quantity demanded of two goods to the proportionate change in their relative prices). Later in this chapter, we will see the concept of timber inventory elasticity that is important in short-term timber supply in a region.

The concept of elasticity has a wide range of applications in economics. In particular, an understanding of elasticity is useful for understanding the dynamic response of supply and demand in a market, to achieve an intended result or avoid an unintended result. For example, a business considering a price increase might find that doing so lowers its profits if the demand for its product is highly elastic, as sales would fall sharply. Conversely, a business might find that a price cut does not increase sale revenues if the demand for the product is price-inelastic. This is an application of elasticity to products pricing.

Elasticity is also an important parameter in investment decisions and can serve as an indicator of industry health. Thus, a firm that is considering building a new plant in a country will want to know the income elasticity in that country for the goods it will be manufacturing. A good forest industry example is the rise and fall of the newsprint industry in North America in the last century. In the 1950s, the price elasticity of demand for local newspapers was very low (around −0.1). This highly inelastic demand was attributed to the lack of substitute news sources. Today this demand elasticity is much higher due to the availability of news source substitutes. In addition, there are many more things to do in one's spare time today than 60 years ago. All this leads to more competition in the local newspaper industry and to consolidation of the industry. This has translated into a significant decline in the demand for newsprint, causing the newsprint industry to shrink drastically in North America in recent decades. Not surprisingly, investors have turned away from the North American newsprint industry in favour of other investment opportunities.

With regard to public policy, elasticities can provide government officials with valuable information about the efficacy and distributive implications of laws and regulations. Tax incidence will be discussed in Chapter 11; for now, it is sufficient to note that the relative size or relative magnitude of supply and demand elasticity determines the distributive impact when supply or demand shifts because of taxes or other reasons – who gains, who loses, and by how much.

Suppose that supply is three times as elastic $(E_s = 0.6)$ as demand $(E_d = -0.2)$ at the equilibrium, a scenario that resembles the softwood lumber market in the United States. In such a situation, consumers will share more of the burden of an excise tax (which is a simple per-unit tax on the sale of a particular item – for example, a $5 tax per thousand board feet of lumber). In fact, consumers will share 75% (0.6/0.8, where 0.8 is the sum of the absolute value of supply and demand elasticity) of the tax burden while producers share the other 25%. If the relative elasticity is the other way around, consumers will share less of the tax burden.

The relative size of supply and demand elasticity also explains why there is a lack of private research and development in certain industries,

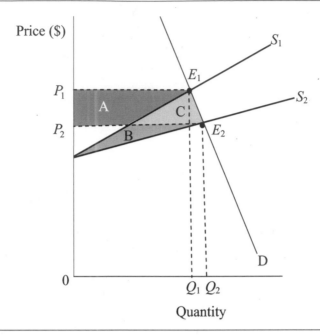

Figure 4.2   Relative elasticity and welfare change resulting from an increase in supply.

such as the softwood lumber industry in the United States. Because the elasticity of supply is greater than the elasticity of demand, the gain from more efficient production in these industries will go to consumers instead of producers. Private research and development often lead to an increase in production, but if increasing production is not in the best interests of an industry, the industry will not invest much in research and development. This conclusion is demonstrated in Figure 4.2.

Figure 4.2 shows that, with supply relatively more elastic than demand, an increase in supply (from S_1 to S_2) will cause a large decline in price and a decline in producer surplus (the net change in producer surplus is area B minus area A); thus, producers are worse off. Consumers, on the other hand, gain as supply increases (the gains in consumer surplus are represented by area A plus area C, the former being a welfare transfer from producers). The total net social benefit – the combination of producer and consumer surplus – increases unequivocally (by the amount of area B plus area C).

From a society's point of view, therefore, an increase in the supply of softwood lumber is desirable. From the producers' perspective, however, a decrease in supply is preferred, because their loss (area A) is more than their gain (area B).

Because demand is less elastic than supply, softwood lumber producers will have an incentive to restrict softwood lumber production and thereby raise its price. How can supply be reduced without triggering US antitrust law? One way is to refrain from investing in research and development in softwood lumber production, as noted earlier. Another way is to restrict foreign imports, thereby reducing the total supply of softwood lumber in US markets. Although trade restrictions do not shift domestic supply, they do lead to an increase in price and quantity supplied by domestic producers. Domestic producers will therefore favour trade restrictions whether the supply is more or less elastic than demand. If the domestic demand for a product is less elastic than its supply, they will have added incentive to restrict foreign imports and gouge domestic consumers. In fact, both strategies have been used by the US softwood lumber industry.

In summary, supply and demand elasticity for a given product largely characterizes the market for that product and determines the welfare gains and losses and their shares associated with shifts in supply and demand because of changes in public policy (such as taxes, tariffs, and environmental regulations), forest certification, research and development, technological advancement, the emergence of substitutes, rising incomes, and so on. These changes or shocks represent a change or displacement of equilibrium, and the associated welfare impacts can often be estimated by using various models, such as the *equilibrium displacement model.* Without going into the details of these models, it is sufficient to mention that one needs the elasticity of supply and demand and the magnitude of the shifts in supply or demand, or both, to use any one model.

We turn now to the forest products sector, its markets, and estimation of supply and demand for forest products, logs, and timber.

THE FOREST PRODUCTS SECTOR

The forest products sector includes industries that produce three groups of products: standing timber, logs, and manufactured products such as paper and paperboard, lumber, plywood, oriented strand board, fibreboard, and wood furniture. These products are traded in three main markets: the stumpage market, the log market, and the market for manufactured products. In the Standard Industrial Classification (SIC) system and the North American Industrial Classification System (NAICS) commonly used in the US and Canada (prior to 1997 and afterwards, respectively), the manufacture of these products is grouped into three industries: the wood products industry, the paper and allied industry, and (part of) the furniture industry. In NAICS, timber growing and logging are in the agriculture, forestry, and fishery sector. In SIC, logging was part of the wood products industry.

Stumpage means standing timber. In recent years, "stumpage" has been used synonymously with "stumpage prices," which reflect the value of standing timber. The stumpage market thus refers to the supply of, and the demand for, standing timber. As standing timber varies with species composition, products (sawtimber, pulpwood), product quality, accessibility, and location, the stumpage market has a strong spatial as well as temporal dimension.

Logs are intermediate products, and log markets bring together the suppliers and demanders of logs, usually delivered to mills, although active log markets are found in parts of the US Pacific Northwest and British Columbia. In the United States, forest products mills often procure timber in woodsheds and bid on timber directly. After winning the bids, they hire logging firms or wood dealers to do the logging and transportation. In recent years, mills have begun inviting bids for logs delivered to their mills and accepting the lowest prices offered. Finally, forest products firms use long-term timber supply contracts, direct timberland ownership, timberland leases, and (short-term) timber storage to supply timber to their mills. In most parts of Canada, stumpage and log markets are integrated as forest products companies obtain a large portion of their timber from public lands, where they hold harvesting rights in the form of licences and leases. These companies pay stumpage fees and typically hire logging firms to do the logging and transportation. In

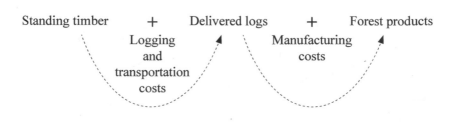

Standing timber   +   Delivered logs   +   Forest products
Logging
and
transportation
costs

Manufacturing
costs

Figure 4.3 Linkage among stumpage, log, and forest products markets.

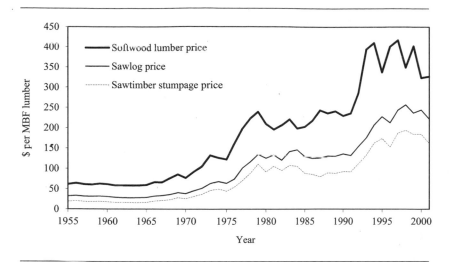

Figure 4.4   Prices for softwood lumber, sawlogs, and sawtimber stumpage in the southern United States, 1955-2001. MBF = thousand board feet. | *Source:* Li and Zhang 2006.

recent years, the amount of timber auctioned through competitive bidding in Canada has increased.

Figure 4.3 demonstrates the linkage of three markets – stumpage, logs, and forest products – in North America. Figure 4.4 shows the (nominal) price relationship among softwood lumber, sawlogs, and sawtimber stumpage in the southern United States from 1955 to 2001. Note that the unit for logs and stumpage is adjusted to the equivalent unit for softwood lumber (thousand board feet, MBF) on an annual basis. Thus, the three nominal prices at any given point in time are directly comparable. Not surprisingly, this figure shows that the three prices are highly correlated, since the three markets are linked.

Table 4.1 shows the contribution of the forest industry to gross state product in the US and to gross provincial product in Canada. The forest industry as a whole contributed about 1% to the gross domestic products in the US and 3% in Canada in 2005, but the industry's importance varies widely among states and provinces and over time. In most cases, as their economies grow and diversify, the value of forest production is expected to increase but the forest industry's share of total economic production gradually declines. This is evident in the recent history of economic development in every US state and Canadian province.

TABLE 4.1	Forest industry's contribution to gross state or provincial product, 2005		
State/province	Forest industry ($ millions)	Total ($ millions)	%
Alabama	4,912	150,582	3.3
Arizona	1,317	215,207	0.6
Arkansas	2,685	86,546	3.1
California	10,102	1,628,599	0.6
Colorado	847	212,582	0.4
Florida	4,061	670,030	0.6
Georgia	5,991	359,521	1.7
Idaho	972	46,584	2.1
Kentucky	2,231	138,592	1.6
Louisiana	2,329	183,022	1.3
Maine	1,824	44,451	4.1
Michigan	4,266	372,009	1.1
Minnesota	3,377	232,802	1.5
Mississippi	2,650	79,521	3.3
Montana	572	29,899	1.9
New York	29,33	956,378	0.3
North Carolina	5,244	348,397	1.5
Oregon	4,707	138,002	3.4
Pennsylvania	6,060	481,957	1.3
South Carolina	3,510	138,614	2.5
Tennessee	3,652	223,784	1.6
Texas	5,627	982,058	0.6
Virginia	3,334	350,897	1.0
Washington	4,541	272,734	1.7
West Virginia	772	52,932	1.5
Wisconsin	6,659	214,821	3.1
United States	120,658	12,339,002	1.0
Alberta	2,598	212,187	1.2
British Columbia	8,894	155,534	5.7
New Brunswick	784	22,514	3.5
Ontario	8,329	497,541	1.7
Quebec	9,571	252,710	3.8
Canada	34,697	1,158,680	3.0

Notes: Dollar amounts are in US and Canadian dollars, respectively. The forest industry in the US includes the following subindustries: forestry, wood products manufacturing, paper manufacturing, and part of the furniture manufacturing industry. As the gross state products (GSP) for forestry are not available (contained in forestry, fishing, and related activities), and furniture manufacturing includes non-wood products, we have assumed that 50% of these industries are forest industry. Thus, forest industry's contribution to GSP includes 50% of the GSP from forestry, fishing, and related activities, 50% of the GSP from the furniture industry, and 100% of the GSP from the wood products and paper manufacturing industry. The forest industry in Canada includes forestry and logging, and wood products, paper, and part (50%) of the furniture manufacturing. Again, half of the furniture manufacturing industry is assumed to be part of the forest industry.

Sources: Data for the US are from US Bureau of Economic Analysis, http://www.bea.gov/regional/gsp/. Data for Canada are from Statistics Canada, "Table 379-0025 – Gross domestic product (GDP) at basic prices, by North American Industry Classification System (NAICS) and province, annual (dollars)," CANSIM (database), 2009, http://cansim2.statcan.gc.ca/cgi-win/cnsmcgi.exe?Lang=E&CNSM-Fi=CII/CII_1-eng.htm.

FOREST PRODUCTS SUPPLY AND DEMAND

Forest products are diverse, and their supply and demand depend on general economic trends as well as product-specific factors. Softwood lumber production, consumption, and prices in the US, for example, are influenced by general economic growth and demographic changes, costs of various production factors, per capita income, housing starts, prices of lumber substitutes, and the amount of net softwood lumber imports. These factors, are either demand-side or supply-side variables.

There is a difference between estimating demand or supply functions as quantity dependent on price and other variables (shifters) on the one hand, and price dependent on quantity and other variables on the other. The former refers to estimating a traditional supply or demand function and is dealt with later under "Estimating Forest Products Supply" and "Estimating Forest Products Demand." The latter can mean two things. One is an inverse demand (or inverse supply) function, where price is a function of quantity and other demand (or supply) factors. Inverse demand or inverse supply is conceptually useful, but is rarely estimated empirically because estimating the traditional demand or supply function is much more intuitive, and converting a traditional demand or supply function to an inverse supply or inverse demand function is a simple matter. The other is a reduced form price equation with both demand and supply shifters on the right-hand side but no quantity variable, presented below in equation 4.7.

Estimating Forest Products Supply

To estimate the supply of a particular forest product, economists often rely on *Hotelling's lemma.* The first part of the lemma states that product supply is a function of output price and input costs:

$$Q = (P_{\text{output}}, C_1, \dots C_n) \tag{4.2}$$

where Q is the product supply, P_{output} is product price, and $C_1, \dots C_n$ are unit costs of various inputs. In the forest economics literature, inputs typically include wood (timber), capital (such as machinery and plants), labour, and energy.

Equation 4.2 is similar to all supply functions for various goods and services. In empirical estimation, the expected sign for the output price variable is positive while those for the cost variables are negative.

Since product price is also influenced by demand, the supply function is often estimated using the two-stage least squares (2SLS) or three-stage least squares (3SLS) technique. In 2SLS, the product price variable is regressed as a function of demand factors in the first stage and, based on the results of this regression, a forecasted price is used in the estimation of the supply function in the second stage. Because the forecasted price is determined by demand factors, the influence of demand on the product price is accounted for. 3SLS is merely an extension of 2SLS where several similar products (such as newsprint, printing and writing paper, tissue paper, and paperboard), which may be subject to the influence of common unknown factors, are jointly estimated.

Estimating Forest Products Demand

Two approaches are often used in estimating products demand. An *aggregated approach* estimates demand (actually apparent consumption, which is production plus imports and minus exports) as influenced by product price, per capita income, and prices of substitutes. Sometimes demand is estimated among a group of countries (or regions) so that a system of equations is used in the estimation process.

This approach has two shortcomings: (1) own-price elasticity and elasticity of substitution are assumed to be the same in all segments of the market, and (2) it may be difficult to select a common appropriate substitute since substitutes can vary across different segments of the market.

The *use-factor approach* estimates a use factor for each end-use sector, which helps identify and understand the demand for the product in each individual segment or end use. Associated with each end-use sector is an indicator of the level of end-use activity (the demand indicator) and a measure of the level of consumption per unit of market activity (the use factor). The estimated use factor is multiplied by the level of each end-use activity or demand indicator to provide an estimate of consumption within each sector.

The demand model for a particular product utilizes information on all end-use sectors. Aggregating all the sectoral end uses gives the total demand for that product in a region or country:

$$D = \sum_{j=1}^{n} UF_j \cdot DI_j \tag{4.3}$$

where D is total demand (domestic consumption), UF_j is the use factor in market segment j (such as new houses, renovation and remodelling, and industrial use in softwood lumber demand), and DI_j is the demand indicator in market segment. This approach allows for an individual demand curve (and elasticity) for each segment (j) of the domestic market.

Using the new house segment of softwood lumber demand as an example (j = new houses), UF_j is the amount of softwood lumber used per newly constructed house, estimated as a function of softwood lumber price, substitute prices, and the size (number of square feet) of houses. DI_j is the number of new houses built in a given year, often taken from estimates from government sources or estimated as a function of per capita or per-household income, interest rates, and perhaps demographic variables such as population between the ages of 20 and 60 (this segment of the population is likely to comprise the main demanders of new houses). The product of UF_j and DI_j is the demand of the jth segment of the softwood lumber market. A summation of all segments (new house, repair and remodelling, and industrial uses) will provide the total demand for softwood lumber.

Estimating Forest Products Price: A Reduced Form
The previous subsections describe how to estimate the supply of and demand for a forest product separately. These procedures will produce a supply function or demand function for a forest product of the kind that we often find in the economics literature. Since the equilibrium price (and quantity) is simultaneously determined by supply and demand factors, it is sometimes more intuitive and convenient to estimate either the price or quantity of a forest product as a function of both its supply and demand factors. Such a function is called the *reduced form* of market supply and demand, which can be used to analyze economic and policy impacts using the estimated coefficients.

For example, the price model for wood pellets (compressed wood particles used as fuel) in the southern United States and/or eastern Canada, two of the largest wood pellet–producing regions in the world, can be jointly derived from the supply and demand function:

$$WP_{supply} = f(P, \text{supply factors}) \tag{4.4}$$

$$WP_{demand} = f(P, \text{demand factors, policy variables}) \tag{4.5}$$

where WP_{demand} and WP_{supply} are the respective demand and supply of wood pellets in a region; P is the market price of wood pellets; supply factors include labour cost, energy cost, biomass or wood cost, and capital cost; and demand factors include national or per capita income and prices of substitutes such as coal.

Here we have included policy variables in the demand equation. This is because the demand for wood pellets in Europe, which consumed 62% of global wood pellets in 2009, and the US, which consumed most of the remaining production, is heavily influenced by government subsidies. The subsidies may reach $60 to $70 per metric ton of wood pellets, which is about 15% to 20% of the market price. These policy variables are used to account for the impacts of government subsidies.

Because price (P) and quantity are simultaneously determined through the interaction of the supply and demand equations, we can use a reduced form equation in econometric estimation. As the quantity supplied equals the quantity demanded, we have:

$$WP_{supply} \ (P, \text{supply factors}) = WP_{demand} \qquad\qquad (4.6)$$
$$(P, \text{demand factors, policy dummy variables})$$

Then solve for the price of wood pellets:

$$P = f(\text{demand factors, supply factors, policy variables}) \qquad (4.7)$$

In empirical estimation, the signs for the supply factors or costs of various inputs used in wood pellet production are expected to be negative. For demand factors, the signs for income and policy dummy variables are expected to be positive. Because a rise in the prices of substitutes will increase the demand for wood pellets, the sign for the price of substitute variables is expected to be positive.

The reduced form can be used in the estimation of timber price or quantity as well. In the next two sections, we turn to the concepts and empirical estimation of timber demand and timber supply, respectively.

THE DEMAND FOR TIMBER

The Derived Demand for Timber

The demand for timber is a *derived* demand, that is, it is derived from the market demand for final goods produced from wood. Final products are those wanted by consumers, and those that depend most heavily on wood are familiar: housing, newspapers, printing and writing paper, toilet tissue, wrapping paper, furniture, and so on. The consumer demand for these products creates a demand for the materials needed to produce them. In this sense, the demand for timber standing in the forest is derived from the demand for consumer products that can be made from wood.

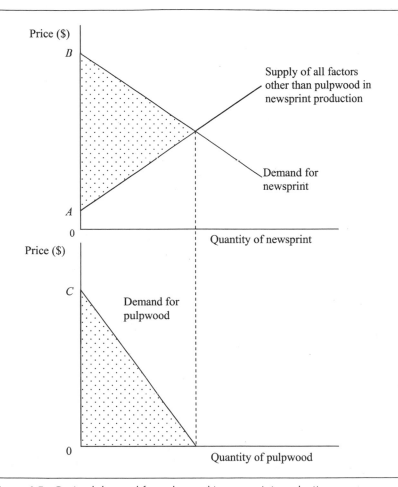

Figure 4.5   Derived demand for pulpwood in newsprint production.

The derived demand for *primary* resources like timber is usually generated through several steps in the production of *intermediate* products needed to produce *final* goods and services. For example, the demand for housing generates a demand for softwood lumber and plywood, which in turn generates a demand for sawlogs and veneer logs suitable for making lumber and veneer, and the demand for sawlogs and veneer logs creates a demand for suitable sawtimber. When price and quantity data are available for intermediate products, the supply of and demand for these intermediate products can be estimated and analyzed in connection with the supply of and demand for final goods and factor markets for these intermediate products.

The derived demand for (delivered) pulpwood, a type of timber used in newsprint production, is illustrated in Figure 4.5. The upper quadrant shows the demand for newsprint, an intermediate product that is used to produce newspapers and that uses pulpwood as a raw material. The supply curve represents the supply cost of all inputs in the production of newsprint except raw timber, including the cost of processing wood into newsprint. To the left of the intersection of the two curves, the excess of the amount that newsprint consumers, often newspaper publishers, are willing to pay for newsprint over the cost of all other factors represents the maximum amount that these producers would be willing to pay for pulpwood. Thus the derived demand for pulpwood, shown in the lower quadrant, reflects the difference between the two curves in the upper quadrant. At the intersection of the two curves in the upper quadrant, the derived demand for pulpwood is nil. Similarly, when these two curves intercept the vertical axis in the upper quadrant, the derived demand for timber is the difference of the two intercepts (A minus B). Consequently, $0C$ in the lower quadrant equals AB in the upper quadrant.

We have shown the derived demand for delivered pulpwood in newsprint production. Aggregating all the derived demand for pulpwood used in producing relevant goods, such as printing and writing paper, engineered wood products, and bioenergy products (such as wood pellets) as well as newsprint, generates a market demand for delivered pulpwood in a region.

What about the demand for pulpwood standing in forests (part of standing timber in forests)? Demand for standing pulpwood is derived from the demand for delivered pulpwood, the lower quadrant of Figure 4.5. Obviously, the cost of supply of all factors other than standing pulpwood in delivered pulpwood production is made up of logging and transportation costs. Thus, subtracting logging and transportation costs from the demand curve for delivered pulpwood gives the demand for standing pulpwood. Put another way, the derived demand for standing pulpwood can be estimated from the demand for delivered pulpwood in a manner similar to the estimation of the derived demand for delivered pulpwood from the demand for newsprint.

In empirical studies, the logging industry is often assumed to be competitive because there are few barriers to entry. Consequently, the short-run supply curve for all factors other than standing pulpwood in delivered pulpwood production – that is, logging and transportation costs – is sometimes assumed to be flat or very elastic. Over time, logging and transportation costs can change considerably due to changes in the price of labour, fuel, machinery, and other inputs. At any point in time, however, the difference between the price of delivered logs and the price of standing timber is accounted for by the prevailing costs of logging and transportation. For example, the difference in any year between the delivered sawlog price and the sawtimber stumpage price in Figure 4.4 is accounted for by the logging and transportation costs, representing the quantity supplied for all factors other than standing sawtimber in log production at the equilibrium point for that year.

Similarly, in lumber production, the difference between lumber price and delivered sawlog price in Figure 4.4 is the manufacturing cost for softwood lumber, representing the quantity supplied for all factors other than sawlogs in softwood lumber production at the equilibrium point of a given year.

In addition to pulpwood, timber stands often contain sawtimber, veneer timber, and pole timber, which are used to produce lumber, veneer and plywood, and poles. Consequently, there is a derived demand for each of these types of standing timber and for delivered sawlogs, veneer logs, and pole logs. The sum of all derived demands for each type of timber is the market demand for (standing) timber in a region. Residues and sawdust, by-products of the wood products industry, are often used as inputs for paper making or energy production. In recent years, many governments have required that a certain amount of recycled paper be used in paper production. This reduces the demand for raw timber.

Figure 4.5 makes clear that the derived demand for pulpwood, like the demand for the final products, follows the law of demand: at lower prices, more will be purchased per period. Thus, demand curves for timber slope downward in the usual fashion, as illustrated in Figure 4.1.

The supply curve for pulpwood slopes upward to the right in the usual fashion, and intersects with market demand to yield the market equilibrium price. This price, added to the cost of all the other factors in newsprint production (which is labelled as "supply of all factors other than pulpwood in newsprint production" in the upper quadrant of Figure 4.5), is the supply function for newsprint. Thus, the market equilibrium price for newsprint is located to the left of the intersection of the two curves in the upper quadrant of Figure 4.5.

For each of the major and/or intermediate products – lumber, plywood veneer, and logs (pulpwood) in this example – there is usually a distinct industry and market, with its own supply and demand relationships. In some cases, however, the wood industries are so integrated that there are no separate markets for intermediate products. Some firms that produce plywood, for example, manufacture it directly from the timber they themselves harvest, eliminating intermediate markets for logs and veneer.

Any change in the demand for final products made of wood will change the derived demand for timber. It is important to note, however, that this relationship is not rigid: a change in the demand for a particular final wood product cannot be expected to generate an identical and proportional shift in the demand for standing timber. There are several reasons for this.

First, a stand of timber is typically best utilized in producing a *mixture* of products, such as plywood, lumber, and pulp, in proportions that can be altered in response to changes in the relative prices of the products. For example, small-diameter "chip-n-saw" sawlogs can be used in making either lumber or pulp chips, and are redirected from one to the other when the relative value of lumber and pulp changes. Thus, if the demand for housing were to double, it would not follow that the demand for all standing timber would double, because only a portion of timber is manufactured into construction materials, and if the demand for other products did not change, the increase in the demand for timber would be less than proportional.

In addition, a doubling of the demand for housing would increase the demand for all of the inputs in housing construction and drive up their prices, but not proportionally. The result would be substitutions and changed proportions of labour, capital, and various kinds of materials used in construction, so the impact on the markets for inputs would vary.

In the long term, the relationship between the demand for timber and the demand for final wood products will change as a result of advances in technology. Consider, for example, the linkages between the demand for timber and the demand for ships, small boats, multiple-unit housing, offices, furniture, sports equipment such as skis and fishing rods, and heating fuel – all these products were formerly made largely from wood, but technological changes have almost eliminated wood as a major component in them. On the other hand, there are the familiar new products made of wood, such as fabrics, paper milk containers, drinking cups, various chemicals, and wood-based bioenergy products such as wood pellets. Moreover, the kinds of raw materials needed to manufacture wood products have changed significantly, often enabling utilization of previously wasted materials, such as the sawmill by-products and low-grade species now used for making pulp and hardboard. Over the long periods involved in producing timber and planning long-term timber supply, discussed in Chapters 8 and 9, such technological changes can dramatically alter the relationship between demands for final products and the input demand for timber.

Technological change can also alter the amount of timber needed to produce the same amount of finished products. For example, the softwood log-to-lumber recovery factor – the ratio between volume of logs consumed by a mill and the volume of its manufactured products – in the US South changed from 5.55 short tons per thousand board feet in the 1970s to 4.75 short tons per thousand board feet in 2002. In Figure 4.4, the prices of lumber, logs, and stumpage are shown to be highly correlated and have similar trends, but the ratio of each to the others changes considerably over the years and are not locked with a ratio of 1 to 1.

Estimating Demand for Delivered Logs
Using the concept of derived demand, we can now demonstrate two methods of estimating the demand (elasticity) of logs (and standing timber in next subsection). One is an econometric approach. The other is called the conversion factor approach.

Under the econometric approach, the supply and demand for logs are estimated jointly using a two-stage least squares (2SLS) or three-stage least squares (3SLS) technique. The demand for delivered logs, when derived from the other part of *Hotelling's lemma,* is a function of log price, final products price, and the prices of all other inputs (such as labour, capital, and energy) used in wood products or paper manufacturing. The demand for logs can also be derived from *Shephard's lemma,* which results in a conditional log demand as a function of all input prices, including the price of logs, and final products output. The supply of delivered logs is a function of log price, logging cost, and standing inventory.

The conversion factor approach uses the conversion factor from logs to final products and directly computes the demand elasticity of logs based on the demand elasticity of final products, and the prices of final products and logs, as well as an estimated processing margin coefficient (β). For example, the demand elasticity of softwood sawlogs is

$$E_{\log} = \frac{E_{\text{lumber}} \cdot P_{\log}}{(1 - \beta) P_{\text{lumber}}} \tag{4.8}$$

where E_{lumber} is the demand elasticity of softwood lumber, P_{lumber} is the price of softwood lumber, $P_{\log}$ is the price of softwood sawlogs, and β is a percentage price spread (also called processing margin coefficient) between logs and lumber.

This approach assumes that lumber is made from fixed proportions of inputs (logs and other inputs) and that the supply function of other inputs is fixed at a particular point in time. Therefore, the price spread (m) between lumber and sawlogs is considered as a combination of a constant amount (α) and a constant percentage (β):

$$m = \alpha + \beta P_{\text{lumber}} \tag{4.9}$$

Consequently, the price of lumber is

$$P_{\text{lumber}} = P_{\text{log}} + \alpha + \beta P_{\text{lumber}} \tag{4.10}$$

And, rearranging equation 4.10,

$$P_{\text{log}} = -\alpha + (1 - \beta) P_{\text{lumber}} \tag{4.11}$$

where α and β are estimated from this equation using historical data.

Equation 4.8 can be derived by differentiating equation 4.11 and using the identity of $dQ_{\text{lumber}}/Q_{\text{lumber}} = dQ_{\text{log}}/Q_{\text{log}}$. The latter is obtained from $Q_{\text{lumber}} = \delta Q_{\text{log}}$, where Q_{lumber} is the quantity of lumber produced, Q_{log} is the amount of sawlogs used, and δ is the conversion factor, which is the inverse of the log-to-lumber recovery factor noted earlier.[1]

Estimating Demand for Standing Timber

The demand for standing timber can be estimated in the same way as the demand for logs. When log markets are active and log prices and the elasticity of demand for logs are available for regions such as coastal Oregon and Washington or countries such as New Zealand, one can use the conversion factor to estimate the demand for standing timber from the demand for logs. If the log markets are not active or log prices and the elasticity of demand for logs are not available, forest economists often use the econometric approach to estimate the demand for standing timber. The demand for standing timber is estimated either as a function of timber price, final products price, and the prices of all other inputs used in wood products or paper manufacturing, or as a function of all input prices (including standing timber prices) and products output. In either case, demand for timber is simultaneously estimated along with supply of timber.

In summary, demand for standing timber can be depicted by a demand curve having the customary downward slope, implying that more
will be purchased per period at lower prices. This demand for timber is
generated by, and thus derived from, the demand for final products that
use wood in their production. The relationship between the demand for
final products and the demand for timber is determined by the relationships described in Chapter 2: the production functions relating inputs
with outputs, the substitutability of inputs in production processes, and
their relative costs.

TIMBER SUPPLY

The meaning of "timber supply" deserves attention because this term is
often used loosely and confusingly to refer to the forest inventory or to
the outlook for future harvests. Here we use it in the conventional economic sense to refer to the quantity of timber that will be supplied in a
market, per period of time, at various prices. Normally, as with other
products, the higher the price, the greater the quantity of timber that
will be supplied. In the case of timber, however, it is especially important
to be clear about the cause and effect and the time period referred to.

Short-Run Timber Supply

A *short-run* supply curve, such as that illustrated in Figure 4.6, indicates
how the quantity of a product supplied in a market will vary in response
to price within a period too short to alter the physical capital used in production, such as timber inventory, logging facilities, pulp mills, and sawmills. In the short run, the physical capital for timber supply consists of
standing timber inventory and capital investment in the logging industry.

In the short run, producers can alter output only by changing their
variable inputs (labour, fuel, and raw materials) while using their
existing capital facilities more or less intensively. To maximize their profits without being able to alter their fixed costs, producers will choose the
level of production at which their marginal revenue is just balanced by
their marginal variable costs, or short-run marginal costs.

The short-run market supply curve is thus the sum of the short-run
marginal cost curves of all the producers serving the market. Because
producers cannot adjust their physical capital within this short period,
short-run supply curves are relatively inelastic with respect to price. In
the case of timber, the physical capital consists of timber inventory.

Thus, the short-run timber supply curve in a region or country is
often characterized as

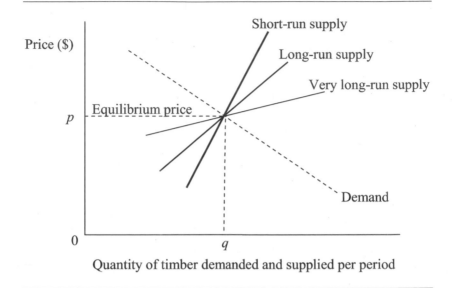

Figure 4.6   Timber demand and supply in the short and long run.

$$Q = (P, I, Z) \tag{4.12}$$

where Q is the annual harvest, P is the current (annual) price, I is the current inventory level, and Z is a vector of other factors (such as interest rate and ownership characteristics) that influence supply.

The supply function for a region is often estimated in conjunction with the short-run timber demand function. Most empirical studies show that stumpage supply elasticity ranges from 0.2 to 0.5 in the US South and 0.1 to 0.4 in the US Pacific Northwest and Canada.

Because timber inventory often changes slowly, it is sometimes not possible to obtain a usable statistical estimate of the inventory variable. In such cases, the inventory variable is omitted from the right-hand side and the supply variable is recast as the ratio of harvest to inventory. This specification implicitly constrains the (timber) inventory supply elasticity to be unity.

Consider a supply function specified as

$$Q/I = f(P, Z) \tag{4.13}$$

The inventory supply elasticity in this model is

$$E_I = \frac{dQ}{dI}\bigg/\frac{I}{Q} - f(P,Z) \cdot \frac{I}{Q} = 1$$

Empirically, the timber inventory supply elasticity is between 0.7 and 1.2.

The above model is an aggregate timber supply model and is often estimated for a region or several regions using many years' data. It is widely used in forest sector analysis.

Forests in many parts of the world are owned or controlled by individuals or families. "Non-industrial private forest (NIPF) owners," the term used in the United States and elsewhere, mainly refers to these individual and family ownerships of forests and forestlands, although it also covers non-forestry private corporations that own forests in some jurisdictions. In contrast, industrial forest owners are those who own both forestland and forest products manufacturing plants. Many studies of timber supply separate these ownerships, because, among other things, they differ in response to changing market conditions and public policy.

Timber supply models built for family-owned or controlled forests are often more detailed and more robust than the aggregate timber supply model for three reasons. First, because these private forest owners "consume" non-timber outputs, researchers have adopted a household production perspective on the timber supply problem. The household makes decisions as though it were maximizing utility derived from income, part of which comes from timber harvesting, and non-timber outputs or amenities. The household production approach lends a better theoretical foundation to the household-level timber supply.

Second, the data used in the household timber supply model often come from landowner surveys or government forest agency records of forest plots. These datasets contain disaggregated, micro-level timber harvesting and inventory data. Finally, because data are disaggregated, individual landowners' objectives and preferences can be modelled explicitly; by grouping landowners with similar objectives and preferences, the impacts of landowner preferences on timber supply and production decisions can be revealed.

The simplest NIPF or household timber supply model considers a risk-averse landowner who maximizes the present value of harvest revenue over two periods in the presence of uncertainty regarding future timber price. The landowner begins with an initial endowment of forest and chooses how much to harvest in the current and future periods. As such, this model is a traditional two-period (life-cycle) model, which has been widely applied elsewhere in economics.

Without getting into detail here, it is worth noting that this two-period model has been developed, in the last two decades, to cover joint production of timber and non-timber outputs (amenities) and extended to problems in which generations of landowners overlap through time. By providing a means of deriving testable hypotheses concerning preferences of NIPF landowners, it has been quite useful in motivating econometric studies of forest landowner behaviour. For more details on this model, interested readers may refer to the readings listed at the end of this chapter.

Long-Run Timber Supply

The dividing line between the short run and the long run depends on how long it takes to change the capital required in producing the product. This will inevitably vary depending on the product and the technical requirements for producing it. The long run might be less than a year for the logging industry compared with several years for the pulp and paper sector, because the capital equipment used in producing logs is relatively mobile and can be obtained and put in place more quickly than capital equipment required for manufacturing pulp and paper. The long run for timber could be a decade or more in North America but only a few years in tropical countries with fast-growing eucalyptus and other species. Whatever the period required for the capital involved, the industry will be able to respond to a change in price more flexibly in the long run than in the short run, and so the long-run supply curve is always more price-elastic, as Figure 4.6 indicates.

Figure 4.7 shows the dynamics of short- and long-run adjustments to an increase in demand for softwood lumber in the United States. Suppose the equilibrium price and quantity of softwood lumber in the US is where the supply and demand curves (S_{2000} and D_{2000}) meet, at P_{2000} and Q_{2000}, respectively, for 2000. As housing starts (construction) increase in the next five years, the demand for lumber increases to $D_{2001\text{-}05}$. For 2001, the resulting higher P_{2001} and Q_{2001} are shown in Figure 4.7 without a supply shift, since the number of sawmills and the total sawmill production capacity are fixed in the short run. By 2005, however, in response to higher price (and higher return to capital), more investment is made in sawmill

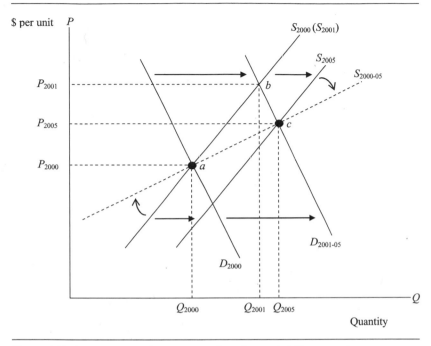

$ per unit P

$S_{2000}\,(S_{2001})$

S_{2005}

$S_{2000\text{-}05}$

P_{2001} .... b

P_{2005} c

P_{2000} .... a

$D_{2001\text{-}05}$

D_{2000}

Q

Q_{2000} Q_{2001} Q_{2005}

Quantity

Figure 4.7   Long-run supply response when demand shifts upward.

production capacity, shifting supply to S_{2005}. At P_{2005} and Q_{2005}, the price is now lower and the quantity higher than the initial (short-run) shock at P_{2001} and Q_{2001}. Thus, the long-run supply curve, connected by points a and c, is flatter than the short-run supply curve S_{2000} (and S_{2001}, which as noted above, is assumed to be identical to S_{2000}), connected by points a and b.

The distinction between short-run and long-run supply response is important in analyzing the impact of price changes in any industry that employs physical capital in production. In timber production, however, there is an even longer run to consider because it involves capital not only in the usual form of plant and machinery but also in the form of forests. Like other capital, the stock of forest capital can wear out through harvesting or be built up through investment. The difference is that the time required for forest management is usually much longer than the time it takes to build a factory.

If an increase in the price of standing timber is expected to be permanent or long-lasting, it will attract more land into industrial forestry, induce more intensive management, and, after many years, increase the market supply of timber. Timber production, however, often requires at least one complete crop cycle to adjust forest inventories to changes in timber prices, unless stand improvement programs are applied to existing forests and the volume growth of existing forests picks up sooner. In this *very long run,* the timber supply curve is even more elastic than in the customary long run. The long-run temporal dynamics of the forest sector, including timber, is presented in Chapter 9.

In summary, three types of supply responses to changes in the price of timber can be distinguished on the basis of the adjustment period to be considered and the range of inputs that producers can manipulate. In the *short run,* producers cannot change any capital, so they are limited to altering variable inputs only and constrained to use existing capital more or less intensively. Over the conventional *long run,* producers can adjust their mechanical production capacities so their response to price changes is more flexible. Over the *very long run,* long enough to grow more timber, producers can adjust all their inputs – variable inputs, physical plant and machinery, and forest capital – and so their supply response to a given price increase is even greater. Figure 4.6 illustrates each of these three supply responses.

Conventional supply curves of the kind illustrated in Figure 4.6 trace points of equilibrium, the quantity of the product that producers will seek to supply at various prices within the relevant period. Each point on the curve indicates the level towards which market supply will tend, but markets are more often in the process of adjustment towards equilibrium than being at equilibrium. The fluctuations we observe in production reflect the ongoing efforts of producers to adjust to changing market opportunities. Their responses are constrained, as we have seen, by the period of time within which they have to adjust.

Timber Supply and the Economic Inventory
Typically, some timber in a market supply region cannot be profitably recovered and used because the cost to harvest is greater than the worth of the recovered wood. This is usually the case in frontier regions with extensive natural forests. In other cases, timber may not be available due to government regulations such as mandatory best management practices (BMPs) or nongovernmental initiatives such as forest certification. An important determinant of the short- and long-run timber supply is the proportion of total forest inventory that can be expected to become available and be worth harvesting. Supply projections of the kind referred to above must therefore be based on estimates, or assumptions, about the economically recoverable portion of the total inventory.

The forest inventory in a market supply region consists of stands of timber that vary in terms of their economic value. On the one hand, differences in the technical characteristics of stands – such as species composition, size, and other qualities – yield varying market prices; on the one hand, their varying distance from the markets, terrain conditions, and so on result in differing production costs. Thus, the timber in each stand has a unique *net value* or *stumpage value* per cubic metre, measured by the difference between the value of the timber that can be recovered from the stand and the cost of harvesting it, including the cost of transporting it to mills.

A forest usually consists of many stands, covering a spectrum of net values. They can be arranged according to their net value to produce a curve like that shown in Figure 4.8. Each point on this curve indicates the quantity of timber, measured on the horizontal axis, recoverable from a stand at the net value per cubic metre equal to or greater than the corresponding point on the vertical axis, while all other variables – such as distance from the market and terrain conditions – are assumed not to be a problem for the timber to be recovered economically. The curve must slope downward, but its curvature and shape may take any irregular form, reflecting the structure and value of the forest inventory.

Figure 4.8 thus portrays the economic recoverability of a forest inventory. Ranking the stands from the left in order of their net value identifies the total volume that has a net value greater than zero. Timber with a zero net value is at the *extensive margin* of recovery. It is marginal insofar as the recoverable values will just cover the recovery costs. The portion of the inventory with a negative net value is beyond the extensive margin, whereas the portion with a positive net value comprises the economically recoverable inventory.

In the US, economic inventory is sometimes regarded as physical inventory on timberland, which is defined as forestland that is capable of producing 20 cubic feet of wood per acre per year (or 1.43 cubic metres per hectare per year) and has not been withdrawn from timber production for legal or administrative reasons. The economically recoverable inventory is a function of timber prices (and costs of recovery), however, and thus should be less than the physical inventory on timberland.

In some regions, economically recoverable inventory may be substantially less than the total physical inventory, which includes the uneconomic or submarginal portion.

As the extensive margin is determined by the balance between values and costs, the economically recoverable inventory is sensitive to any changes in these economic variables as well as to government regulations.

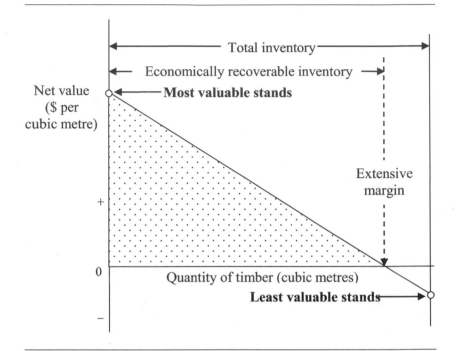

Figure 4.8   Relationship between net value of timber and economically recoverable inventory.

This is particularly important in making projections of timber supplies over long periods, where we need to estimate not only how much timber has a positive net value today but also how the extensive margin is likely to shift over time as a result of changes in prices, costs, and technology. Over the long periods common in forest yield planning, such changes can be substantial, placing a heavy burden on assumptions about future trends in costs and prices.

The shaded area in Figure 4.8 represents the total net value of the entire forest inventory, of which only a small portion is usually harvested in any year. Over time, the inventory is diminished through harvesting and natural losses, shifting the curve in Figure 4.8 to the left, while increase through growth shifts the curve to the right. Thus, the net effect on the harvestable inventory will depend on the balance between these reductions and additions, both of which are influenced by forest management decisions.

Timber Harvesting Sequences

The foregoing explains how the merchantable inventory of timber available to a particular timber market at any moment is constrained by its cost of recovery and its recovered value, or price. It now remains for us to examine the most economically efficient pattern of harvesting merchantable inventory.

To define the most efficient pattern of harvesting over time, we must consider three related issues: the *sequence* in which timber will be removed from the economic profile of the inventory, the *level* at which harvesting will initially take place, and the *change* in the level of harvesting over time. The objective, we assume, is to generate the highest possible return from the inventory, which, as explained in detail in Chapter 3, requires maximizing the present worth of harvests.

Initially, let us assume that the merchantable stock of timber is fixed (that is, it does not grow) and can be depicted in terms of its range of net value, as illustrated in Figure 4.8. We will also assume that costs and prices are not expected to change in the future.

The first question is the *sequence* in which timber of varying value should be harvested. As was shown in Chapter 3, the goal of maximizing the total value of a future stream of harvests requires that the greatest net values be discounted least. This means that the timber with the highest net value should be harvested first, depicted graphically by removing successive slices from the left-hand portion of the inventory in Figure 4.8 until the extensive margin is reached.

As harvesting progresses in this way, the net value of the timber removed declines because it brings lower prices, or it costs more to harvest, or both. Especially in developing forest regions, harvesting costs are likely to rise as operations advance into more remote areas. Progressively higher costs will shift the short-run supply curve upward in successive production periods.

The optimal *level* of harvests is determined by comparing present and future net returns, as suggested earlier. It is the level at which the marginal net revenue (the difference between the price received for the timber and its marginal recovery cost) is just equal to the marginal user cost of the harvested timber (the present worth of the net revenue that could be realized by harvesting the timber in a future period). In a perfectly competitive market, this level is achieved by expanding production until the short-run marginal cost plus the marginal user cost of production rises to the price of timber.

The final issue is *change* in the level of harvesting over time. Under our assumptions of a fixed resource stock and constant prices and costs, present worth maximization calls for a declining rate of harvest for two reasons. First, production must lean towards the present in order to equate marginal net revenue with marginal user cost in all periods. Harvests in the more distant future must, because of the effect of discounting, generate a higher net return at the margin in order to equate with present marginal net revenues, implying a lower rate of future production. (Note that this would apply even if the entire merchantable inventory was of uniform value.) Second, whenever depletion involves a progression into timber that is more costly to harvest, the short-run cost curve shifts upward, resulting in reduced market supply at any given price.

LONG-RUN WOOD SUPPLY PROJECTIONS

In much of the forestry literature, long-run timber supply refers to the quantity of wood that will become available over time, usually over many decades. This type of supply projection is depicted in Figure 4.9, which shows the volume of timber expected to be supplied per period – usually per year – over future years. Such projections may relate to the supply of timber in a region, the supply in a particular market, or the production from a particular forest.

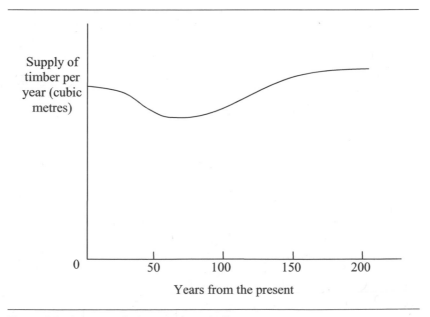

Figure 4.9   Long-run timber supply projection.

For example, in 2005 the Canadian Council of Forest Ministers provided wood supply projections for Canadian provincial Crown land over the next 50 to 100 years. In such projections, the trajectory of timber production over future decades is estimated on the basis of the age composition, rate of growth, and other characteristics of the forest inventory, often in the context of some harvest regulation policy designed to sustain yields.

It is important to recognize the difference between such supply projections and the economic concept of supply reflected in the supply curves discussed above as well as the timber supply projections based on the economic concept of supply. These long-run timber supply projections depict the quantity supplied as a function not of price but of time. The economic assumptions embedded in them are often unclear, so the results must be interpreted cautiously.

Market forces will generate a supply of timber over time according to such projections only if every point along the trajectory represents an equilibrium of supply and demand at the corresponding time. If the projection is not based on these market forces, or ignores likely trends in prices, costs, and technology that influence timber producers, the projected results can be realized only through governmental control. Indeed, projections of this kind are often based on artificial yield controls or policies (discussed in Chapter 8) and, in that context, they may represent only some regulated upper limit on timber production.

Most forests cannot be realistically characterized as consisting of a fixed merchantable stock of timber, as assumed previously. Over time, the inventory is reduced not only by harvesting but also by natural and often unpredictable losses such as forest fires, insect and disease outbreaks, and hurricanes. It also increases with forest growth, which can be enhanced by increasing silvicultural investment and technological advances such as genetically modified seedlings. Changes in prices and costs of timber, and in other forest values, will shift the extensive margins of production and lead to additions or withdrawals of forestland from timber production, as explored in more detail in Chapter 6.

Moreover, when growth is taken into account, it may not always be most efficient to harvest the most valuable timber first. For example, the timber with the highest net value may also be growing rapidly in value, so its high user cost means that its harvest should be postponed. As we shall see in Chapter 7, the optimum time to harvest a stand depends not only on its current value but also on its rate of growth in value and the productivity of the forest site it occupies.

When all these variables associated with actual forests are accounted for, the problem of determining the optimum pattern of harvesting (and thus timber supply) over time is considerably complicated. Recently, modern computer programs have facilitated the development and use of sophisticated dynamic models to enable detailed projections and assessments of forest growth and yield over time. Combined with economic information or estimates relating to costs and prices, these models can identify the pattern of harvesting that will best meet a specified objective. These projections of long-run timber supply are based on economic models of forest products supply and demand at a regional, national, or global scale. They also have projections for prices, and their detailed trajectory can be quite different from that shown in Figure 4.9.

DETERMINANTS OF STUMPAGE PRICES

At the beginning of this chapter, we noted that the determining factors for the value of timber are the same as for other goods and services, namely, supply and demand. We have heard non-industrial private forest landowners questioning why the stumpage prices they received for their timber were much less than what they had expected several years ago, and less than what their neighbouring landowners had just received for their timber sold roughly at the same time. Indeed, if one looks at the stumpage prices for the same species and timber sizes reported by various sources, one will notice that stumpage prices differ in different parts of a state and vary among states in the United States, sometimes very substantially.

The first part of the landowners' question relates to changes in supply and demand over time (mostly demand in the short run), while the second part relates to variations in supply and demand (mostly supply) in space (or different sites). In this section, we focus on the second part of the question: the spatial variations in the stumpage prices at any given time. In this case, demand is set at any given location, and all price variations can be attributed to supply factors. In Chapter 9, we present the long-term dynamics of supply and demand and the ensuing changes in timber prices in the US.

Foresters know that timber stands are often heterogeneous across the landscape. They may differ widely in species composition; size, quality, and density of timber; terrain; and accessibility of the standing timber. The locations of these stands, or their distances to market or mills, vary. Even though the amount of timber in a given stand at a given time is fixed, a timber stand is not like a standardized good such as TV sets that we buy at stores. When the demand is set at a given location and time and there are no price distortions in the local timber market (see next section), stumpage prices received by landowners can still vary because the inherent supply factors associated with their timber stands differ. Again, these determinant factors are:

- species composition
- size and quality of timber (products composition)
- density of standing timber (which affects logging costs)
- terrain, accessibility, and size of the tract (which affects logging costs)
- location (distance to market or mills).

Landowners may also require loggers to comply with mandatory or voluntary regulations (such as best forest management practices) and stipulate additional conditions for timber harvesting when they sell their timber. This can affect the bidding prices they receive. Finally, empirical studies have shown that sealed-bidding timber sales may get landowners better prices than negotiated sales, and often the bidding price spread narrows when the number of bidders for their timber increases.

PRICE DISTORTIONS
In any timber market, the interaction of supply and demand generates an equilibrium price. Both supply and demand change more or less continuously, however, causing the shifts and trends in prices observed in the markets for timber. Suppliers and demanders are constantly adjusting, with lagged responses, to these changes. As a result, the equilibrium or market-clearing price of timber in a particular market at a particular moment is the price towards which the market is adjusting.

Timber markets are seldom perfectly competitive, and various imperfections impede the interplay of supply and demand. Monopolies and oligopolies may prevail in local timber markets, distorting costs and prices, artificially restricting timber supplies, and maintaining prices above marginal revenue and cost. Geographic monopsonies are also common, where suppliers face only one buyer and both price and supply are depressed. And wherever firms manufacture into forest products all the timber they harvest, there is no market at all for unmanufactured timber.

Governments also influence market supply, demand, and prices through taxes, royalties, subsidies, and other fiscal or regulatory policies. Sometimes these interventions are aimed at correcting market failures or redistributing income. Moreover, governmental agencies such as forest practices acts are often deeply involved in managing forest inventories directly, especially on public lands. They sometimes require that the timber inventory be harvested in a sequence governed by the age of its constituent stands. Examples are the "oldest first" and "old-growth first" rules applied to some public forests. More common are sustained yield policies that prescribe more or less constant rates of harvesting, as examined in Chapter 8. Such harvesting regimes deviate from the economically most efficient pattern because they take no direct account of interest rates, costs, and prices.

Whenever obstacles to competition or other market imperfections exist, the market price of timber is likely to differ from its social value, or marginal social benefit, as defined in Chapter 2. Moreover, the prices actually paid for timber may differ from those that would have resulted from purely competitive market forces as a result of barriers to competition of the kind noted in Chapter 2. Public timber, which accounts for a major share of the total supply in many regions of the world, is often made available to private users through licensing arrangements at prices that are not determined by market competition. The payments take a variety of forms, including rentals, licence fees, royalties, special taxes, stumpage fees, and other levies. These matters, and the distinction between the economic value of timber and the price actually paid, are examined in Chapter 11.

NOTE

1 The conversion factor from logs to lumber (δ) is defined as

$$\delta = \frac{Q_{\text{lumber}}}{Q_{\text{log}}} \tag{4.11a}$$

This implies that

$$dQ_{\text{log}} = \frac{dQ_{\text{lumber}}}{\delta} \tag{4.11b}$$

Differentiating equation 4.11 in the main part of this chapter yields

$$dP_{\text{log}} = (1 - \beta)dP_{\text{lumber}} \tag{4.11c}$$

Dividing equation 4.11b by equation 4.11c yields

$$\frac{dQ_{\log}}{dP_{\log}} = \frac{dQ_{\text{lumber}}}{\delta(1-\beta)dP_{\text{lumber}}} \qquad (4.11d)$$

Using equation 4.11a, equation 4.11d can be changed to

$$\frac{dQ_{\log}}{dP_{\log} \cdot Q_{\log}} = \frac{dQ_{\text{lumber}}}{(1-\beta)dP_{\text{lumber}} \cdot Q_{\text{lumber}}} \qquad (4.11e)$$

Finally, multiplying $P_{\log}$ by the numerator and denominator of the left-hand side of equation 4.11e, and multiplying P_{lumber} by the numerator and denominator of the right-hand side of equation 4.11e, and using the definition of elasticity, one can get equation 4.8.

REVIEW QUESTIONS

1 Why is demand (and supply) elasticity important to investors, consumers, and policy makers? In what sense is the demand for timber a "derived demand"? Describe the connection, if any, between the demand for newspapers in New York and the demand for timber in Quebec or South Carolina.

2 Why does the full market supply response to a change in demand take longer for timber than for most other products?

3 Use a diagram of supply and demand curves to illustrate how improvements in manufacture of artificial Christmas trees will affect the demand, price, and sales of natural Christmas trees.

4 The diagram below illustrates how total logging costs and total revenue from the harvest increase with the amount of timber recovered

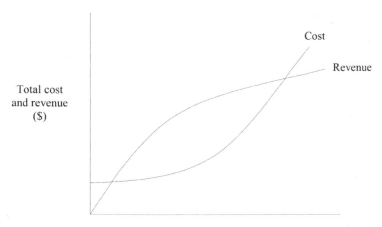

Volume of timber recovered per hectare (m³/ha)

from a hectare of forest, assuming that the harvest always consists of the most valuable timber. Identify in this diagram (a) the fixed costs of logging, and (b) the break-even levels of utilization. Draw the corresponding marginal cost and revenue curves to show the most profitable level of utilization, or the intensive margin of recovery. How would an increase in the price of timber affect this intensive margin?

5   What is the difference between a long-term projection of timber production for a region and a conventional market supply curve for timber?

6   What is the "extensive margin" of timber production? Give examples of changes in prices and costs that would shift the extensive margin.

FURTHER READING

Amacher, Gregory S., Markku Ollikainen, and Erkki Koskela. 2009. *Economics of Forest Resources.* Cambridge, MA: MIT Press. Chapter 4.

Binkley, Clark S. 1988. Economic models of timber supply. In *The Global Forest Sector: An Analytical Perspective,* edited by M. Kallio, D. Dykstra, and C. Binkley. Chapter 6. London: John Wiley.

Canadian Council of Forest Ministers. 2005. *Wood Supply in Canada, 2005 Report.* http://www.postcom.org/eco/sls.docs/Can%20Forest%20Ministers-2005%20Wood%20Supply.pdf.

Duerr, William A. 1960. *Fundamentals of Forestry Economics.* New York: McGraw-Hill. Part 3.

–. 1988. *Forestry Economics as Problem Solving.* Blacksburg, VA: Author. Parts 2 and 5.

Gregory, G. Robinson. 1987. *Resource Economics for Foresters.* New York: John Wiley and Sons. Chapters 9 and 10.

Haynes, Richard W. 1977. A derived demand approach to estimating the linkage between stumpage and lumber markets. *Forest Science* 23: 281-88.

Li, Yanshu, and Daowei Zhang. 2006. Incidence of the 1996 US-Canada Softwood Lumber Agreement among landowners, loggers, and lumber manufacturers in the US South. *Forest Science* 52 (4): 422-31.

Max, Wendy, and Dale E. Lehman. 1988. A behavioral model of timber supply. *Journal of Environmental Economics and Management* 15 (1): 71-86.

Nautiyal, Jagdish C. 1988. *Forest Economics: Principles and Applications.* Toronto: Canadian Scholars Press. Chapters 4 and 5.

Newman, David H. 1987. An econometric analysis of southern softwood stumpage markets: 1950-1980. *Forest Science* 33 (4): 932-45.

Rideout, Douglas B., and Hayley Hesseln. 2001. *Principles of Forest and Environmental Economics.* 2nd ed. Fort Collins, CO: Resource and Environmental Management, LLC. Chapters 3-6.

Sedjo, Roger A., and Kenneth S. Lyon. 1996. Timber Supply Model 96: A Global Timber Supply Model with a Pulpwood Component. Report prepared for Resources for the Future. Discussion Paper 96-15. http://www.rff.org/documents/RFF-DP-96-15.pdf.

Stier, Jeffery C., and David N. Bengston. 1992. Technical change in the North American forestry sector: A review. *Forest Science* 38 (1): 134-59.

Chapter 5   Unpriced Forest Values

Chapter 4 considered commercial timber as the product of forest management, but a wide variety of other products and services are produced from forests, ranging from livestock forage and water to recreational, aesthetic, and environmental amenities. The importance of these non-timber benefits varies widely; in some forests they are insignificant, whereas in others one or more of them constitute the dominant value. Almost everywhere, the demand for non-timber benefits is growing, their value is rising accordingly, and they are becoming increasingly influential in forest management.

To provide appropriately for these other forest products and services, forest managers and planners must take account of their values and costs of production. Frequently, more of one can be produced only with some sacrifice of another, or of commercial timber production, implying a need for careful compromises in order to maximize the total value generated by a forest. We discuss this problem of trade-offs and the economics of multiple use in the next chapter. First, however, we must deal with the methods for evaluating each type of benefit, in order to identify efficient levels of production and investment in them. This requires measuring different types of benefit consistently, including benefits that are not priced or sold in markets.

UNPRICED VALUES: A PROBLEM OF MEASUREMENT

Some non-timber forest products and services are priced and marketed much like industrial timber. For example, livestock forage is sometimes sold to ranchers by means of grazing rights issued at prices determined by competition. In such cases, there is no special problem in assessing the value of the benefits other than accounting for market imperfections that may cause prices and social values to diverge.

Other forest values are made available to users without charge. These unmarketed benefits include services such as outdoor recreation, amenity, and flood control; and products such as game, wild berries, and fuel wood. The range of forest products and services available to people without charge, in contrast to those that are priced and sold, varies greatly among countries and regions, partly because of their varying forest conditions and partly because of their differing systems of property rights. In almost all cases, however, forest managers must be concerned with a mixture of marketed and unmarketed products.

There are two main reasons why some forest benefits are not priced and sold, one technical and one political. The technical reason is that certain forest values are difficult to price and market in the usual way. As noted in Chapter 2, the aesthetic value of a forest landscape, for example, would be difficult to parcel up and sell to individual consumers, and it would be difficult to exclude those who were unwilling to pay for it. In any event, it would not be desirable to price it. A view of a forest landscape is thus an example of a true *public good* in the sense that its consumption by one consumer does not reduce its availability to others; thus, it involves no user cost, so any positive price would inefficiently ration its consumption and reduce the value it generates. Other public goods are associated with the contribution of forests to the quality of the natural environment, such as the quality of air and water, and carbon sequestration.

The political reason is that some forest products and services are not marketed because of public choice. For example, in contrast to the view of a forest landscape, access to recreational areas and campgrounds, though often provided free, present no technical obstacle to pricing. Indeed, they are often priced by private owners and sometimes by governments as well. In North America, governments often charge a licence fee for a general privilege to hunt or fish, but the licence fee is usually unrelated to any specific resources consumed, or even the amount taken. It is typically a nominal administrative fee rather than a market-determined price. The use of resources is often rationed not by market pricing but by regulatory measures such as bag limits and closed seasons.

Governments resort to regulatory measures rather than market pricing to control demands on public parks, forests, and recreational resources because of pressure from voters who argue that they have a right to free use of the public domain, that they pay for such use through taxes, and that prices are unfair to people with low incomes, among other things. Whatever the reason, the absence of prices to indicate the value of these resources to consumers raises the problem of measuring their value by other means.

The fact that they are not priced does not mean that these forest products and services are valueless, of course – only that there are no market indicators of their value. The issue in this chapter is how to estimate the value of a forest product or service where users cannot express their evaluation of it by paying a price. In the absence of the usual market indicators of value, we must resort to indirect evidence about the demand for the product or service. As long as the full benefits accrue to individual consumers, the problem is one of finding and analyzing other information in order to estimate how much consumers would be willing to pay for the good or service even though they are not required to pay for it.

The economics of producing any good or service call for attention to both the benefit (value) of the product and the costs of producing it. The costs of producing unpriced goods and services are usually not very different in kind from those associated with the production of commercial products. In both cases, the costs are incurred in the expenditures for labour, land, and capital needed to manage the forest for particular purposes. The difficulty lies benefit measurement. The following sections concentrate on the special problem of evaluating benefits wherever they are not priced.

CONSUMER SURPLUS AS A MEASURE OF VALUE

The value or utility that consumers gain from a good or service is reflected in their willingness to pay for it. Their willingness to pay indicates their willingness to give up other things, or income, in order to obtain that particular good. Thus, willingness to pay is a measure of their relative evaluation of the good in monetary terms.

The willingness of consumers to pay for some articles in a particular market is indicated by a market demand curve of the kind described in Chapter 4, so the problem of quantifying the value of a good or service focuses our attention on its demand curve.

Figure 5.1 shows a typical downward-sloping demand curve, dd'. If the product were sold in a market at a price p, the quantity purchased would be q and the total amount paid by purchasers would be the price multiplied by the quantity consumed, represented by the rectangle $0pp'q$. The value that consumers gain from consuming this good is equal

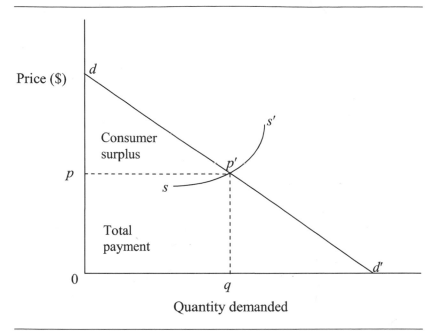

Figure 5.1   Market demand and consumer surplus.

to the price only at the margin, however. The triangle *pdp'* indicates that some consumers would be willing to pay more than the price they actually pay, *p*. As noted in Chapter 4, this amount – the amount that consumers would be willing to pay in excess of the amount that they actually pay – is referred to as *consumer surplus*. It is obviously part of the total value that consumers gain from the good or service.

If a product or service is available at zero price, then *all* the value accruing to consumers is in the form of consumer surplus. In Figure 5.1, if the price were zero, consumption would increase to *d'*, the value gained by consumers would be the entire area under the demand curve 0*dd'*, and all this value would be in the form of consumer surplus. So, to establish the value of a good or service provided without charge, we must estimate the consumer surplus it generates, which in turn requires some way of defining its market demand curve and measuring the area under it.

As long as the product or service is sold to all consumers at the same price, at least part of the total benefit accruing to them will be in the form of consumer surplus. Sellers could eliminate consumer surplus altogether only if they could charge each buyer a different price, extracting from each the maximum amount that buyer was willing to pay for each unit purchased. Sellers rarely have the opportunity to discriminate among buyers, however, so they cannot capture all the benefits in sales revenues in this way.

The existence of consumer surplus obviously depends on a downward-sloping demand curve. If the demand were perfectly elastic, so that the demand curve was horizontal at price level p in Figure 5.1, there would be no consumer surplus. This is important, because it means that the need to estimate consumer surplus arises only when a significant change in the quantity of a good or service, or in its price, is being considered. If the quantity is marginal, the change in consumer surplus can be ignored. For example, if we want to estimate the value of all the product consumed in a market, q, in Figure 5.1, the consumer surplus pdp' is a significant component of that value. On the other hand, if the task is to estimate the value of a marginal increment to the supply of a good or service, such as that provided by one of many producers or that supplied from one of many tracts of land, the change is not likely to alter much the equilibrium price or the consumer surplus.

The latter is sometimes the case facing forest managers. They often want to know the benefit of providing for more recreation or some other value in a particular forest when this would add a relatively small increment to the total supply available to the relevant consumers. In such cases, the potential change in consumer surplus may be negligible. The larger the increment of supply relative to the total, however, or the more unique and distinctive the character of the good or service, or the less elastic the demand curve, the more the change in consumer surplus needs to be considered.

Other cases may involve a significant change in either the supply or demand, or both, of the unpriced good or service, sufficient to cause a change in the equilibrium price and the consumer surplus.

To the extent that changes in consumer surplus or producer surplus are significant, the equilibrium displacement model, noted in Chapter 4, can be used to evaluate the welfare change for consumers and suppliers, respectively. Sometimes an all-or-nothing decision must be made about the preservation of a unique natural feature or wilderness area. In these cases, the total demand for the site, including the consumer surplus, must be evaluated.

EVALUATING UNPRICED RECREATION

Let's consider a particular type of unpriced benefit that is increasingly important in forest management: unpriced outdoor recreation. More specifically, let's consider the recreational value of a particular *site,* which forest managers and planners often need to know in order to compare the values of alternative uses for land-use planning purposes. Other unpriced goods and services can be evaluated using similar techniques.

Outdoor recreation is pursued in many forms, but most recreational experiences are a combination of anticipation, travel to and from the site, on-site activity, and recollection afterwards. The demand for a particular recreational site is derived from the demand for the recreational experience that the site provides, just as the demand for timber is derived from the demand for final products made of wood, as explained in Chapter 4 and shown in Figure 4.5. The demand curve for recreation at a site, therefore, is as shown in Figure 5.1. If access is free, however, we can observe only the point d', the quantity demanded at zero price. Our task is then to estimate the amount of recreation that would be consumed over the range of prices above zero.

Recreationists who are not required to pay a price for access to a forest for recreational purposes are still likely to incur costs when they take advantage of this opportunity. They usually incur travel costs to reach the site, and they may have to purchase supplies and equipment, among other things. Such expenditures do not measure the value they derive from a particular recreational opportunity or their willingness to pay for access to it. These expenses are analogous to the costs one might incur in attending a movie, such as the cost of gasoline, parking, and babysitting services. What is needed is an estimate of the amount that people would be willing to pay in the form of a toll to enter the particular recreational site, analogous to the theatre ticket.

The distinction between consumers' evaluation of something and the costs they incur in consuming it is important because the two are often confused. The expenditures of tourists and other recreationists are sometimes misrepresented as the value of the facilities that attract them, but recreationists' spending represents only the costs they incur, not the benefits they enjoy. The benefits of a recreational activity are reflected in the demand for the activity itself, not the demand for ancillary goods and services.

The expenditures of recreationists undoubtedly affect business activity, incomes, and employment in the region where the spending takes place, and therefore constitute useful information for assessing the economic impact of recreational opportunities on a region or on the local population. To measure the value of particular recreational sites or facilities to the people who use them, however, we must estimate the demand curves for them. These reflect only the willingness of consumers to pay for access to specific recreational opportunities, which is quite separate from the incidental expenses they incur in travelling to the site.

Direct Techniques for Estimating Consumer Surplus

Direct techniques for estimating the value of unpriced recreational opportunities involve directly asking recreationists for information that is needed to specify relevant demand curves. Recreationists are asked to declare the maximum amount they would be prepared to pay to visit the site and participate in the recreation.

This is usually done through mail or on-the-spot surveys of a sample of recreationists. Then, by arraying the responses from highest to lowest, the demand curve for the recreational opportunity can be drawn. The total consumer surplus is the sum of these responses after adjustment for sample size, represented by the area under the demand curve. This technique is called the *contingent valuation* method of estimating the value of unpriced benefits because it is based on the consumers' expressed willingness to pay, contingent on specific information about the recreational site and their own economic and social circumstances. This method has been frequently used to estimate the value of recreational sites such as parks and sport fisheries, as well as some of the non-use values of environmental goods and services.

If, for example, a thousand recreationists per year indicate an average willingness to pay $50 for access to the site, the annual consumer surplus is estimated at $50,000. Using techniques described in Chapter 3, the present value of such an annual amount can be calculated to indicate the gross capital value of the recreational facility.

An alternative to asking about recreationists' willingness to pay for a recreational opportunity is to ask them the minimum payment they would accept to refrain from using it. Theoretically, this would yield a similar result as long as the values involved are insignificant in terms of the consumers' total income. The *income effect,* however, which explains that a dollar reduction in income is weighted more heavily than a dollar increase in income, ensures that the minimum acceptable bribe to abstain, or willingness to accept (WTA), will never be less than the maximum willingness to pay (WTP), and some surveys have found it to be much greater.

These two approaches therefore yield different measures of consumer surplus. The maximum willingness to pay in addition to what a consumer already pays for a good is called the *equivalent variation* measure, whereas the minimum acceptable compensation to do without is the *compensating variation* measure. Which one is the more appropriate measure of consumer surplus depends on the reference level of the good to be valued or the property rights. If individuals (consumers) have a right to the original level of consumption or original level of utility, compensating variation is the more appropriate measure. On the other hand, if they want to have a higher level of consumption or a higher level of utility, then equivalent variation is more appropriate.

In other words, if individuals must purchase a good or want more of it, the appropriate measure is equivalent variation – the maximum amount they will pay (WTP). If they already own a good that may be taken away (for example, if they have the right to a clean environment but this right is threatened because of pollution), the appropriate measure is compensating variation – the minimum compensation (willingness to accept, WTA) they require to keep their utility at the same level as before they lost the good.

The theoretical underpinning of both compensating variation and equivalent variation is rooted in individuals' expenditure functions, which are thus the starting point of empirical estimation of either compensating variation or equivalent variation.

Although the theoretical basis for the contingent valuation method is straightforward, the design of a survey questionnaire for use with this method demands careful consideration. For example, to elicit willingness to pay for a recreational site in a questionnaire requires, at the minimum, a detailed description of the site to be valued, the mechanism by which the site will be provided, and the method of payment (such as an increase in the respondents' utility bill). The questionnaire needs to be pretested.

Another difficulty with these direct techniques lies in obtaining rational and consistent expressions of value from recreationists in response to hypothetical questions about their willingness to pay to use a recreational resource that they currently use without charge. The emotion attached to freely accessible recreation, coupled with suspicions about the purpose of such questions, is likely to produce answers that are deliberately or subconsciously distorted (such as "it is priceless" or, at the other extreme, "I would refuse to pay anything"). Moreover, if the value declared by any recreationist were actually charged, the recreationist would probably alter somewhat the quantity of the recreation he or she consumes, which also complicates the interpretation of results. Thus, the contingent valuation technique must be used with care.

More sophisticated survey methods and econometric techniques have recently been developed to overcome some of the limitations of the contingent valuation method. At minimum, a validity testing for internal consistency of responses is required for contingent valuation studies. Specifically, one can use regression analysis to determine the manner in which the socioeconomic and demographic characteristics of respondents influence willingness-to-pay responses, and to test whether or not these influences coincide with economic theory. For example, those with higher income levels should have a higher willingness to pay than those with lower income levels.

The contingent valuation method is referred to as a *stated preference* method, because it depends on respondents' answers to questions about what they would do in hypothetical circumstances, in contrast to a *revealed preference* method, which is based on observations of their actual choices. This dependence on hypothetical questions and answers is a source of weakness and possibly biased information. Nevertheless, contingent valuation is one of the few ways to assign dollar values to many unpriced environmental values that do not involve market purchases and may not involve direct participation. These values are sometimes referred to as "passive use" values, which are typically described as consisting of option, bequest, and existence values, noted in Chapter 1. They include everything from the basic life support functions associated with ecosystem health and biodiversity, to the enjoyment of a scenic vista or a wildness experience, to an appreciation of the option of fishing or bird-watching in the future, or to the right to bequeath those options to grandchildren. They also include the value that people place on simply knowing that the giant pandas or whales exist.

Previous contingent valuation studies of forest-related goods and services include protection of ponderosa pine from mountain pine beetle infestation, reduction of fire hazard to old-growth forests, restoration of old-growth longleaf pine forests for Red-cockaded Woodpecker (an endangered species) habitats, and creation of national parks and protected area to preserve 10% of tropical rainforests.

Another method, called the *contingent choice method* or *conjoint analysis,* is similar to the contingent valuation method. The contingent choice method was developed in the fields of marketing and psychology to measure individuals' preferences for different characteristics or attributes of a multi-attribute asset. Like the contingent valuation method, it is a hypothetical method – it asks people to make choices based on a hypothetical scenario. It differs from contingent valuation, however, because it does not directly ask people to state their values in dollar terms. Instead, values are inferred from the hypothetical choices or trade-offs that people make. Thus, inferring values from individuals making choices or trade-offs among different alternatives (that is, the asset in different forms and with different prices or costs) in simulations – rather than directly assigning pecuniary values to the asset, as is done in the contingent valuation method – is the characteristic feature of the contingent choice method.

Indirect Measures of Consumer Surplus

The practical limitations of direct techniques have led to the development of econometric methods for inferring recreationists' willingness to pay from indirect evidence drawn from their observed behaviour. These techniques, including the travel cost approach and hedonic pricing method, reveal willingness to pay. These approaches can be best explained by first sketching their theoretical underpinning.

Theory of Recreation Behaviour

Some of the costs of participating in a recreational experience are *fixed* costs that do not vary with the quantity of recreation consumed as measured in recreation days spent at the site. Fixed costs include the cost of travelling to and from the site and any fee that must be paid for access. Other costs, such as the cost of food and supplies, are *variable* costs because they depend on the duration of the recreational experience.

The fixed and variable costs incurred by an individual recreationist are depicted in Figure 5.2, where the recreationist's income is measured along the vertical axis and the number of on-site recreation days consumed is indicated on the horizontal axis. Suppose that the recreationist's total income is equal to the vertical distance $0Y$, and that to participate in the relevant recreational opportunity at all the recreationist must incur fixed costs equal to T_XY, reducing his remaining income to $0T_X$. The recreationist's variable or on-site costs of recreation are represented by the slope of the line T_XR_X (which is shown as a straight line on the assumption that the marginal cost of a recreation day is constant). The kinked line YT_XR_X traces the recreationist's consumptive opportunities for dividing his income between this recreational activity and other things.

The various combinations of recreation and income that will yield the recreationist the same utility or satisfaction can be represented by *indifference curves,* which show consumption bundles that give the consumer the same level of satisfaction. One such curve, curve I in Figure 5.2, is just tangent to the line tracing the recreationist's consumptive opportunities, and thus represents the highest level of satisfaction that he can attain by

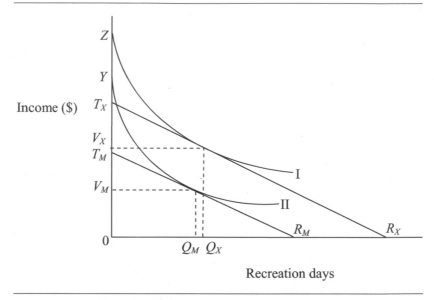

Figure 5.2   Equilibrium level of recreation consumption at two levels of fixed cost.

consuming $0Q_X$ recreation days. By doing so, the recreationist will spend V_XY on the recreation, of which T_XY are fixed costs and V_XT_X are variable costs. Obviously, this level of satisfaction, reaching the vertical axis at point Z, is higher than spending his income, $0Y$, on all other things without participating in any recreational activities.

Clearly, this recreationist would be willing to pay more for access to the recreational opportunity than the fixed cost T_XY that he incurs. At most, the recreationist would pay fixed costs of T_MY, which would give him the same level of satisfaction as if he had consumed no recreation at all. This is shown by indifference curve II, which runs through point Y, indicating the level of satisfaction without any recreation.

Now, the total amount that the recreationist would have spent on recreation has three components: fixed costs as described above (T_XY), variable costs (V_MT_M), and the maximum amount that the recreationist would be willing to pay for entry (T_MT_X), which becomes part of his total fixed costs. Note that if the recreationist had to incur this extra T_MT_X amount of fixed cost, he would spend a total of V_MY on recreation and reduce his consumption of it to $0Q_M$.

This depiction illustrates the two measures of consumer surplus identified earlier. The amount T_MT_X is the *equivalent variation*, the maximum that the recreationist would pay to gain access to the site in addition to the costs that he actually incurs, which include T_XY (fixed costs) and V_MT_M (variable costs). The amount YZ is the *compensating variation,* the minimum amount that the recreationist would need to have added to his income in order to leave him equally satisfied without participating in the recreation. In the following paragraphs, we adopt the former, more traditional definition of consumer surplus, equivalent variation, and show ways of estimating it.

The Travel Cost Approach

The value of a recreational resource enjoyed by a recreationist under free access – that is, his consumer surplus – can be expressed as the maximum access fee or toll that the recreationist would be willing to pay in addition to his original fixed costs. As long as he would react to a toll in the same way that he would react to fixed costs, the maximum toll he would pay is T_MT_X in Figure 5.2. The sum of these maximum hypothetical tolls for all participating recreationists is the total consumer surplus generated by the recreational opportunity, or the area under the demand curve.

The simplest way to estimate this willingness to pay is based on survey information about the travel costs incurred by the participating recreationists in travelling to and from the recreational site, which depend mainly on the distance they travel. By assuming that all participating recreationists have the same willingness to pay, and that the one who incurs the highest travel cost is marginal in the sense that she would not be willing to pay anything more, the consumer surplus of all the rest can be expressed as the difference between the travel cost each incurs and the travel cost of the marginal participant. Ranking these estimates of individual willingness to pay from the highest to the lowest can be expected to reveal a downward-sloping demand curve like that in Figure 5.1.

Thus, if a survey of the users of a camping area were to reveal that the camper who incurred the highest travel cost spent $40 travelling to and from the site and that the average of all campers was $15, the average consumer surplus would be estimated at $25 by this method. Multiplying this figure by the total number of campers who visit the site each year provides an estimate of the annual value of the recreational facilities.

This procedure involves some tenuous assumptions: that all the recreationists are equally willing to pay for the recreation regardless of differences in their incomes and other characteristics; that their cash outlays for travel fully represent the costs they incur to obtain access to

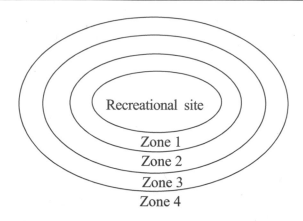

Figure 5.3   Zones of origin of travel to a recreational site.

the site; that they would respond to a toll in the same way that they respond to travel costs; and that the participant who incurs the highest travel cost is marginal. These assumptions are not likely to conform to reality, and they obviously limit confidence in such estimates.

The assumption that all recreationists are equally willing to pay is particularly questionable because we normally observe larger numbers participating at relatively low costs and smaller numbers at high costs, consistent with the law of demand. To assume that all would be prepared to pay the same amount as the one who paid the most almost certainly exaggerates the consumer surplus.

An alternative technique that avoids making this assumption involves measuring the sensitivity of recreationists to the costs they must incur to reach the site, and using this information to estimate their response to an access fee. The method is illustrated with a simple example in Figures 5.3 and 5.4 and Table 5.1.

The geographic area from which recreationists are drawn to a recreational site is divided into concentric zones (Figure 5.3), within each of which travel costs to the site are approximately uniform. For each zone, the total population, the average travel cost to and from the recreational site, and the number who visit the site are assessed, as shown in columns 2 to 5 of Table 5.1. The numbers in these columns provide the information for establishing the relationship between the participation rate and travel costs for the entire population, as shown in Figure 5.4A.

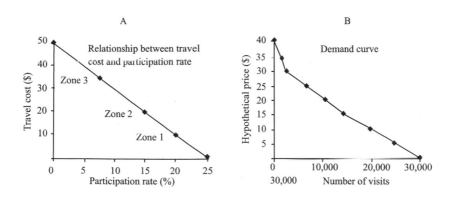

Figure 5.4   Derivation of the demand curve for a recreational site from travel costs.

The remaining columns in Table 5.1 use this relationship to estimate the reduction in the number of visitors to the site that would result from adding an access fee, of varying amounts, to the travel cost of all visitors. The table shows the calculation for a hypothetical access fee, or price, of $5, $10, $15, $20, $25, $30, $35, and $40. The relationship in Figure 5.4A gives the lower participation rate that could be expected from each zone when the price is added to the travel cost. That rate is applied to the total population of the zone to yield the number of participants who could be expected to participate at that price. These estimates are then used to plot the demand curve for the recreational site, as shown in Figure 5.4B.

We use zones 1 and 2 as an example to simulate the total visits (demand) by comparing the behaviour of the populations in zones 1 and 2 when price changes. First, as we have assumed, the difference in participation rates between zones 1 and 2 attributed to travel cost when price is equal to zero is given by

$$(15\% - 20\%)/(\$20 - \$10) = -0.5\%/\$1$$

This shows that for a $1 increase in price, there is a 0.5% reduction in the number of visits. The same is true, in our example, if we compare the behaviour of the populations in zones 2 and 3.

Second, as the total cost per visit in zone 1 increases from $10 to $15 ($10 travel cost plus $5 hypothetical access fee or price), the participation rate from zone 1 declines from 20% to 17.5%, a 2.5% net reduction

TABLE 5.1 Visitation data for a recreational site and total visits per year with simulated increases in travel cost

Zones	Population (number)	Travel cost ($)	Price = 0 (observed)		Price = $5 (simulated)		Price = $10 (simulated)	
			Participation rate (%)	Participation (number)	Participation rate (%)	Participation (number)	Participation rate (%)	Participation (number)
Zone 1	50,000	10	20.0	10,000	17.5	8,750	15.0	7,500
Zone 2	100,000	20	15.0	15,000	12.5	12,500	10.0	10,000
Zone 3	60,000	35	7.5	4,500	5.0	3,000	2.5	1,500
Zone 4	20,000	50	0.0	0	0.0	0	0.0	0
Total	230,000			29,500		24,250		19,000

Zones	Price = $15 (simulated)		Price = $20 (simulated)		Price = $25 (simulated)	
	Participation rate (%)	Participation (number)	Participation rate (%)	Participation (number)	Participation rate (%)	Participation (number)
Zone 1	12.5	6,250	10.0	5,000	7.5	3,750
Zone 2	7.5	7,500	5.0	5,000	2.5	2,500
Zone 3	0.0	0	0.0	0	0.0	0
Zone 4	0.0	0	0.0	0	0.0	0
Total		13,750		10,000		6,250

Zones	Price = $30 (simulated)		Price = $35 (simulated)		Price = $40 (simulated)	
	Participation rate (%)	Participation (number)	Participation rate (%)	Participation (number)	Participation rate (%)	Participation (number)
Zone 1	5.0	2,500	2.5	1,250	0.0	0
Zone 2	0.0	0	0.0	0	0.0	0
Zone 3	0.0	0	0.0	0	0.0	0
Zone 4	0.0	0	0.0	0	0.0	0
Total		2,500		1,250		0

(0.5% reduction × $5 increase in cost). Multiplying this number by the total population generates 8,750 visits from zone 1.

Finally, using the same method to calculate visits from zones 2 and 3, and adding all visits, generates a total of 24,250 visits from all zones when price (access fee) is increased to $5. This information is provided in column 7 of Table 5.1.

The same procedure can be used to simulate scenarios for all price increases to generate the participation rates and visitations reported in Table 5.1. Plotting the number of visits against price generates a typical downward-sloping curve, as shown in Figure 5.4B.

In this simplified example, the total consumer surplus represented by the area under the demand curve can be calculated at $458,750:

$$\$458,\!750 = (29,\!500 + 13,\!750) \times 15/2 + (13,\!750 + 2,\!500) \times 15/2 \\ + \; 2,\!500 \times 10/2$$

Note that the demand curve is kinked at a price equal to $15 and $30.

Because this technique depends on estimates of participation rates from whole populations, it becomes less reliable when applied to recreational sites that attract visitors from a wide area that encompasses large populations with different recreational alternatives.

Other techniques have been devised to recognize dissimilar circumstances that influence recreationists' willingness to pay. For example, using data obtained through surveys, those who utilize the relevant recreational resource can be classified by income, which can be expected to be a primary determinant of their willingness to pay for it. Then it should be safe to assume that the recreationists who have similar incomes and participate in the same recreational activity have equal willingness to pay.

Within each income group, they can be ranked according to the fixed costs (that is, travel costs) they incur. If there is a sufficient sample in each income group, it can be assumed that the recreationist who incurs the highest travel cost (corresponding to the one who incurs a fixed cost of $T_M Y$ in Figure 5.2) is at or near the margin and enjoys no consumer surplus.

Given these assumptions, the consumer surplus enjoyed by each recreationist is equal to the difference between the travel cost the recreationist incurs and the travel cost of the marginal (highest-cost) recreationist in the same income group. The sum of these differences for all recreationists in all income groups is the total consumer surplus generated by the recreational opportunity.

Using this approach, the demand curve for a recreational facility can be estimated by obtaining the number of recreationists who would participate at any given price. Based on the difference between a recreationist's travel cost and the travel cost of the marginal recreationist in his income group, a recreationist will participate if his consumer surplus exceeds the hypothetical price.

This technique is capable of recognizing differences in income and possibly other characteristics of recreationists – such as education, age, and family circumstances – that are likely to influence their willingness to pay. Some of the assumptions on which the estimates are based may give rise to error, however, especially the assumptions that all recreationists within a defined group are equally willing to pay for the activity, that the one who incurs the highest cost is marginal, and that all recreationists have similar alternatives.

The Hedonic Method

A rather different approach to evaluating recreational resources is based on recreationists' responses to the characteristics of different sites. In contrast to the techniques discussed above, which attempt to evaluate a recreational resource in isolation, the *hedonic method* uses information about the characteristics of a variety of sites to estimate the value that recreationists ascribe to them.

Each site is viewed as a bundle of characteristics important to recreationists. For example, the characteristics for a sportfishing site may be fish abundance, fish size, privacy, cleanliness of the water, scenic beauty, and so on. Every sportfishing opportunity accessible to a consumer has a different combination of these qualities. The hedonic method uses *observations* about recreationists' choices among alternative sites to estimate the value they attach to particular characteristics. The observations are of such things as the number of visits the recreationists make to particular sites, the time they spend at them, and the expenses they incur to reach them.

The technique involves analyzing the relevant characteristics of a variety of recreational sites and attaching a value to each characteristic using a numerical scale, similar to our comparable sales approach used in valuing pre-merchantable timber stands, discussed in Chapter 3. The costs incurred by recreationists who visit the sites are regressed against the site attributes to yield a "shadow value" for each. The result is an estimate of recreationists' willingness to pay for each attribute. Any particular site can then be evaluated in terms of its quality characteristics.

An attractive feature of the hedonic method is its specific recognition of the quality of recreational opportunities, and its ability to explain differences in the value of different sites on the basis of their characteristics. This can provide valuable information for management planning and land allocation. The technique presents a number of practical difficulties in scaling the quality of site attributes and in calculating the values attached to them, however.

The hedonic method, like other methods of estimating consumer surplus that draw inferences from recreationists' responses to costs, implies an assumption that the recreationists' sole purpose in incurring their costs is to participate in the recreational opportunity in question. This is probably a realistic assumption for certain types of recreation such as hunting or fishing. Campers and tourists are likely to be less single-minded about their recreational objectives, however, and if they casually visit a particular area in the course of a wide-ranging tour, it would obviously be inappropriate to attribute all their travel costs to that single experience. When costs are incurred for multiple purposes, some method of prorating them is called for.

A related implication of these techniques is that consumers' travel costs are fully measured by their cash outlays – specifically, that they does not consider as a cost the time they must expend in travelling to and from the site. In reality, these circumstances vary. A family on a camping trip may consider the travelling to be part of the recreational experience and so the time involved is appropriately ignored in accounting for costs. In contrast, hunters or fishermen are likely to regard time spent travelling as an encroachment upon their on-site recreation, so some cost should be ascribed to it. Some of these problems can be elucidated by careful attitude surveys of recreationists.

Expenditure-Based Methods
As noted in Chapter 3, an asset can be valued through the replacement cost approach, which is an expenditure-based method described here. Often, estimating the cost or expenditure of providing an unpriced asset is simpler than estimating the value of the benefits it provides and, in the absence of reliable estimations of value, costs offer useful guidance for some purposes. For example, if a government has made a decision that the supply of a particular unpriced forest benefit must be maintained or increased, the value of an asset that already provides this benefit can be assumed to be worth at least the cost of replacing it by the lowest-cost alternative means. Thus the value of a forest in carbon sequestration, for

example, is at least equal to the cost of achieving the same result by another method.

The cost or expenditure can be measured in three possible forms, and there are three cost-based methods for valuing an asset. One is the cost of replacing an asset if it were lost or were going to be lost (*replacement cost*). Another is the cost or sacrifice that people are willing to incur to avoid the damages associated with the actual or potential loss of the asset (*damage avoided cost*). The third one is the cost of providing a substitute that generates the same benefits as the original asset (*substitute cost*). Thus, the value of some environmental goods and services generated from forests can be measured by estimating the cost of actions that people are willing to take to replace lost goods and services (replacement), or to avoid damages to the goods and services (damage avoided), or to replace the lost services with a substitute (substitute cost). All these methods produce an imputed willingness to pay for an unpriced asset.

For example, the value of an otherwise unpriced 100-acre wetland is, arguably, at least equal to the cost of providing a 100-acre substitute wetland. Thus, the no net loss wetlands policy in the United States, in fact, puts a minimum price on the wetlands of the United States at their substitute cost, because anyone who wants to develop a piece of wetland must go through the government permitting process and, if approved, provide a substitute wetland that has similar size and, hopefully, ecological function.

All three costs are closely related, and the expenditure-based methods are based on the assumption that if people incur costs to avoid damages, or replace services in case they are lost, those services must be worth at least what people paid to maintain or replace them. When there are good reasons to believe that costs of damage avoidance or replacement of an environmental good/service differ substantially from the benefits it provides, the expenditure-based methods should be avoided.

Cost-Effectiveness Analysis
The expenditure-based methods are closely related to another concept in benefit/cost analysis: cost-effectiveness. Again, for some purposes, forest managers find it sufficient, instead of attempting to estimate the value of certain unpriced forest benefits, to compare the cost of providing them in alternative ways or places.

This is an expedient technique where the resource planner's objective is to produce a certain quantity of fish or game or camping opportunity in a region. She can compare the cost of all the alternative ways of contributing to the goal, and by choosing the least-cost means she will meet her objective at the lowest possible aggregate cost.

Obviously, such analyses of cost-effectiveness reveal nothing about the value of the benefits produced, which is the purpose of all the techniques discussed earlier in this chapter. It is nevertheless a helpful technique for ensuring consistency and efficiency in meeting predetermined forest management objectives.

OTHER CONSIDERATIONS IN VALUING RECREATIONAL RESOURCES

Intermediate Products in Forest Recreation

Outdoor recreation is consumed directly by consumers, so its value is properly assessed in terms of consumers' willingness to pay for it, as described above. Forest managers, however, are often concerned with providing less direct human benefits in the form of wildlife, fish, aesthetics, biodiversity, and general quality of the natural environment. Such benefits add another dimension to the problem of evaluation.

For purposes of economic evaluation, it is essential to identify clearly the nature of the human values created when forests are manipulated for a particular purpose. Fish and wildlife, for example, are usually valued for their contribution to the recreational quality of a forest, either as objects of viewing or of hunting and fishing. There may be exceptions, where wildlife is managed for commercial, scientific, or purely environmental purposes, but for present purposes we shall assume that the value generated is in the form of recreational fishing, hunting, or viewing, as is often the case.

Wildlife, valued by hunters, fishermen, and sightseers, is thus an intermediate product in the production of recreation, in the same sense that timber was described as an intermediate product in the production of housing and newspapers in Chapter 4. This observation is important because the issue is often confused in attempts to ascribe values directly to fish and game or to the harvest of hunters and fishermen. It must be kept in mind that the ultimate product of fish and game management is not the fish and game themselves but the recreation they support. More fish and game will only increase the quantity or quality, and hence value, of the recreational opportunity, but a bagged bird, deer, or fish must be regarded primarily as a by-product of a recreational experience. A fish may or may not have consumptive value to the sportfisher (who may release it, eat it, or give it away). The essential characteristic of an attractive recreational fishing opportunity is a reasonable chance of catching a fish; actually catching one enriches the experience, but the fact that fishermen can enjoy fishing without catching anything indicates that the harvest is only one of many dimensions of the quality of a fishing experience.

Additional fish or game can increase the recreational value of a forest in either or both of two ways. One is by enhancing the quality of the recreational experience. More wildlife will usually result in more sightings by nature lovers, and more fish or game will increase the success rate of sportfishers and hunters. Such improvements in the quality of the experience will increase the recreationists' willingness to pay for it, shifting upward the demand curve for the recreational opportunity.

The other effect, resulting from the increased attractiveness of the recreational opportunity, is that more recreationists will be attracted to it. This implies a shift of the base of the demand curve to the right, again increasing the area under the demand curve. A sufficient increase in the number of users could forestall any improvement in the average harvest that would otherwise result from the enhanced resources. In any event, through either or both of these effects, the demand curve will shift, and the value of the additional fish or wildlife is reflected in the increase in the area under it, which is the increment in consumer surplus enjoyed by hunters or fishermen.

Recreational Capacity, Quality, and Crowding

These considerations illustrate the interdependence between the quality of a recreational experience and its level of consumption. Both affect the value of the recreational experience.

One dimension of the quality of a recreational opportunity is congestion, or the degree of crowding. If the demand for an unpriced recreational opportunity increases sufficiently, and pricing or other means of rationing access is unavailable, the site will become crowded, reducing the quality and value of the recreational experience.

The effect of crowding is illustrated in Figure 5.5. At zero price, increased numbers of recreationists shift the base (the intercept to the horizontal axis) of the demand curve to the right. The lower quality of the recreational experience, however, means that the recreationists will not be willing to pay as much for it, hence the demand curve (reflected in the intercept to the vertical axis) is lower. Whether such a change will increase or decrease the total value generated, represented by the area under the curve, will depend on its shape and the degree of these

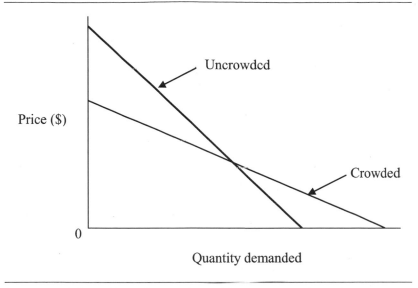

Figure 5.5   Effects of crowding on demand for a recreational opportunity.

opposing effects. In Figure 5.5, the lower triangle between the two curves measures the gain in consumer surplus from additional numbers of consumers; the upper triangle indicates the loss resulting from the lower quality of the more crowded site. The net impact of crowding on total consumer surplus is the difference in the areas of these two triangles.

It should be noted that crowding is not always detrimental; a "good crowd" can enhance the appeal of recreational facilities such as ski resorts or holiday camps. This may be related to the so-called *bandwagon effect* or *herd instinct*, which means that people often do and believe things merely because many other people do and believe the same things. For most forms of forest recreation, however, solitude enhances quality and, conversely, crowding diminishes the value of the experience.

Both the quality and the capacity of a recreational site can usually be enhanced through investment. The *quality* of a recreational resource can be improved by such measures as grooming campsites and trails, enlarging wildlife populations, or refraining from unsightly timber-harvesting practices. In terms of the conceptual framework above, the effect of such quality improvement is to shift the demand curve for the recreation to the right, indicating larger numbers of demanders at zero price, and upward, indicating their greater willingness to pay for the opportunity.

The *capacity* of a recreational site can be expanded by investment in additional campsites, trails, and so on, so that more recreationists can be accommodated without diminution of the quality of the experience. Expansion of capacity can therefore accommodate a shift to the right of the base of the demand curve without reducing participants' willingness to pay, represented by the height of the curve.

In all these cases, the benefit resulting from the investment is the increase in the area under the demand curve, which can be estimated by the techniques described earlier in this chapter. It is this benefit that must be compared with the cost of the improvement in order to measure the net gain.

If a recreational site were operated by a profit-maximizing owner who was free to price access, given the capacity and quality of the site in the short term, the owner would select the combination of price and number of recreationists that would generate the greatest net revenue. In the long term, he would invest in the site's capacity and quality whenever the cost of doing so is less than the resulting increase in revenue. This behaviour would ensure that crowding would not diminish the value of the resources. Where access is free and uncontrolled, however, this sort of optimization is not possible, and crowding may erode the value of the recreation. It is therefore common, where a policy of free access to recreational resources applies, to find examples of campsites, wilderness areas, and fishing and hunting opportunities becoming so crowded that their attractiveness to potential users is diminished.

In the absence of pricing, the erosion of recreational values through crowding can be prevented through other means of rationing access, such as by allowing a limited number of users on a first-come, first-served basis, by drawing lots, or by some other scheme for admitting only some of the demanders. These other techniques are less effective in generating the maximum value from the recreational opportunity because, unlike pricing, they cannot ensure that those who value it most highly will have access to it. They can, however, protect the quality of recreational sites from deteriorating through crowding.

In short, crowding, which can diminish the value of recreational sites, can be controlled by a variety of means, but they have quite different implications for efficiency and equity. Pricing, or an auction system, can ensure that the maximum value will be realized, but it may be perceived to disadvantage people with low incomes. Drawing lots or utilizing a first-come, first-served system will not generate maximum value because access will not be provided to those who value it most. This is a limitation of all regulatory methods.

EXTERNALITIES AND INTRINSIC VALUE

Externalities

So far, we have been referring to benefits that accrue to consumers, so that the value of a good or service can be assessed in terms of the potential willingness of consumers to pay for it. In some cases, however, benefits or costs accrue to people other than the consumers, giving rise to *externalities* of the kind described in Chapter 2. These external values are sometimes significant, and it is necessary to estimate them to supplement the values accruing to consumers.

Sometimes external costs and benefits can be observed in monetary terms. For example, a forest or park may not only provide benefits to its users but also increase the value of adjacent private properties. In this case, the external benefit accruing to local landowners is reflected in their enhanced property values, which may be estimated using the hedonic pricing method described earlier.

Often, external benefits are not reflected in any market prices (such as travel cost or enhancement of adjacent property values). Estimating their value is problematic, even though, in theory, they can be valued by applying the contingent valuation method to an inclusive sample of population. An inclusive sample includes all people who potentially receive positive or negative externalities associated with an asset (such as a forest or a park) that is being valued. This is the only way to value all external benefits and costs, but it is often impractical to value all the externalities of an asset or a forest management action across a widely dispersed population. Thus, it is helpful to identify and distinguish several kinds of such external effects that are relevant in forest management.

Public goods. As noted in Chapter 2, one type of externality is public goods. Forests sometimes yield true *public goods,* benefits that suppliers cannot parcel up and sell to those who are willing to pay for them while excluding others, and that consumers can consume without diminishing the supply available to others, such as the amenity of a landscape and mitigation of climate change.

Option value. People who do not participate in a recreational experience or seek out a particular amenity may nevertheless value the opportunity to do so. They are thus willing to pay something for preserving the option for themselves or their children, even though they are not currently among the active consumers. As noted in Chapter 1, the value that people place on preserving such opportunities is the *option value.*

Option value is particularly relevant to decisions about unique features of nature where irreversible decisions are being considered. For example, evaluation of a proposal to build a hydroelectric reservoir that would eliminate a unique wilderness should take account of the option value associated with the wilderness, in addition to any current recreational or aesthetic value it generates. Where the relevant resources are not unique, or can be replaced, which is more often the case in managed forests, this special value is less important.

A closely related concept is *preservation* or *existence value,* which is the value that people place on something regardless of any interest in direct consumption – and often people have no interest in direct consumption or enjoyment of the resource now or ever. For example, many people are willing to pay something to preserve the whooping crane, the wood buffalo, and a rainforest even though they have no expectation of seeing these things. Their support for public expenditure and private efforts to protect them is evidence of their willingness to pay simply to maintain their existence.

These various types of externalities are sometimes difficult to distinguish. The value of maintaining a forest may be a mixture of consumer surplus, option value, preservation value, and public goods and may generate values in all these forms. They are all exceedingly difficult to evaluate except in the rare cases where market indicators of value are available. Estimates of their value must rely on subjective assessments or on public surveys designed to establish the willingness of people to pay for them rather than go without them.

Intrinsic Value

Economics deals with human values, production, and consumption as well as conservation and preservation. A forest, a tract of land, or a recreational site is often valued in monetary terms or in terms of instrumental value to human beings, and it should be conserved or preserved if its net present value of continuing existence is greater than the consumptive value today. As noted in Chapter 1, however, some philosophers and conservation biologists view all living organisms and works of nature as having their own intrinsic value, independent of human perceptions of their usefulness or of their aesthetic and spiritual appeal. They advocate conservation and preservation strictly on ethical grounds. Whatever the appeal of this approach, it denies that economic analysis of the value of these resources has any relevance.

OTHER PRACTICAL COMPLICATIONS

The unmarketed products and services of forests are often difficult to define and measure, and they often generate emotional responses that can confuse or distort the information needed to assess them. Thus, in any attempt to measure the value of such benefits, it is particularly important to clarify precisely the nature of the benefits being generated, the beneficiaries, and how the chosen valuation technique captures these values.

It is worth noting that the problem of evaluation discussed in this chapter is not due to the intangibility of natural values, as is often suggested. The enjoyment people derive from a painting or a record album or a book of poetry is intangible in the usual sense, but the value of these things is nevertheless observable in market prices. The problem of evaluation is difficult not because certain forest values are intangible but because they are unpriced.

The difficulties of estimation suggest that analysts should take advantage of market indicators of values wherever they are available. For example, external costs and benefits of the kind encountered in forestry often become capitalized in property values. And sometimes the value of unpriced recreational facilities such as campgrounds can be estimated with reference to the prices charged by comparable facilities that are priced.

In this chapter, we have considered a variety of unpriced values that may be produced from forests, and various approaches to evaluating them. Each of the direct and indirect techniques for estimating consumer surplus and other unpriced values has its special strengths and weaknesses that make it most suitable for use in particular circumstances.

None of these methods provide more than rough estimates of values. Even so, they can provide useful guidance in deciding how forests can best be used and developed, which otherwise must be arrived at through guesswork, and can at least provide some consistency in decision making. Techniques for evaluating unpriced goods and services continue to be developed, and cumulative experience in applying them is leading to improvements in the reliability of estimates. They are rarely precise, however, and the results must be used cautiously.

Finally, it must be realized that there are conceptual and practical differences in estimating the value of unpriced and market goods. When the supply of an unpriced good is ignored due to measurement problems, its value is estimated based on the total area under its demand curve. As such, the estimated values are overstated because the cost to supply or the opportunity cost of producing the good is ignored. In contrast, the value of a market good or service is the area below its demand curve and above its supply curve. We have seen some empirical studies that place the value of the environmental or ecosystem services from certain ordinary or working (not unique) forests several times higher than the market value of the forests. These results may very well reflect the real value of environmental services from these forests. Closer examinations, however, often reveal that these values are inflated, either because they are estimated using the total area below the demand curve or because there are issues with the study designs; consequently, the results of these studies are controversial. Anyone making a trade-off between producing an unpriced good and producing a market good needs to know whether the values of both goods are estimated in the right way.

REVIEW QUESTIONS

1   The right to fish in a particular stream or to hunt in a particular area is usually priced in Europe but not in some parts of North America. What are the reasons for this difference? How does it affect the problem of evaluating these recreational resources?

2   What is "consumer surplus"? With reference to a typical demand curve, explain the relationship between consumer surplus and the total amount consumers are willing to pay for a product when the price is (a) positive, and (b) zero. Is the consumer surplus large for unique natural resources such as Banff National Park, Old Faithful geyser, or the Himalayas (Mt. Everest)? Is there any consumer surplus associated with one of several local campgrounds or picnic sites?

3   Why is it inappropriate to use the expenditures of visitors to a park as a measure of the value of the park?

4   Recalculate the total consumer surplus that would be attributable to the recreational site depicted in Figures 5.3 and 5.4 if the participation rate for each of the three zones were doubled.

5   How does the number of game animals in a forest affect the demand for hunting? What other factors are likely to affect the demand for and value of hunting in the forest?

6   Why are "option value" and "existence value" likely to be more important in assessing the economic implications of harvesting a virgin stand of old-growth Douglas-fir than the harvesting of a Douglas-fir plantation?

FURTHER READING
Bishop, Richard C. 1982. Option value: An exposition and extension. *Land Economics* 58 (1): 1-15.

Bowes, Michael D., and John V. Krutilla. 1989. *Multiple-Use Management: The Economics of Public Forestlands.* Washington, DC: Resources of the Future. Chapter 7.

Brown, Gardner Jr., and Robert Mendelsohn. 1984. The hedonic travel cost method. *Review of Economics and Statistics* 66 (3): 427-33.

Clawson, Marion, and Jack L. Knetsch. 1966. *Economics of Outdoor Recreation.* Baltimore: The Johns Hopkins University Press, for Resources for the Future. Part 2.

Davis, Lawrence S., K. Norman Johnson, Pete Bettinger, and Theodore E. Howard. 2005. *Forest Management: To Sustain Ecological, Economic, and Social Values.* 4th ed. New York: Waveland Press. Chapter 8.

Kurtilla, John V., and Anthony C. Fisher. 1985. *The Economics of Natural Environments: Studies in the Valuation of Commodity and Amenity Resources.* Rev. ed. Washington, DC: Resources for the Future.

Pearse, Peter H. 1968. A new approach to the evaluation of non-priced recreational resources. *Land Economics* 44 (1): 87-99.

Sills, Erin O., Karen Lee Abt, eds. 2003. *Forests in a Market Economy.* London: Kluwer Academic Publishers. Section 3.

Smith, V. Kerry, and Yoshiaki Kaoru. 1987. The hedonic travel cost model: A view from the trenches. *Land Economics* 63 (2): 179-92.

Walsh, Richard C., John B. Loomis, and Richard A. Gillman. 1984. Valuing option, existence, and bequest demands for wilderness. *Land Economics* 60 (1): 14-29.

van Kooten, G. Cornelis, and Henk Folmer. 2004. *Land and Forest Economics.* Northampton, MA: Edward Elgar. Chapter 4.

Chapter 6   Land Allocation and Multiple Use

The most fundamental decisions in forest management relate to the allocation of land among alternative uses. The economic challenge of this issue is to identify the use or combination of uses that will generate the greatest value. This chapter reviews the economic and technical relationships that determine the optimum pattern of land use.

Land can usually be utilized for a variety of purposes. Choosing the most productive form of land use calls attention to economic efficiency at two levels. One is efficient management when land is used for any particular purpose – that is, how to apply labour and other inputs to generate the maximum return under a given land use. The other is selection among the alternative uses: to find the use, or the combination of uses, that yields the highest net return. We turn first to the problem of maximizing the return under a particular form of land use.

INTENSITY OF LAND USE

Efficient production involves combining various factors, such as land and labour, in a way that generates the maximum net return. Where one factor in the production process is fixed in supply, maximizing the aggregate net return involves maximizing the return to the fixed factor, after the costs of all the variable factors have been met. In forest production, land is usually the fixed factor.

The problem of efficient use of a tract of forestland is therefore one of identifying how much of the other productive factors, such as labour and capital, can be advantageously applied to it in the forest production process – that is, how much labour and capital must be used to generate the maximum *land rent*. This is, in other words, the question of the optimum

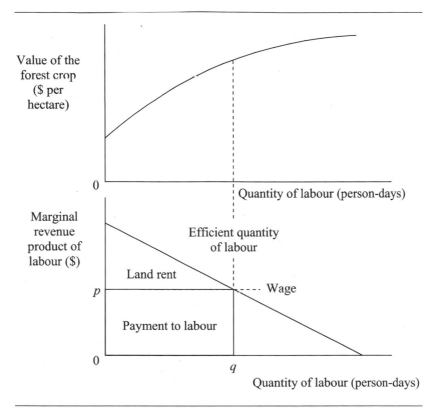

Figure 6.1   Efficient application of labour to a forest site.

intensity of forest management. We will assume that the forest manager uses the available factors of production in the most productive way that technology allows. The task is to identify the quantities of each factor that combine to generate the maximum return to the forestland.

The solution to this problem lies in the conditions for economic efficiency reviewed in Chapter 2. There we noted that each factor in production should be employed up to the point at which the return it generates at the margin declines to equal its cost.

This condition is illustrated in Figure 6.1 in terms of the efficient amount of labour to employ on forestland of specific quality. The curve in the upper quadrant shows how the value of a forest crop can be increased if more labour is applied to land used in cultivating the crop, increasing the intensity of silviculture. The curve in the lower quadrant shows the corresponding *marginal revenue product* of labour – that is, the additional value generated for each increment of labour over the range of labour use. As explained in Chapter 2, this incremental value declines as more labour is added because of diminishing returns, reflected in the diminishing slope of the curve in the upper quadrant of Figure 6.1.

Given the price of labour, or the wage rate, p, additional labour can be advantageously employed, contributing to the return to the land, or the land rent, up to the quantity q, at which point its marginal revenue product and cost are equal. This point defines the *intensive margin* of land use. At this level of employment, labour makes its maximum contribution to the land rent, as illustrated by the upper triangle between the curve and the price line in the lower quadrant of Figure 6.1. This is therefore the efficient combination of labour with land of this quality and other characteristics, and a corresponding relationship applies for all other variable factors of production.

Each forest site has a unique combination of fertility, location, and terrain that determines its productivity and its responsiveness to other factors of production. Land of lower productivity will usually yield a lower total return to labour, resulting in a lower curve in the upper quadrant of Figure 6.1. The *marginal revenue product* of labour, reflected in the *slope* of the curve in the upper quadrant of Figure 6.1, is not directly correlated with the inherent productivity of the land, however. As shown in Chapter 2, the marginal revenue product of labour is equal to output product price times the marginal physical product of labour.

A lower or more rapidly diminishing marginal revenue product of labour would indicate that a smaller amount of labour could be employed advantageously. Its contribution to the land rent would be smaller as well. Similar relationships apply to other inputs in forest cultivation, so the lower the responsiveness of the forest to silviculture, the lower the potential land rent.

These relationships determine the optimal intensity of forest management. Generally, the higher the productivity of the land, the higher the potential land rent, and the greater the quantities of other factors of production needed to generate it. This explains why highly productive lands usually warrant more intensive silviculture and other forms of forest management than land with low productivity.

EXTENSIVE MARGIN OF LAND USE
In Chapter 4, we examined how the supply and demand for timber determine its equilibrium market price, and noted that an increase in demand would result in a higher price and greater production. The higher price

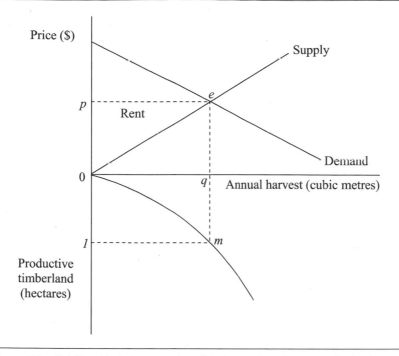

Figure 6.2   Relationship between price of timber and productive timberland.

provides incentive for producers to cultivate forestland more intensively, as explained above, and also to bring more land into forest production. Both more intensive production on existing forestland and expansion of forestry onto additional land add to total production. The former is referred to as an expansion in the *intensive margin* of production, and the latter, bringing new land into production, as an expansion in the *extensive margin* of production.

Figure 6.2, a slightly modified version of Figure 4.1, illustrates how the extensive margin is governed by the price of the product. The upper quadrant shows the equilibrium price and annual production of timber resulting from the interaction of market supply and demand. The lower quadrant shows, for any possible change in demand and equilibrium price, how much land can profitably be used for timber production.

If the land has no other productive use, the most productive land will be employed in timber production at low timber prices, and at higher prices progressively more and more of the less productive land will be drawn into production. At the price *p* in Figure 6.2, the equilibrium level of production, q, is produced using l amount of land, which can generate a positive net return in timber production. Of this land, the most productive will earn a rent, as illustrated in Figure 6.1. The poorest, or marginal land, will earn no rent, which can be illustrated in Figure 6.1 by a lower curve of marginal revenue product that intersects the vertical axis at point p. Figure 6.2 can be used to show how an increase in demand will increase both the equilibrium price and level of production and thus attract more land into timber production.

ALLOCATION AMONG USES

Efficient allocation of land to timber production becomes more complicated when the land can be used in a variety of productive ways. Choices often have to be made about allocating land to agriculture, forestry, recreation, or other uses when two or more of these uses can alone, or in combination, generate a positive return. Thus, the allocation of land among its alternative uses is the second dimension of efficient land use.

We have already discussed how other factors of production are applied to land in order to generate maximum land rent. If, for each possible use of the land, the rent-maximizing combination of other factors is known, then the task of efficient allocation of land among uses is simplified and it becomes a matter of selecting the use that will generate the greatest rent in any particular time and place.

The capability of land to generate economic returns depends on a variety of factors: fertility, distance from markets, topography, accessibility, and so on. The importance of each characteristic differs for each use. The quality of land, or the determinants of its economic potential, can thus be viewed as a bundle of characteristics that have varying importance for different uses.

The importance attached to certain characteristics is relative to the purpose for which the land is used. To illustrate the problem of allocating land among competing uses, let us consider one such characteristic in isolation: the proximity of the land to an urban centre. This usually has a significant influence on the value of land, and we commonly observe that land close to urban centres is used more intensively than land at great distances from them. Proximity to an urban centre is a more important dimension of the quality of land for some purposes than for others, however: it contributes more to the potential productivity of land for commercial purposes than for forestry, for example. So land at greater distance from urban centres is not only utilized less intensively in any particular use but is also utilized for different purposes.

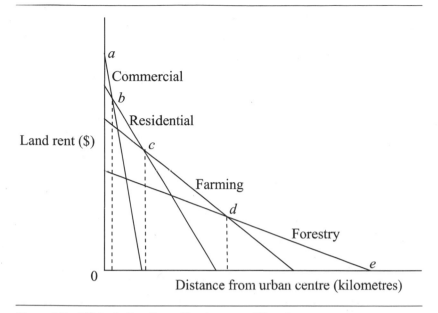

Figure 6.3   Efficient allocation of land among different uses.

Figure 6.3 illustrates the pattern of land allocation as a function of distance from an urban centre. As all other characteristics of the land are assumed to be identical, the figure shows how the potential land rent under various uses declines at progressively greater distance from the urban centre. In all uses, the value the land can generate is greater the closer it is to the urban centre, but the use that generates the highest rent varies over the spectrum. Commercial use, which is the most intensive, yields a higher return than all others at the urban centre. Forestry, the least intensive activity, makes the most productive use of the remotest lands. Farming generates the highest rent over an intermediate range (*cd*).

Thus we observe concentric patterns of land use around urban centres. Distance from an urban centre is only one of many characteristics of land that determine the land's capacity to generate rent, however. Its fertility, topography, and many other dimensions similarly influence its relative value in different uses, and all these qualities blended together result in complicated patterns of efficient land allocation.

This illustration in Figure 6.3 also helps explain a number of other important concepts about land use. First, it shows that land can often generate returns in more than one use. The use that generates the highest returns is the most efficient, whereas the rent in its next-highest use represents the *opportunity cost* of the land, or the value it could generate in its next-best use. The *differential rent* is the rent that the land earns in excess of its opportunity cost. Unlike the case in which land has no alternative use, land capable of earning rent in other uses can be efficiently allocated to timber production only if the rent it earns in timber production exceeds the rent under the alternative uses.

Second, it is not always efficient to allocate land to a particular use merely on the basis of its productivity in that use. Consider a tract of land closer to the urban centre that is more fertile and that can earn a higher return in farming than one within the range *cd*. Based on productivity, this tract of land would be suited for agricultural use, but Figure 6.3 demonstrates that its proximity to the urban centre gives it an opportunity to fetch even higher returns in residential and commercial uses.

Third, the allocation of land to its highest use ultimately depends on the value of the outputs and the cost of the inputs in each alternative use, and because these constantly change, so does the efficient allocation of land. Changes in technology, costs, and prices continuously shift the boundaries between highest uses, illustrated by points *a, b, c, d,* and *e* in Figure 6.3. These points mark the *extensive margins of land use* for the four uses indicated where differential rents are eliminated. Land allocation often changes near these boundaries, shifting with market conditions, but tends to be more stable on land well within the boundaries of extensive uses. For example, the forestland most subject to change is typically that closest to farmland (at point *d* in Figure 6.3).

Finally, competition in land markets tends to channel land to buyers who put it to its best use and are thereby able to pay the most for it. The last section of this chapter notes how institutional investors have found ways to enhance the values generated from forestlands by putting some of them to their best uses in the United States. Markets function only imperfectly in allocating land among uses, however. We have referred to land rent as though it were a continuous or annual return to the land, but land rent is more often capitalized in the value of the land, in accordance with the relationship between annual returns and their present worth (described in Chapter 3). The land-use allocation process responds only slowly to changing economic conditions and perhaps also slowly to government policies, and it is faulty whenever externalities in land use prevent social benefits or costs from being reflected in market prices.

COMBINATIONS OF USES

So far, we have considered the economics of producing various goods and services from a forest one good or service at a time. We must now turn to the opportunities for multiple use – the joint production of two or more goods or services.

Multiple use is a popular idea, frequently extolled as a means of reconciling the growing and often conflicting demands on land and natural resources. It is also a vague concept, however, difficult for resource managers to apply. When, and to what extent, is it *technically feasible* to accommodate two or more uses? In those cases where it is technically feasible, when is it *desirable* to do so on economic or social grounds? And when multiple use is desirable, *how much* of one use should be sacrificed for another? This section examines these questions.

Interdependence and Production Possibilities

As already noted, the pattern of demands on land varies widely and so does the capability of land to produce various goods and services. The most productive use, or combination of uses, must be considered separately for each tract of land in light of its particular circumstances. It is useful, however, to identify the range of conditions with respect to the possibilities for multiple use.

At one extreme, land may be subject to no demand for any use. Some remote and inaccessible forestlands fall into this category. In Figure 6.3, they lie to the right of point *e*, beyond the extensive margin of forestry where rent-earning capacity declines to zero. For these lands that are located too far away, it is uneconomical to harvest even naturally growing timber. These lands may contribute in a general way to the natural environment, but if they are perceived as having no value in any specific use, the decision-making problem does not arise. They may, of course, take on some value in the future, but it is not until then, or at least until a future value is anticipated, that decisions must be made about their allocation.

A second category comprises lands that are demanded for only one form of use. Many productive forests, rangelands, and remote recreational resources fall under this heading. They are illustrated in Figure 6.3 by the range of land over which only forestry is capable of generating a positive rent, near point *e*. This case raises the issue of the appropriate intensity of use, but there is no problem of allocation among competing uses.

It is worth noting that the absence of competing demands on land does not mean that the land is technically incapable of other forms of production; it only means that it cannot generate an economic rent in other uses. The physical or biological capability of land to produce various goods and services presents an allocation problem only when two or more uses can yield a positive net return.

The remaining categories refer to lands that are capable of generating rent under two or more forms of production, raising the problem of choosing the best use or combination of uses. The efficient choice depends heavily on the technical interdependence of the uses, which can take several forms. Most common are *competing uses,* where the production of one product requires some compromise in the other. For example, forestland used for timber production can also be used for recreation, but recreational capacity can usually be expanded only by sacrificing some timber production, and vice versa.

This is like the well-known trade-off between "guns and butter" used to illustrate production possibilities in economics textbooks: if all factors of production in an economy were devoted to producing guns or butter, more of either of these products could be produced only by producing less of the other. The versatility of factors of production effectively enables one product to be transformed into another, according to a transformation or *production possibilities curve* like that shown in Figure 6.4A. The curve joins up all the possible combinations of timber and recreation that can be produced on a fixed tract of forest with the same amount of labour and other variable inputs.

The production possibilities curve in Figure 6.4A shows that without any provision for recreation, it is possible to produce T timber, measured in cubic metres of wood per year, and without any timber production, R recreation days could be accommodated. The points along the curve between these extremes indicate the possible combinations of the two products that can be produced with the same inputs.

The nearly horizontal slope of the curve near its intersection with the vertical axis indicates that it is possible to produce some recreation with very little sacrifice in timber production, but the more recreation produced, the greater the sacrifice in timber required to produce another unit of recreation. Similarly, the more timber produced, the greater the sacrifice in recreation required to produce another unit. This gives the curve its concave shape, reflecting an *increasing marginal rate of transformation* of one product for the other. Thus, the curvature of the production possibilities curve reflects the degree of competitiveness between

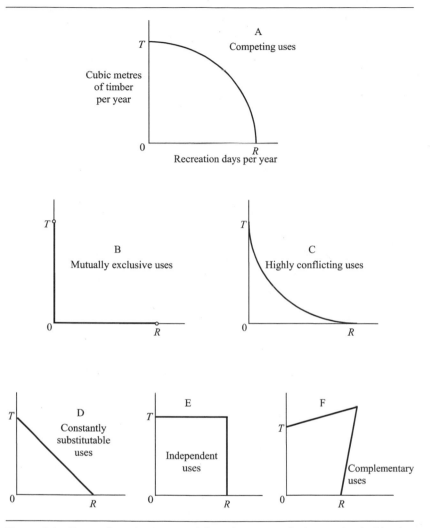

Figure 6.4   Types of production possibilities for two products on a tract of land.

the two outputs, and this degree of competitiveness varies over the range of possible combinations.

Any point inside the curve indicates a possible combination of the two products, but it implies that the resources available are not being fully utilized, or are being used inefficiently, because there are points on the curve showing that more of both products can be produced with the given inputs. Thus, the curve is a frontier of production possibilities.

Production possibilities curves may take other shapes, as illustrated in the other portions of Figure 6.4. They are encountered less frequently in forest management, but they sometimes have important implications for land-use decisions.

- *Mutually exclusive uses* are uses that are entirely incompatible. An example is timber production and preservation of a virgin forest for its scientific value. Thus, Figure 6.4B shows the quantities of each of two products that can be produced, either *T* or *R*, but no combinations of the two.
- *Highly conflicting uses* are those where successive increments in the output of one product can be accommodated with progressively smaller sacrifices of the other. This relationship is unusual, but in some circumstances the trade-off between timber production and amenity values can take this form. For example, a small amount of industrial forestry can have a significant aesthetic impact on an otherwise undisturbed landscape, but successive increments of timber production would have smaller effects. This relationship is illustrated by the convex production possibilities curve in Figure 6.4C, which implies a *decreasing marginal rate of transformation* of one output for the other.
- *Constantly substitutable uses* are those for which the trade-off between the two products remains the same throughout the full range of production possibilities. Two such products produced on a tract of forest might be fuel wood and industrial timber, although there are few examples in forestry of such a *constant marginal rate of transformation*. This case is illustrated by the straight-line production possibilities curve in Figure 6.4D.
- *Independent uses* are uses that have no effect on each other. Managing a forest for purposes of watershed protection may have no impact on recreational values, for example. Independent production possibilities are illustrated by the right-angled curve in Figure 6.4E, which indicates that either product can be produced without impinging on the other.
- *Complementary uses* are found where one form of production enhances another. For example, management of a forest for timber production may benefit wildlife or range values in some circumstances. The production possibilities curve in Figure 6.4F illustrates how the production of one product increases the capacity to produce the other. This is a rare case, as the only relevant point is at the kink. Any other point along and inside the frontier as drawn is technically inefficient, since one output can be increased without decreasing the other.

These two-dimensional diagrams illustrate the relationship between only two products. To illustrate similar relationships among three products, it would be necessary to add a third axis, at right angles to the other two, and the production possibilities curves would take the form of three-dimensional curved surfaces and planes.

The impact of a marginal change in the production of one product on the capacity to produce a joint product varies with the intensity of land use. Most importantly, low intensities of use often permit non-conflicting uses, whereas highly intensive management of land for any purpose often gives rise to conflict with other values. For example, low-intensity timber management and low-intensity recreation may be able to exist together, but intensive production of either timber or recreation requires reduction in the other.

Relative Values and Optimal Combinations

Even if two products or services can be produced on a tract of land and both can generate rent, it is not always advantageous to produce both. Accommodation of a second use might impinge so heavily on the first that the aggregate net value generated by both may be reduced. Moreover, even when two or more uses can be served beneficially, there remains the question of how much one ought to be sacrificed for another – that is, what is the appropriate compromise between the two?

As long as the objective is to generate the maximum land rent, account must be taken of the relative values of the interdependent products. Graphically, this can be represented by the slope of an *exchange line,* such as V_tV_r in Figure 6.5, where the value of timber, V_t, is equal to the value of recreation, V_r. Most important here is the *relative* value of the two products, which governs the slope of the exchange line. In the figure, this line is positioned so that it is just tangent to the production possibilities curve, *TR,* for timber and recreation, at point *E.*

This point, *E,* indicates the optimum combination of the two products – *y* timber and *x* recreation – since no other possible combination will yield as high a total value. To the left of *E,* the more gradual slope of the production possibilities curve compared with the exchange line means that additional recreation is worth more than the necessary sacrifice in timber. To the right of *E,* total value can be increased by sacrificing recreation for timber. Thus, value can be added by shifting production in favour of one product as long as the resulting gain is greater than the sacrifice in value of the other product. The optimal combination of the two products is found at the point where the trade-off in physical possibilities is just equal to the trade-off in value between the two products –

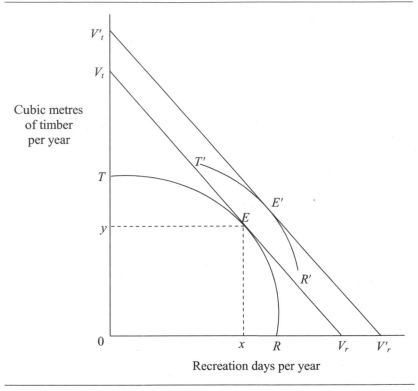

Figure 6.5   Optimal combination of two products.

that is, where the marginal rate of transformation of one product for the other is just equal to the ratio of their marginal values. Therefore, the highest total value of combined production occurs at the point where the two curves are parallel, which is the combination indicated by *E* in Figure 6.5.

With other forms of production possibilities, illustrated in Figure 6.4, the corresponding solutions are simpler. Independent and complementary uses should always be accommodated if both yield a net return, because doing so will increase the aggregate rent to the land. The point at which a sloping exchange line would be tangent to the production possibilities curves in Figures 6.4E and 6.4F is sharply defined and it indicates combined production of both products. For mutually exclusive, highly conflicting, and constantly substitutable uses, the solution will always be to produce only one product, whichever can generate the greatest rent. Thus, in Figures 6.4B, 6.4C, and 6.4D, a sloping exchange line would intersect the production possibilities curve where the latter intersects either the vertical or horizontal axis, indicating that the maximum rent can be generated by producing only the product measured on that axis.

Expanded Possibilities with Additional Inputs

The preceding representation of joint production possibilities is based on an assumption of fixed inputs. If the inputs available for production are variable, the frontier of production possibilities is not constrained to a single curve as described above. With a bigger land base or more labour and capital, the production possibilities curve will shift outward, indicating that, with more inputs, more of either or both products can be produced, as illustrated by the curve $T'R'$ in Figure 6.5.

On the expanded frontier, $T'R'$, the new optimal combination of products is E'. This expansion of production will be advantageous if the value of additional outputs produced at E' exceeds the cost of the additional inputs required to shift production from E to E'.

The relationships depicted in Figure 6.5 closely parallel those used in Chapter 2 to demonstrate efficient factor proportions in production and returns to scale.

Joint and Separated Forms of Production

The relationships described in the preceding paragraphs and figures are general and simplified; in practice, the most efficient forms of multiple production are more complicated and vary widely. In some circumstances, two or more products may be produced most efficiently if they are produced together throughout the forest, as might be expected, for example, in a forest devoted to both timber production and sequestration of greenhouse gases. In others, perhaps involving recreation and industrial use, it will be more efficient to produce each separately, in different parts of the forest. In still other cases, the uses may be separated by time, such as where forests support livestock grazing when they are young and wood production when they are older. The possibilities for efficient combinations of production are almost infinite, depending on economic conditions and the natural character and capabilities of the forests themselves.

Competing Land Use: Industrial Timber Production and Environmental Service

Industrial timber production and environmental services provided by forests are, by and large, competing land uses. Even conflicting land uses at a small scale are competing land uses at a larger, regional or national scale. The Northern Spotted Owl is a good example. At a small scale, protecting nesting habitat for an owl pair and engaging in timber production may be conflicting uses, but protecting an owl population and timber production can be viewed as competing uses at the regional scale.

As noted earlier, competing uses often implies that it is optimal to find and produce a combination of outputs rather than one output or another. This does not mean, however, that a combination of outputs need be produced in every acre of land. Some scholars have argued that it is possible and perhaps desirable to allocate some or a small portion of forestland to intensive management and industrial timber production so that the rest of the forest can be used more towards providing environmental services.

In essence, they would advocate the implementation of a division of land theory in forestland uses, similar to the division of labour theory in production and international trade. Forestlands in various parts of a region, a country, or the world have different productivities in timber production and in producing environmental services. Thus, in theory, making some forestlands "specialize" in timber production to meet society's demand for timber and other forestlands "specialize" in providing environmental services can be beneficial to society.

Indeed, we have found that, in many countries, some forestlands are under very intensive management and are primarily, if not exclusively, used in timber production, whereas others are used exclusively for environmental services. Between these two extremes is a continuum of multiple uses resulting from a combination of natural factors, economic forces, and government policies.

Perhaps this division of forestland is mostly an economic phenomenon that evolves with human values and natural conditions. Whether a government should pursue it as a goal on large scale is another question, since advancing the division of forestland with government policies may lead to undesirable consequences in efficiency and geographic distribution of income.

PRACTICAL DIFFICULTIES

If market economies functioned perfectly, as described in Chapter 2, financial incentives would induce landowners to allocate land to its highest use. Most western industrial economies rely heavily on private landowners driven by market incentives to determine how land is used, but markets for forestland often cannot be relied upon to allocate resources efficiently. For one thing, forestland is often publicly owned and thereby withheld from market processes. For another, externalities in land use cause governments to use zoning and other controls to determine how private owners use land. In addition, taxes distort economic incentives, and unpriced costs and benefits distort market signals and decisions.

The limitations of markets as a means of determining the best pattern of land use make empirical analyses of efficient solutions, based on the general theory described in this chapter, increasingly important. These economic analyses, and techniques for incorporating them into integrated resource management plans, have developed rapidly in recent years and are now familiar to practising foresters.

This discussion reveals two dimensions of the problem of optimizing land use where more than one output is involved. One is the identification of the most advantageous combination of products or services that can be produced at a given level of management intensity – that is, the point on a production possibilities curve, as discussed above. The other is to determine the optimal intensity of use, or how much to expand the production possibilities. Both of these questions require that attention be paid to the economic values involved as well as to the technical possibilities of production.

A simple example can elucidate these two issues. Suppose a forest produces a continuing yield of 1,200 cubic metres of timber annually, and accommodates 1,000 recreation days. The net value of a cubic metre of timber is $20, and that of a recreation day is $25.

Managers estimate that without increasing expenditures they could produce an additional 800 cubic metres per year by eliminating half the recreational capacity. The potential gain in the value of timber produced is 800 m^3 × $20/m^3 = $16,000. This value exceeds the loss in recreational value of 500 recreation days × $25/day = $12,500. The difference, a gain of $3,500, implies that the original combination was to the right of E on the curve TR in Figure 6.5, and the proposed change is closer to the economic optimum.

Suppose that this new combination was precisely the optimum, and that resource planners estimated that with additional annual management expenditures of $15,000 they could increase the production of both products by 50%, that is, by 1,000 cubic metres and 250 recreation days. Clearly, the extra value generated – (1,000 m^3 × $20/m^3) + (250 days × $25/day) = $26,250 – exceeds the cost, and expansion of the production possibilities is advantageous.

The practical application of the theoretical solution to optimal land use described in this chapter is rarely as simple as this illustration. Although the theory is straightforward, the data required to establish the relationships involved raise many difficulties. Two types of data are needed: the technical interdependencies of the products that underlie their joint production possibilities, and the economic information about their relative values. The crucial information about their interdependence is their marginal rate of transformation, or the trade-off in increasing one at the expense of another over the range of possibilities. Earlier, we noted the complicated ways in which the production of different goods and services may be integrated in managing a forest. Relatively little empirical research has been directed towards these technical relationships, and the findings cannot be readily transferred from one forest situation to another.

The extent to which more intensive management can expand production possibilities is also likely to be more complicated than our diagrams suggest. Although it is relatively easy to identify the joint production of an established forest management regime, information about the potential outputs from more intensive or less intensive management is often sketchy, and is complicated by the fact that the rate of transformation between joint outputs is likely to differ at different levels of management intensity, causing the shift in the production possibilities frontier to be asymmetric and irregular. These are major challenges in analyses of multiple land-use opportunities.

Corresponding difficulties surround the economic data. The problem calls for information about marginal costs and benefits of each product. The cost of producing most products and services can usually be estimated, but where two or more products are produced jointly, it is often difficult to separately identify the costs to be ascribed to each. For example, if reforestation of a site increases both recreational values and timber values, the cost must be prorated between them so that the benefit of each can be compared with its cost.

Measurement of benefits is often difficult also. Market prices are often unreliable and must be corrected for market distortions. Most problematic are the values of unpriced benefits such as outdoor recreation and amenity, discussed in Chapter 5.

For all these reasons, the optimal solution may be difficult to calculate. Forest managers often know, however, or can estimate, possible alterations in their management regimes and the changes in output and value they would yield, or at least whether one possible alternative is superior or inferior to another.

AN ILLUSTRATION: LAND-USE CHANGE ASSOCIATED
WITH THE RISE OF INSTITUTIONAL TIMBERLAND OWNERSHIP
IN THE UNITED STATES

So far in this chapter, we have discussed spatial land use allocation and multiple uses of land at a given time. Although the vast majority of land may be allocated for a particular use for a long time – even though the management intensity changes along the way – some land-use change occurs in certain parts of the world all the time. We conclude this chapter with an example of changing use of forestland in the United States in recent years, which is fuelled in part by ownership changes that have much to do with government policies. A full treatment of temporal dynamics in the forest sector in the long term, including forest land base, will be provided in Chapter 9.

Before the mid-1990s, industrial forest landowners owned some 66 million acres or 13% of all US timberland, which contributed nearly 30% of the nation's timber supply. Since then, however, almost all large forest products companies have either become Real Estate Investment Trusts (REITs) themselves or sold most of their timberland to institutional investors. These institutional investors include pension funds, endowments, insurance firms, and foundations that favour diversified investment portfolios. To manage their lands, institutional investors hire Timberland Investment Management Organizations (TIMOs). Most industrial firms now focus on production of manufactured forest products and have timber supply contracts with the institutional investors that bought their timberland.

Interestingly, empirical studies have found that timberland ownership can enhance the profitability of forest products firms and lower their levels of risk. All this industrial restructuring therefore raises two questions: On the supply side, why do these industrial firms choose to sell their timberland? On the other hand, why do institutional investors want to buy them? Obviously the land is worth more to institutional investors than it is to industrial forest owners, but why would TIMOs (and REITs) be able to capture greater benefit than industrial owners?

A possible explanation lies in financial pressure on forest companies. Between 1980 and 2000, low earnings put pressure on integrated forest products companies to liquidate or sell their timberlands. In addition, some forest products firms had incurred heavy debts in purchasing other firms, and the need to reduce their debts added to the pressure to sell assets.

Besides these financial and market circumstances, however, institutional and tax arrangements played a role. US accounting rules (GAAP, or generally accepted accounting principles, which apply to all publicly traded companies) undervalue timberland because tree growth is not considered an increase in asset value. This lowers the value of industrial owners' assets, leaving them subject to potential buyouts.

The demand for timberland by investment companies was also spurred by the 1974 Employee Retirement Income Security Act (ERISA), which required pension plans to diversify their holdings. Previously, many pension plans held only portfolios of bonds, which often left them underfunded during periods of high inflation. ERISA encouraged pension plan managers to think about other assets, including timberland.

Another significant factor in the shift in forestland ownership has been the tax treatment of the revenue generated from timberland. The US Internal Revenue Code provides capital gains tax preference for individuals as well as organizations (such as pension funds) that generate passive income from timberland. This means that individuals or organizations selling trees they have owned for at least a year are subject to a tax on their income of only 15%, compared with 30% to 40% income tax paid by industrial forest companies on such income.

Moreover, most forest products firms are classified as *C corporations,* so they also pay dividend tax on earnings distributed to their shareholders. A recent study shows that the difference in timber harvest profits between a double-taxed C corporation and a tax-advantaged financial ownership such as a REIT or a TIMO can be as high as 39%. As a result, the same timberland yields a higher after-tax return to owners organized as REITs and TIMOs than to those organized as industrial corporations. Although owning timberland may otherwise be advantageous, such a high tax margin makes it virtually impossible for vertically integrated forest products companies to compete with REITs and TIMOs.

Finally, and equally relevant to the subject of this chapter, because they are not focused on timber production to feed forest products manufacturing plants, TIMOs and REITs have been more active than industrial owners in seeking out higher returns by reallocating timberland to more valuable residential, commercial, and recreational uses.

To summarize, the ownership and use of forestland in the US have changed dramatically over the last couple of decades, as large industrial timberland ownership, mostly for timber production purposes, has been replaced by institutional timberland ownership aimed at maximizing profits through timber production and non-timber uses and through conversion of forestland to other uses. Changing economic conditions as well as government laws, regulations, tax policy, and accounting rules have much to do with these changes.

In Chapter 7, we turn to optimal rotation age, a management intensity issue when land is used for forestry purposes. Another management intensity issue, silvicultural investment, is considered in Chapter 9.

REVIEW QUESTIONS

1 How does an increase in the wage rate affect the amount of labour that must be employed to generate maximum returns to a tract of forestland? How do diminishing returns to labour influence this effect?

2 Why can more intensive silviculture be justified on highly productive land but not on land of low productivity?

3 Explain how an increase in the price of timber would lead to (a) more intensive cultivation of forestland, and (b) a shift in the extensive margin of timber production.

4 Some land that finds its highest use in timber production is less productive in this use than some land devoted to agriculture. Does this mean that the land devoted to agriculture would be better used in timber production?

5 A forest produces 2,000 cubic metres of timber annually, valued at $50 per cubic metre, and livestock forage that supports 500 cattle, worth $100 per head. The livestock carrying capacity could be increased by 25% by reducing the forest cover, but this would reduce the annual harvest of timber by 300 cubic metres. Would such a change improve the efficiency of land use?

6 A forest produces 1,200 cubic metres of timber annually and accommodates 1,000 recreation days annually. The value of timber is $20 per cubic metre. The forest could produce an additional 400 cubic metres of timber annually with the same input costs, but only if recreational use were restricted to 100 days annually. We don't know how much recreation is worth, but we can calculate the break-even value for this decision. What is the break-even recreational value that would make it worthwhile to increase timber production to 1,600 cubic metres and decrease recreation to 100 days? If recreation is worth more than your break-even value, would you produce at 1,200 cubic metres and 1,000 visits, or at 1,600 cubic metres and 100 visits? This exercise shows that the manager often does not need exact information to make a good decision. Often he can make good decisions if he has most of the necessary information, and an idea about whether the remaining values are greater or less than some critical value.

7 Why can't market forces always be relied on to ensure that land will be allocated to its highest use or combination of uses?

8 Explain the rise of institutional timberland ownership (and the fall of industrial timberland ownership) in the United States in the last two decades.

FURTHER READING

Binkley, Clark S. 2007. *The Rise and Fall of the Timber Investment Management Organizations: Ownership Changes in US Forestlands.* Pinchot Distinguished Lecture. Washington, DC: Pinchot Institute for Conservation.

Bowes, Michael D., and John V. Krutilla. 1989. *Multiple-Use Management: The Economics of Public Forestlands.* Washington, DC: Resources for the Future. Chapter 3.

Clawson, Marion. 1978. The concept of multiple use forestry. *Environmental Law* 8: 281-308.

Gregory, Gregory R. 1955. An economic approach to multiple use. *Forest Science* 1 (1): 6-13.

Hof, John G., Robert D. Lee, A. Allen Dyer, and Brian M. Kent. 1985. An analysis of joint costs in a managed forest ecosystem. *Journal of Environmental Economics and Management* 12 (2): 338-52.

Hyde, William F. 1980. *Timber Supply, Land Allocation, and Economic Efficiency.* Baltimore: The Johns Hopkins University Press, for Resources for the Future. Chapter 4.

Li, Yanshu, and Daowei Zhang. 2010. *Industrial Timberland Ownership and Finacial Performance of U.S. Forest Product Companies.* Working Paper. Auburn, AL: School of Forestry and Wildlife Services, Auburn University.

Pearse, Peter H. 1969. Toward a theory of multiple use: The case of recreation versus agriculture. *Natural Resources Journal* 9 (4): 561-75.

Vincent, J., and C.S. Binkley. 1993. Efficient multiple use may require land use specialization. *Land Economics* 69: 370-76.

PART 3

The Economics of Forest Management

Chapter 7   The Optimal Forest Rotation

One of the most important economic questions in forestry is the age at which trees should be harvested, or the *crop rotation period.* This decision governs how long the capital tied up in the crop must be carried before it is liquidated, and it also governs the size of the forest inventory that must be carried to maintain a given level of production. It is a problem that calls for analysis of biological as well as economic relationships over time and, like the aging of wine, it has intrigued economists for over a century as a classic problem in investment analysis.

Foresters have developed a variety of criteria for selecting the age at which to harvest forest stands, some of which take no account of the economic variables involved. Such criteria include the age at which the trees reach a size best suited for making certain products, the age at which the volume in the stand is greatest, and the age at which the rate of growth in volume is maximized. These technical criteria are likely to prescribe widely divergent rotation periods, but none can be expected to yield efficient results in an economic sense. In this chapter, we are concerned with finding the age to which forests must be grown to yield maximum *economic* returns, taking account of how the volume of wood they contain and its value change with age.

Determining the economically most advantageous rotation period is an investment problem of the kind discussed in Chapter 3 as well as a management intensity (efficient capital allocation) problem associated with a particular land use, discussed in Chapter 6. To maximize the net benefits from growing a crop of timber, we must examine how its value and the costs of producing it vary with its age. Since the costs and benefits associated with a forest crop accrue over time, they must be discounted to their present equivalent values so they can be compared consistently. The rotation age that will generate the greatest difference between the present value of benefits and the present value of costs, and therefore yields the greatest net benefit, is the most efficient, or optimal, economic rotation.

This chapter demonstrates how this economically optimal rotation age is determined, and how it depends on the biological growth and economic characteristics of the forest. The solution is presented using both discrete interest rates and continuous interest rates, whose relationship was presented in Chapter 3. Later discussion compares the economic rotation age with other rotation criteria and examines some important influences on the optimal solution, including the effect of values other than timber.

OPTIMAL ROTATION AGE IN DISCRETE FORMAT

Some Initial Simplifications

To begin, we can simplify the problem by assuming that the only benefit of concern to the forest manager is commercial timber. Recreational benefits, wildlife, livestock forage, and the host of other non-timber values that are important considerations in particular circumstances will initially be set aside, to be considered later.

We can further simplify the presentation by assuming that the appropriate management regime involves clearcutting the entire crop when it reaches harvest age. This implies an even-aged forest, where the rotation period is the number of years between harvests of all the trees on the site. As we will see later in this chapter, the analytical framework nevertheless applies to partial harvests as well.

Finally, to minimize complications, we will initially assume that the crop is established without cost and without delay; that only one crop is to be considered, after which the land will be valueless; that no taxes or management costs will be incurred as the crop grows; and that all costs and prices will remain constant during the growing period. Later, we will relax these assumptions in order to examine more realistic conditions.

Stumpage Value and Stand Age

As mentioned in Chapter 4, the value of the timber in a forest stand is referred to as its *stumpage value*. This is the maximum price that competitive buyers would be prepared to pay for the timber standing in the forest. Accordingly, the stumpage value, S, is equal to the revenue, R, that an efficient producer could expect from harvesting the timber and selling it in the best available market, minus his expected costs, C_h, for harvesting the timber and delivering it to market:

$$S = R - C_h \qquad\qquad (7.1)$$

The costs must include both capital and operating costs and a normal profit to the producer. Thus, stumpage value is the *net* value of the forest crop, which an owner can expect to realize whether he harvests it himself or sells it to competitive logging contractors, dealers, or wood products manufacturers.

Stumpage prices are usually expressed in dollars per board foot (or per cubic foot, cubic metre, or ton) of standing timber. Thus, if sold as standing timber, the value of a forest stand, $S(t)$, is the stumpage price multiplied by the volume of timber that the stand contains:

$$S(t) = P(t) \cdot Q(t)$$

where t refers to the age of the stand, $S(t)$ is the stumpage value of the stand, $P(t)$ is the stumpage price, and $Q(t)$ is the standing timber volume, all at age t.

At a constant stumpage price, $P(t) = P$, a forest stand growing on a hectare of land generally increases in stumpage value as it increases in age, following the pattern indicated by the curve $S(t)$ in the upper quadrant of Figure 7.1. Even when we assume that the stumpage price is constant, the stumpage value increases with stand age for at least three reasons. First, the volume of merchantable timber on the hectare increases as the trees grow. The growth in volume of the stand is illustrated by the dashed sigmoid curve $Q(t)$ in the upper quadrant of Figure 7.1. Its slope increases up to an inflection point then decreases, a growth pattern frequently observed in biology. In the case of a forest stand, the volume continues to increase as long as the (diminishing) annual increment of growth exceeds the (increasing) losses due to insects, disease, and natural mortality in the stand.

Another reason is that the trees become bigger, and more valuable products can be manufactured from larger timber. For example, large-dimension lumber and high-quality veneer can be milled only from large logs, and larger logs have a larger proportion of clear grain. Such quality effects usually cause the value of a stand of timber *per cubic metre* to rise as the trees grow larger with age.

A third reason is that larger timber can usually be harvested at lower cost *per cubic metre*, reflecting economies of log size. Large logs contain

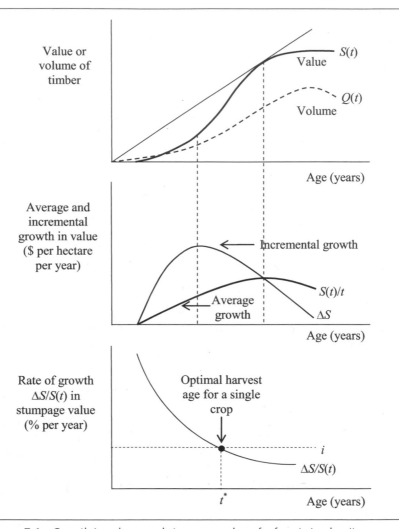

Figure 7.1   Growth in volume and stumpage value of a forest stand as it increases in age.

more volume than small logs and therefore require less handling per cubic metre than small pieces, so logging costs per cubic metre are lower for older and larger trees. There are exceptions to this generalization; for example, uniformity in size of the logs is often as important an influence on handling costs as their average size.

The effect of these influences on the value of a forest stand is that growth in stumpage value usually follows a pattern similar to the forest stand's growth in volume, but at a faster rate and over a longer period, as shown in the upper quadrant of Figure 7.1.

With the information about stand value at various ages, reflected in the curve $S(t)$, the *average rate of growth* in value to any particular age can be readily calculated: it is simply the value of the stand at that age divided by the number of years to that age – that is, $S(t)/t$. Geometrically, this growth rate is the slope of a ray drawn from the origin of the upper quadrant in Figure 7.1 to the point on the total stumpage value curve corresponding to the relevant age. The middle quadrant of Figure 7.1 shows the progression of this average value over the corresponding range of stand age.

The *incremental growth* (ΔS) in value is the increase in stumpage value from one year to the next – that is, $\Delta S = S(t + 1) - S(t)$. It can also be expressed as $dS(t)/dt$, which is the derivative of $S(t)$ with respect to t. It represents the incremental gain in value of the stand if the harvest is delayed an additional year. This incremental gain rises then falls with the age of the stand.

The relationship between the curves of average and incremental growth in value corresponds to the relationship between average and marginal cost curves in the economic theory of the firm. As long as the increment in value from one year to the next is greater than the average growth in value to that age, the average curve must continue to rise. At its maximum, the average growth in value is equal to the incremental growth in value and declines over the range where the incremental growth is less than the average.

Before we move on to optimal economic rotation age in the next subsection, we need to look at two terms that foresters use: *mean annual increment* (MAI) and *current annual increment* (CAI). MAI is the total increment of a stand divided by the age of the stand, $Q(t)/t$. Graphically, MAI is similar to the average growth in value, $S(t)/t$, in the middle quadrant of Figure 7.1. CAI is the incremental growth of a stand in a year, expressed as $\Delta Q = Q(t + 1) - Q(t)$, or $Q_t = dQ(t)/dt$, which is the derivative of $Q(t)$ with respect to t. Graphically, CAI is similar to the incremental growth in value, ΔS, in the middle quadrant of Figure 7.1. As we shall see, choosing the forest rotation at the age when these two terms are equal will yield the maximum possible average annual amount of timber that can be produced on the land in perpetuity.

Optimal Economic Rotation

The *incremental rate of growth* in the value of the stand, expressed as a percentage of the current stumpage value, $\Delta S/S(t)$, follows the pattern illustrated in the lower quadrant of Figure 7.1. It declines as the stand ages, because the denominator increases and the increment in value growth declines over a wide range.

Forest owners seeking the most advantageous harvest age must consider the marginal benefit and marginal cost of carrying the crop from one year to the next. More specifically, in any year, owners must weigh the expected return on their capital from growing the crop for another year, $\Delta S / S(t)$, against the cost of doing so. Ignoring for the moment the cost of the land, the owners' cost in carrying the crop is the interest they could earn on the capital tied up in the crop if they liquidated it and invested the proceeds at the going rate of interest, i. So, to maximize their gains, they will carry the crop only so long as the rate of return on growing the stand exceeds the interest rate, which is to say they will carry the crop as long as the marginal benefit from doing so exceeds the marginal cost, and no longer. Therefore, the age at which the return on the stumpage value falls to the interest rate is a first approximation of the optimal rotation age, denoted by t^* in the lower quadrant of Figure 7.1.

In other words, under our simplifying assumptions, the optimal rotation age is the age at which the marginal benefit of carrying the forest capital for another year, expressed in terms of the percentage increase in stumpage value, is just equal to the opportunity cost of the capital:

$$\frac{\Delta S}{S(t)} = i \tag{7.2}$$

Because it is the age at which the incremental rate of return from another year's growth falls to the opportunity cost of capital, the optimal harvest age is longer *the higher and more prolonged the rate of stand growth in value and the lower the rate of interest.*

This solution is consistent with maximizing the present worth of the harvest according to the formulation in Chapter 3. There (in equation 3.2), the present net value (V_0) of the harvest, discounted to the beginning of the crop rotation, is

$$V_0 = \frac{S(t)}{(1 + i)^t} \tag{7.3}$$

The optimal economic rotation is the age at which this value is greatest. It is also the age at which the annual increase in present value of the stand falls to zero:

$$\Delta V_0 = 0$$

or

$$\frac{S(t+1)}{(1+i)^{t+1}} - \frac{S(t)}{(1+i)^t} = 0$$

This equation is equivalent to

$$S(t+1) = (1+i)S(t)$$

or

$$\Delta S = iS(t)$$

which is the same as the rule in equation 7.2.

Optimal Rotation with Successive Crops
The solution above takes into account only one cost, namely, that of carrying the forest capital. In forestry, however, there are always at least two factors of production involved, capital and land. Hence, there are two categories of costs to be considered, and the cost of the land must be incorporated into our analysis of the optimal economic rotation.

Let us assume for the moment that the land is suitable only for growing timber, or finds its most productive use in this activity, and that each successive crop will involve identical values and costs. The net present value, V_0, of an infinite series of future harvests, $S(t)$, expected at regular intervals, t, can be expressed as the geometric series

$$V_0 = \frac{S(t)}{(1+i)^t} + \frac{S(t)}{(1+i)^{2t}} + \frac{S(t)}{(1+i)^{3t}} + \cdots + \frac{S(t)}{(1+i)^\infty}$$

where each successive term on the right-hand side represents the present net worth of another crop after an additional rotation period of t years. As noted in Chapter 3, this expression can be simplified to

$$V_0 = \frac{\dfrac{S(t)}{(1+i)^t}}{1 - (\dfrac{1}{1+i})^t}$$

or its equivalent:

$$V_0 = \frac{S(t)}{(1+i)^t - 1}$$

which is identical to equation 3.9.

This present worth of an infinite series of future harvests, net of all costs of producing them, is sometimes referred to as the "land expectation value," "site expectation value," "soil rent," or "bare land value." We will use the term *site value*, V_s. If there are no costs involved in producing the crop, the site value can be expressed as

$$V_s = \frac{S(t)}{(1+i)^t - 1} \tag{7.4}$$

This formulation of the site value applies under the assumptions noted above, including the assumption that the land will be used for continuous forestry, that the evaluation is made when the land is in a bare state at the beginning of a rotation, and that reforestation or afforestation is costless. (The effect of reforestation costs on the rotation is examined below.)

The optimal economic rotation for a continuing succession of crops is the rotation that generates the highest site value. This is the age at which the present net worth of the forestry enterprise cannot be increased by extending the rotation age by another year; that is, $\Delta V_s = 0$, which means that

$$\frac{S(t)}{(1+i)^t - 1} = \frac{S(t+1)}{(1+i)^{t+1} - 1}$$

which can be simplified to[1]

$$\frac{\Delta S}{S(t)} = \frac{i}{1 - (1+i)^{-t}} \tag{7.5}$$

This equality will be satisfied at the optimal forest rotation, t_F. Once more, this implies balancing the marginal benefit, expressed as the percentage increase in stumpage value to be gained by carrying the crop another year (the term on the left-hand side of equation 7.5) with the marginal cost, including the annual cost of holding the land (the term on the right-hand side). At any rotation age less than t_F, it is advantageous to

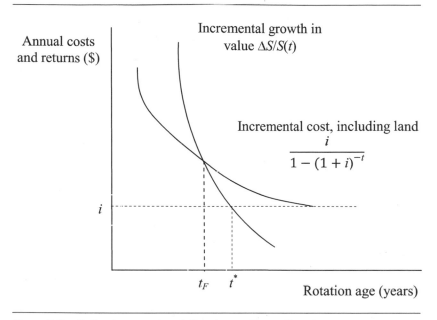

Figure 7.2   Optimal economic rotation for continuous forest crops.

postpone harvesting because the incremental gain in value exceeds the incremental cost. At any greater age, the incremental cost exceeds the gain in value, as shown in Figure 7.2. This expression for the optimal economic rotation for continuous forestry is known as the *Faustmann formula,* after the German forester Martin Faustmann, who derived it in 1849.

Figure 7.2 illustrates the relationships between the terms on the two sides of equation 7.5, and the optimal rotation at age t_F, where they are equal.

Taking account of successive crops shortens the optimal rotation age from t^* to t_F. Algebraically, this is because the incremental cost under successive crops, $i \div [1 - (1 + i)^{-t}]$, is greater than the incremental cost, i, for a single crop (because the denominator in the former is less than 1). Graphically, this means that the curve of incremental cost under successive crops is higher, and therefore intersects the curve of incremental growth in value at an earlier rotation age. As Figure 7.2 shows, the incremental cost $i \div [1 - (1 + i)^{-t}]$ exceeds the interest rate, i, and converges with it at longer rotation ages.

Logically, the shorter rotation under successive crops is explained by the addition of the cost of the second factor of production in forestry,

TABLE 7.1 Value and costs of growing a forest to various harvesting ages

Age t	Volume per hectare[a] Q(t)	Value per cubic metre[b]	Stumpage value per hectare S(t)	Annual incremental change in stumpage value ΔS(t)	Annual interest on stumpage value[c] iS(t)	Site value[d] V_s	Land rent[e] A	Annual incremental cost[f] iS(t) + A
Years	m³	$	$	$	$	$	$	$
10	2	0	0		0	0	0	41
15	14	0	0	0	0	0	0	41
20	51	0	0	0	0	0	0	41
25	124	0	0	0	0	0	0	41
30	232	2	464	93	23	140	7	64
35	366	4	1,464	200	73	324	16.2	114
40	513	6	3,078	323	154	510	25.5	195
45	662	8	5,296	444	265	663	33.2	306
50	802	10	8,020	545	401	766	38.3	442
55	929	12	11,148	626	557	817	40.9	598
				680				

58	995	12	13,187		659	827	41.4	700
				690				
60	1,039	14	14,546		727	823	41.1	768
				713				
65	1,132	16	18,112		906	793	39.6	947
				730				
70	1,209	18	21,762		1,088	739	37.0	1,129
				732				
75	1,271	20	25,420		1,271	672	33.6	1,312

Notes:

a Volume for Douglas-fir in coastal British Columbia, site index 40 at reference age 50 years. Figures rounded to nearest cubic metre. *Source:* BC Ministry of Forests, Inventory Branch. 1982. *Variable Density Yield Tables and Equations for Coastal Douglas-fir in British Columbia.* Forest Inventory Report No. 2. Victoria: Ministry of Forests.

b Values based on assumed stumpage values increasing by $2 per cubic metre every five years beginning at age 25.

c Using an interest rate of 5%.

d Where $v_s = \frac{S(t)}{(1+i)^t - 1}$

e Where $A = iV_s$ and the maximum value in this column, $41.40, reached at age 58, is the land rent for this site.

f That is, the sum of the annual interest on stumpage value, $iS(t)$, and the maximum annual opportunity cost of land, A ($41.40 at age 58), rounded to the nearest dollar.

land. The opportunity cost of the land is the value it could generate if any existing crop were removed and a succession of new crops were initiated. It increases the incremental cost of carrying the present crop and causes it to intersect with the incremental growth in value at an earlier age – at t_F rather than t^* in Figure 7.2.

The extent to which the optimal rotation is shortened by taking account of subsequent crop values in this way may be small, however, because the force of discounting reduces the value of later harvests, especially over long rotation periods and at high discount rates. For example, a single crop worth $10,000 in 60 years has a present value, discounted at 5% (using equation 7.3) of $535. The present value of an infinite succession of crops every 60 years, each valued at $10,000, discounted at the same rate (using equation 7.4) is only $30 greater. This is the present value of all the harvests following the first one. Consequently, the economic rotation is only slightly shorter when more than one crop is considered, but the effect is greater the lower the discount rate and the shorter the rotation age. The effect may be significant where, for example, fast-growing species such as eucalyptus and radiata pine are produced in the southern hemisphere on rotations as short as 5 to 10 years.

Since the annual opportunity cost of the land, or the land rent, A, is the annual equivalent of this site value, iV_s, equation 7.5 can also be expressed as[2]

$$\Delta S = iS(t) + A \tag{7.6}$$

This implies that forest owners will continue to grow their timber crop as long as the increment in value exceeds the increment in costs of holding the timber and the land, but no longer. We will use equation 7.6 to demonstrate the optimal economic rotation age in the following example.

An Illustration
The way the economic variables converge to define the optimal economic rotation can be illustrated with reference to Table 7.1, which shows how a particular forest type grows in volume and value over time. The stumpage values of the stand, $S(t)$, in column 4 enable us to calculate the site value at different ages using equation 7.4. Using a discount rate of 5%, the maximum site value of $827 per hectare occurs at a rotation age of 58 years, based on straight-line interpolation between the values for 55 and 60 years.

The annual opportunity cost of the land, or the land rent, A in this case, is calculated as the annual equivalent of this site value:

$$A = iV_s$$
$$= 0.05(827)$$
$$= \$41.40$$

This is the highest value appearing in column 8 ("Land rent") of Table 7.1. This maximum site value and its equivalent annual land rent help identify the optimal economic rotation.

It is economically advantageous to continue to grow a forest crop as long as its increase in value, from one year to the next, exceeds the incremental cost of doing so. The incremental growth in stumpage value of the stand is shown as ΔS in Figure 7.3. As equation 7.6 shows, the incremental cost has two components: the cost of the land and the capital, or more precisely the land rent, A, and the interest on the stumpage value of the forest, $iS(t)$. Thus, the optimal rotation age, t_F, is the age at which ΔS is equal to $iS(t) + A$.

These relationships are illustrated in Figure 7.3 with data from Table 7.1. The figure shows the incremental growth, ΔS, exceeding the incremental costs, $iS(t) + A$, up to 58 years, which is the optimal economic rotation age.

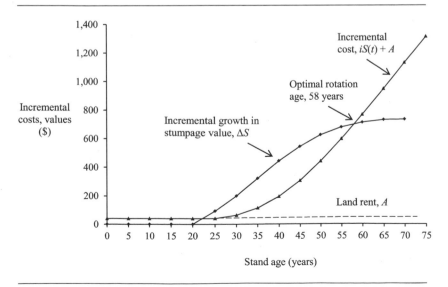

Figure 7.3   Incremental growth in value and costs with stand age.

In this example, the optimal economic rotation age occurs before the incremental growth in stumpage value, ΔS, reaches its maximum, but in other cases it may be later. In any event, it will usually differ from rotation age determined by using other criteria discussed later in this chapter.

Planting Costs

The preceding examples are based on the assumption that no costs are incurred in establishing forest crops. Afforestation (planting on land that has no forest on it) and reforestation (planting following a harvest) are not always costless, however. If a reforestation cost, C, must be incurred for planting or other measures to re-establish the forest whenever it is harvested and no afforestation cost is needed, the value of the harvest, $S(t)$, must be reduced by that amount in calculating the site value:

$$V_s = \frac{S(t) - C}{(1 + i)^t - 1} \tag{7.7}$$

The rate of incremental growth in stumpage value, $\Delta S/S(t)$ in Figure 7.2, must also be revised, to $\Delta S/[S(t) - C]$.

Another possibility is that the land, to begin with, is bare, so the planting cost must be incurred at the beginning of the first rotation as well as at the end of it and all subsequent rotations. In this case, the site value must be further modified by subtracting the planting cost from the site value in equation 7.7:

$$V_s = \frac{S(t) - C}{(1 + i)^t - 1} - C$$

$$= \frac{S(t) - C(1 + i)^t}{(1 + i)^t - 1} \tag{7.8}$$

$$= \frac{S(t)(1 + i)^{-t} - C}{1 - (1 + i)^{-t}}$$

Note that for simplicity we assume that afforestation cost and reforestation cost are equal and remain constant. The incorporation of planting costs introduces an important general point in forestry: the values that forests generate depend not only on time, as implied by the term $S(t)$, but also on silvicultural effort and investment. The economic implications of silvicultural investments are further considered below and in Chapter 9, where timber volume is dealt with as a function of time as well as silvicultural effort, using continuous discounting.

OPTIMAL ROTATION AGE IN CONTINUOUS FORMAT

The formulas in the previous section can be applied more flexibly using continuous discounting. Other than derivational flexibility and clarity by explicitly treating timber volume as a function of time as well as silvicultural efforts, the continuous form of Faustmann rotation age does not add new knowledge. It only helps us better understand the economics of optimal rotation. Readers who choose to skip this section but understand the last section well will not miss much. On the other hand, readers who go through this section will be prepared to access most of the current forest economics literature that uses the continuous form of rotation age. Again, the continuous version of the discrete discounting factor $(1 + i)^{-t}$ is e^{-rt}, where t is time and r represents the continuous interest rate that is equivalent to the annual interest rate, i, as noted in Chapter 3.

Conceptual Model

Let us begin with the productive process of a representative hectare of forestland that first becomes available for afforestation in year zero and will be reforested at the end of each rotation. The production function for a given hectare is

$$Q = Q(t, E)$$

where Q is timber volume, which is a function of time, t, and the level of initial period silvicultural effort, E. In the absence of silviculture, land lies idle until it regenerates naturally, and only time affects the production level, $Q = Q(t)$. On the other hand, when a positive amount of silvicultural effort is applied, it can affect the biological production level at a given time indirectly, such as by decreasing reforestation lag, as well as directly, by changing annual timber growth. Therefore, the level of silvicultural effort affects both the level and the rate of production.

Some silvicultural efforts, such as site preparation, planting, and fertilization, are applied in the initial period. Others, such as second fertilization, herbaceous weed control, and prescribed burning, are applied in later years. To simplify the analysis, we convert all of these efforts, weighted by their unit cost and appropriate discount rates, to the beginning of the first period to come up with a total silvicultural cost (which is the reforestation or afforestation cost, C, of the previous section):

$$C = wE$$

where w is the unit factor cost of silviculture and E is a quantity variable and represents total silvicultural efforts.

In other words, we can think of w as the unit cost of labour in silviculture. If a landowner spends a total of $750 per hectare in site preparation and tree planting in the initial period (year 0) and an additional $200 per hectare in year 5 in herbaceous weed control, and his discount rate is 6%, then the total silvicultural cost in the initial period is

$$\$750 + \$200/(1 + 0.06)^5 = \$750 + \$200/1.338 = \$899.50/\text{hectare}$$

Thus, if the prevailing unit labour cost is $15 per hour, we can think that the landowner has used 60 hours of labour ($899.50/15) in the initial period. Thus, our w is $15 per hour, and E equals 60 hours. In this way, all silvicultural costs, including materials and machinery, can be treated as labour cost.

Another important economic variable is stumpage price. As noted earlier, the landowner's total revenue, $S(t)$, is equal to stumpage price times timber volume. As revenue occurs only when timber is sold in some future period, we discount it to come up with a present value. If the continuous discount rate $r > 0$, then the present value of the total revenue stream for each hectare, at the moment that idle land becomes available for timber production, is

$$PQ(t, E)e^{-rt}$$

where P is stumpage price, which is assumed to be constant, and $PQ(t, E)$ is thus identical to the stumpage value, $S(t)$, discussed in the previous section. Again, we assume here that timber volume is a function of time as well as silvicultural effort, whereas in the previous section timber volume is expressed as a function of time only.

If timber production is the highest-value use of the land, both now and in the future, we need to look at the case of successive rotations on the same land, or perpetual timber production. The net present value of the first rotation (V_1) is

$$V_1 = PQ(t, E)e^{-rt} - wE$$

Maximizing the net present value of the land or site value, V_S, means maximizing the summation of net present value in all the rotations, V_j, from 1 to infinity. This will generate a site value:

$$V_s = \sum_{j=1}^{\infty} V_j$$

$$= [PQ(t, E)e^{-rt} - wE](1 + e^{-rt} + e^{-2rt} + e^{-3rt} + ...) \qquad (7.9)$$

$$= \frac{PQ(t, E)e^{-rt} - wE}{1 - e^{-rt}}$$

This formula is equivalent to equation 7.8, although Q in equation 7.8 is responding only to time, not reforestation and afforestation efforts. Again, reforestation and afforestation costs (C) in equation 7.8 are decomposed to w and E in equation 7.9.

Obviously, if it is not economical for landowners to invest in silviculture at all (that is, if natural regeneration is most advantageous and there are no reforestation, afforestation, or other silvicultural efforts), the term wE in equation 7.9 disappears and $Q(t, E)$ becomes $Q(t)$. Thus, we have

$$V_s = \frac{PQ(t)e^{-rt}}{1 - e^{-rt}}$$

$$\qquad (7.10)$$

$$= \frac{PQ(t)}{e^{rt} - 1}$$

which is equivalent to equation 7.4.

If there are reforestation costs, as demanded by some governments, but not afforestation costs, equation 7.9 will become

$$V_s = \frac{[PQ(t, E) - wE]e^{-rt}}{1 - e^{-rt}}$$

$$\qquad (7.11)$$

$$= \frac{PQ(t) - wE}{e^{rt} - 1}$$

which is equivalent to equation 7.7.

If trees are already growing on the site, equation 7.9 also needs to be modified. If the trees are already merchantable, then the net return from timber sales today must be added to the right-hand side of this equation. If the trees are not yet of merchantable age, the value of the pre-merchantable trees must be added on the right-hand side as well. In addition, the right-hand side must be multiplied by e^{-rx}, where x is the number of years until the trees currently growing on the site are sold.

TABLE 7.2   Faustmann formula in discrete and continuous formats

Costs	Discrete	Continuous
No afforestation and reforestation costs $(C = wE = 0)$	$V_s = \dfrac{S(t)}{(1+i)^t - 1}$	$V_s = \dfrac{PQ(t)e^{-rt}}{1 - e^{-rt}}$
	$= \dfrac{PQ(t)}{(1+i)^t - 1}$	$= \dfrac{PQ(t)}{e^{rt} - 1}$
With reforestation costs only	$V_s = \dfrac{S(t) - C}{(1+i)^t - 1}$	$V_s = \dfrac{[PQ(t,E) - wE]e^{-rt}}{1 - e^{-rt}}$
	$= \dfrac{PQ(t) - C}{(1+i)^t - 1}$	$= \dfrac{PQ(t,E) - wE}{e^{rt} - 1}$
With both reforestation and afforestation costs	$V_s = \dfrac{S(t) - C}{(1+i)^t - 1} - C$	$V_s = \dfrac{PQ(t,E)e^{-rt} - wE}{1 - e^{-rt}}$
	$= \dfrac{S(t) - C(1+i)^t}{(1+i)^t - 1}$	
	$= \dfrac{PQ(t)(1+i)^{-t} - C}{1 - (1+i)^{-t}}$	

The formula for determining the value of land when trees are already growing on the site is then

$$V = \frac{1}{e^{rx}}\left[PQ(t,E) + \frac{PQ(t,E)e^{-rt} - wE}{1 - e^{-rt}}\right] \tag{7.12}$$

This formula can be used to determine the expected value of forest-land with pre-merchantable timber stands.

Table 7.2 shows the similarities and differences when the Faustmann formula is expressed in discrete and continuous formats.

Faustmann Rotation Age

In the foregoing expressions, t is an arbitrary rotation age. Differentiating equation 7.9 with respect to t and setting the resulting equation equal to zero, we find the necessary conditions for a maximum of the site value or land expectation value:

$$V_t = [PQ_t(1 - e^{-rt}) - rPQ + rwE]e^{-rt}(1 - e^{-rt})^2 = 0$$

where V_t and Q_t are the derivatives of site value, V_s, and timber volume, Q, respectively, with respect to t. This expression can be simplified as

$$\frac{Q_t}{Q - \frac{wE}{P}} = \frac{r}{1 - e^{-rt}} \tag{7.13}$$

If trees regenerate naturally and $wE = 0$, then equation 7.13 can be rewritten as

$$\frac{Q_t}{Q} = \frac{r}{1 - e^{-rt}} \tag{7.14}$$

which is equivalent to equation 7.5. Note that the left-hand side of equation 7.5 is expressed in stumpage value. When the term for stumpage price (P) in the numerator and denominator of the left-hand side of equation 7.5 is cancelled out, the left-hand side of equation 7.5 is the same as the left-hand side of equation 7.14.

Again, the optimal economic rotation age that meets the condition expressed in equations 7.13 (with afforestation and reforestation cost) or 7.14 (without reforestation or afforestation cost) is the aforementioned Faustmann rotation age (t_F).

CUTTING CYCLES FOR AN UNEVEN-AGED STAND
Uneven-aged forest stand management is popular in many parts of the world. Although we postulated even-aged stands in the foregoing discussion, the Faustmann formula is equally applicable in determining the optimal cutting cycle and growing stock for uneven-aged stands. In fact, if we treat the merchantable volume left after each partial harvest of an uneven-aged stand as a cost of regenerating the forest stand, the theoretical basis for maximizing the land and forest value in an uneven-aged stand is similar to that for maximizing the land expectation value, or site value, of an even-aged stand. Thus, we have a unified approach towards the management of even-aged and uneven-aged timber stands.

The economic interpretation of the optimal stocking, the amount that should be harvested and retained, and the optimal cutting cycle can be presented in the usual fashion. At the optimal cutting cycle, the extra revenue that can be earned by delaying the cutting cycle for one year must equal the cost of holding both the land and the timber for one year. Similarly, if the extra revenue that one can earn later in a cutting cycle by leaving one more unit of growing stock in the stand now is greater than the compounded value of that unit of growing stock, that unit should be left on the stand.

COMPARISONS WITH OTHER ROTATION CRITERIA

The Faustmann formula maximizes the economic rent to the fixed factor of production, land, and thereby maximizes the present net value of the future stream of forest crops. It follows that any other rotation age will yield lower net returns. Nevertheless, a variety of other criteria for determining the rotation age have been proposed and often adopted, and deserve mention here.

Maximum Growth of Wood

A widely accepted criterion for establishing the rotation age in traditional forestry regulation is the age that maximizes the average annual growth in volume of wood, or the forest's *mean annual increment* (MAI). This rule takes no account of any economic variables. As a result, it usually indicates rotations significantly different from the economically optimal age described above.

Nevertheless, this criterion has deep roots in forest policies for public lands in the United States and Canada and some private lands in other countries, where it has been a cornerstone for pursuing the objective of *maximum sustained yield* (MSY). In the simplest case, the maximum sustained yield of a forest is achieved by growing all the trees, or stands of trees, to the age at which their average annual growth of wood (MAI) is greatest, and harvesting the total *growth* or *increment* of the forest every year or at regular periods in perpetuity (as discussed in the following chapter).

The age of maximum average annual growth is usually (but not always) greater than the economic optimum because it takes no account of the (usually faster) increase in the forest's opportunity costs or the costs of postponing harvests. The economic return is thus diminished, and may be eliminated altogether, but the effects vary widely.

Maximum Growth in Stumpage Value

Another criterion for choosing the rotation is the age at which the *average rate of growth in stumpage value, $S(t)/t$,* is greatest, which occurs when the ΔS curve intersects the $S(t)/t$ curve in the middle quadrant in Figure 7.1. This rule has the familiar tone of the MSY criterion, which can be readily seen in Figure 7.1. This rule is sometimes referred to as the "forest rent," which must not be confused with the land rent discussed here.

The significance of this rule lies in the popularity of "forest rent" as a guide to the optimal rotation age. It is argued that by maximizing the average annual growth in stumpage, the revenue from the forest is maximized. "Forest rent" ignores the opportunity costs of capital and land, however, and maximizing "forest rent" almost always lowers the net value of forest crops.

Maximum Value of a Single Crop

Earlier in this chapter, we considered the optimal rotation for a single forest crop, where the land has no opportunity cost, noting that the economic return will be maximized by extending the rotation as long as the rate of return on growing the forest exceeds the rate of interest. The optimal rotation is the age at which this incremental benefit and cost are equal, that is, where $\Delta S = iS(t)$ in the lower quadrant of Figure 7.1.

This formulation of the optimal rotation, sometimes referred to as the *Fisher rotation age*, does not recognize the opportunity cost of the land, and so is inadequate whenever the land has an alternative productive use, including subsequent forest crops. Because it fails to consider this cost, it always results in a longer rotation than the optimum.

Summary

In short, the optimal economic rotation age is likely to differ, sometimes substantially, from the age determined by using one of these other three criteria where some or all economic variables are missing.

As implied by its name, the objective for maximum sustained yield (MSY) is to find the forest rotation age that leads to the maximum possible annual output that can be maintained in perpetuity. To find a rotation age that maximizes sustained yield in perpetuity, we need to use the concepts of mean annual increment (MAI) and current annual increment (CAI), noted earlier.

Recall that MAI is the average volume growth, $Q(t, E)/t$, in a stand (or district). In the parlance of production economics, the MAI is simply average product. Graphically, it is given by the slope of a straight line from the origin to a point on the growth curve, $Q(t, E)$. The current annual increment (CAI) is analogous to marginal product, and is given by the slope of line tangent to the growth curve. Mathematically, CAI is defined as $Q_t = dQ(t, E)/dt$.

It is well known from microeconomic theory that the marginal product curve intersects the average product curve from above and at the point where the average product attains its maximum:

$$\frac{Q(t, E)}{t} = Q_t$$

Rearranging this equation gives the usual relationship of finding the MSY rotation age:

$$\frac{Q(t, E)}{Q_t} = \frac{1}{t}$$

As price and interest rate are not in the equation that determines the MSY rotation age, this MSY rotation age is a biological criterion. The single rotation age model does not consider the opportunity cost of land in growing successive crops. The "forest rent" rotation age model does not have an interest rate component and thus ignores the incremental cost of carrying the forest capital (in both land and timber). Only the Faustmann model incorporates both biological and full economical (prices, costs, and interest rates) factors.

Note that the relationship between the Faustmann rotation age and the maximum sustained yield rotation age is not unidirectional. In most cases, the Faustmann rotation age is shorter than the maximum sustained yield rotation age. In rare circumstances, the Faustmann rotation age is longer than the maximum sustained yield rotation age. The latter situation arises mostly with short rotation and fast-growing species, and when the interest rate is less than the inverse of the maximum sustained yield rotation age.

OTHER IMPACTS ON THE OPTIMAL ROTATION
The optimal forest rotation is influenced by the productivity of the land, the value of the timber produced, the costs of harvesting it, taxes and other costs of management, the interest rate, non-timber benefits, and other conditions. These vary widely in different circumstances. This section considers how some of these factors affect the optimal forest rotation.

Interest Rate
The importance of the interest rate should be emphasized because it determines the incremental cost of carrying the forest capital. Equation 7.5 (or equations 7.13 and 7.14) shows that the incremental cost on the right-hand side of the equation must increase with any increase in i (or, in its continuous form, r). Thus, *the higher the interest rate, the shorter the optimal rotation.*

The interest rate is never as low as zero, for reasons explained in Chapter 3, but it is instructive to consider such an extreme case. Then in the absence of the force of discounting, the optimal rotation would simply be the age at which the average annual rate of growth in stand value, $S(t)/t$, is maximized (referred to earlier as the age of maximum "forest rent"). This is why we said in the previous section that the "forest rent" rotation age ignores the opportunity cost of capital (tied in timber) and land.

Reforestation Costs and Stumpage Prices

As noted earlier, when reforestation costs are incurred in establishing new crops following harvests, the rate of incremental growth in stumpage value, $\Delta S/S(t)$ in Figure 7.2, must be revised to $\Delta S/[S(t) - C]$. This reduces the value of the crop by the cost, C, of replanting. The effect will be to shift the incremental growth curve upward to the right, lengthening the optimal rotation age.

This effect can also be seen in equation 7.13, where an increase in reforestation cost (wE) will increase the value of the left-hand-side of the equation, which is the marginal revenue of carrying the forest capital for an additional year. *The greater the reforestation cost, the longer the optimal rotation.*

It follows that a price increase that is perceived as permanent will decrease the value of the left-hand-side of equation 7.13. Thus, *a one-time price increase will cause a decrease in optimal rotation age.*

Where planting is necessary but no other costs are involved, growing successive crops will be advantageous if the value of the harvest, discounted to the beginning of the rotation, exceeds the planting cost. If this is true for one crop, it will be the case for successive crops, notwithstanding the impact of discounting.

Note that if the reforestation cost were equal to the present worth of the harvest, that is, $C = S(t)/(1 + i)^t$, so that the present net worth of growing a new crop were zero, it can be seen by substituting $C = S(t)/(1 + i)^t$ into equation 7.7 that the latter equation simply collapses into equation 7.3, and that the optimal rotation age would be the same as in the case of a single crop, derived from equation 7.2. This is because the land would yield no net value in subsequent crops. This land, to use a term from Chapter 6, is at the extensive margin of continuous plantation forestry. Natural regeneration is the only rational choice.

Land Productivity

Not only reforestation costs but anything that raises the cost of growing the crop has the effect of lengthening the optimal forest rotation. The same is true of anything that lowers the value of the harvest or reduces the productivity of the land in terms of its capacity to generate value. All these reduce the value of the crop, and hence also reduce the two incremental costs in equation 7.6, the interest on the forest capital, $iS(t)$, and the land rent, A. They also reduce the incremental growth in value, ΔS, although by a smaller amount. This result can be seen in Figure 7.3. The incremental costs and benefits will intersect later, indicating a longer optimal rotation age, t_F. In other words, as long as other influences remain the same, *the more productive the forest site, the shorter the optimal rotation.*

Annual Costs

Forestry usually involves some continuing costs of management, protection, administration, and taxes. If these costs are a constant annual amount, m, they can easily be introduced in the above formulas. As we saw in Chapter 3, the present worth of such an infinite future series is simply m/i, which can be subtracted from the right-hand side of equation 7.4, reducing the site expectation value by this amount.[3] Constant annual costs will not alter the optimal rotation, however, because the age that maximizes the site value also maximizes the site value minus a constant m/i.

This illustrates a general rule: *the optimal rotation is not affected by any cost that is independent of the way the forest is managed.* Any such cost reduces the net returns from the forestry enterprise, of course, but as long as the burden cannot be shifted or lessened by changing the management regime, it will not affect the rotation age.

Property Taxes

Forest taxes and other charges are dealt with in detail in Chapter 11, but we must note here their impact on the optimal forest rotation. Taxes on forest properties usually consist of a percentage rate applied against the tax base, which is typically the value of the land, the value of the timber, or both of these combined.

Tax on Land Value

The simplest property tax is one based on the bare land, usually consisting of a percentage tax rate applied annually to the site value. This produces a fixed annual charge and has the same effect as other annual costs noted above. Although it will lower the land's value to the owner, *a tax on the bare land value will not alter the most advantageous harvest age.* It will simply divert some of the value generated by the land from the owner to the government.

Tax on Timber Value

A tax rate applied annually to the value of the standing timber will result in an annual charge equal to a constant fraction of the stumpage value.

The forest owner will thereby incur an annual cost, increasing each year with the stumpage value of the stand, and cumulating at compound interest until the stand is harvested. The return from growing the forest is the stumpage value when it is harvested minus the accumulated tax payments.

Such an annual *ad valorem* (or percentage of the value) tax on growing timber has the same effect on the owner's choice of the rotation age as an increase of that percentage in the rate of interest in the Faustmann formula. As already seen, increasing the discount rate shortens the optimal rotation.

Correspondingly, the longer the crop is grown, the greater the accumulated taxes that must be charged against the harvested timber. This means that *an annual tax on the value of the standing timber provides an incentive to shorten the rotation.*

Tax on Harvest Value

A yield tax, in the form of a percentage rate applied against the stumpage value, is sometimes levied on timber when it is harvested. A tax of this kind has the effect of reducing the value of the harvest in proportion to the tax rate.

Like higher reforestation costs that must be incurred when the forest is harvested, a yield tax can be postponed, and thereby reduced in present value, by postponing the harvest. The effect of this, and the reduced stumpage return to the owner, means that a *yield tax on the harvest lengthens the preferred rotation.*

Trends in Costs and Prices

The discussion so far has implied that the values for stumpage and the costs that enter into the calculation of the optimal rotation are constant over time. In reality, they are likely to change. Although it is often difficult to predict the direction and magnitude of future changes in these economic variables, we must incorporate our expectations, or best guesses, into our calculations.

Values and costs fluctuate constantly with changing economic conditions, but for purposes of planning forest crops over long periods, the concern is not with short-term swings but with long-term trends. The costs and values used in the Faustmann formula must be adjusted to reflect expected trends over the long periods for which the calculations are made.

For example, if stumpage values are expected to rise over time at some regular percentage rate, an approximation of the present net worth and the rotation age can be obtained by reducing the discount rate applied to the Faustmann formula by that percentage.[4] As stumpage value equals stumpage price times volume, this means that *when stumpage price increases at a constant rate over time, the rotation age will increase.*

This appears to contradict the comparative static result, reached in the subsection "Reforestation Costs and Stumpage Prices" above, of the negative effect of stumpage price on rotation age. What we must remember, however, is that the earlier result is based on a single change in price level, whereas the result described here relates to a continuous change of price at a constant rate.

This simple procedure of adjusting the discount rate to account for trends in values and costs can be used only when the trends are expected to be steady. If they are expected to change in some irregular fashion, each term in the geometric series that leads to equation 7.4 will differ, so they cannot be reduced to a simple expression. Instead, each term must be treated separately, making the calculation of the optimal rotation much more cumbersome.

There are a variety of other costs, values, and taxes associated with growing timber. All such additional charges and values affect the optimal rotation except those that are independent of the value of the forest, and most can be assessed by modifications of the Faustmann formula.

THE HARTMAN ROTATION AGE

So far we have considered only timber benefits in our optimal rotation age. Other forest products and services are often affected by the way timber is managed and harvested, however. As we noted in the preceding chapters, efficient forest management must take into account impacts on values other than timber. We can demonstrate this by drawing on the 1976 formulation of economist Richard Hartman for incorporating non-timber values in the Faustmann solution for the optimal rotation.

Here non-timber values are broadly defined and include benefits derived from a forest's scenic amenities, watershed functions, carbon sequestration services, biodiversity, and non-timber products such as wildflowers and mushrooms. Each of these non-timber values is often related to timber volume or the age of the forest, but the exact relationship varies and may be estimated only by empirical investigations. Complicating the matter is that we have to aggregate all these benefits, correlate them with timber value at various ages, and analyze them along with commercial timber values to determine the optimal rotation age.

Hartman Rotation Age in Discrete Format
We first present the Hartman rotation age in discrete format. This enables readers who do not want to deal with calculus, especially integration, to skip the next subsection, "Hartman Rotation Age in Continuous Format," without loss of understanding of the Hartman rotation age concept. Let us start with a single rotation model and then move on to the rotation age with successive forest crops.

Recall that the optimal condition for the single rotation age is represented by equation 7.2, which can be rearranged as

$$\Delta S = iS(t)$$

The left-hand side is the marginal benefit of timber growth and the right-hand side is the marginal cost of holding timber for an additional year. Now, assume that $n(t)$ represents marginal non-timber benefit at age t. Adding $n(t)$ to equation 7.2 yields

$$\Delta S + n(t) = iS(t) \tag{7.15}$$

which states that timber should be harvested only when the sum of marginal timber and non-timber benefits is equal to the marginal cost.

Similarly, the optimal condition for a perpetual series of crops is represented by equation 7.6:

$$\Delta S = iS(t) + A$$

or by

$$\Delta S = iS(t) + iV_s$$

where A is the land rent and V_s is the site value. Similarly, adding $n(t)$ to this equation yields

$$\Delta S + n(t) = iS(t) + iV_s \tag{7.16}$$

Again, this equation implies that the forest crop should be allowed to grow until the marginal benefits (consisting of both timber and non-timber benefits) are equal to the increment in costs of holding the timber and land. Note that in equation 7.6, the site value (V_s) and consequently the land rent (A) include only the value of timber, whereas in equation 7.16, these terms include non-timber benefits as well. The rotation age that meets the requirement of equation 7.16 is the *Hartman rotation age*.

Equation 7.16 shows that as long as no timber benefits are positively related to age, the Hartman optimal rotation age is longer than the Faustmann rotation age. In some cases, the marginal benefits may always exceed the marginal costs, implying that the timber should never be harvested.

Hartman Rotation Age in Continuous Format

Let us put the timber value aside for a moment and consider maximizing only non-timber value in our optimal rotation model.

The difference between commercial timber and non-timber benefits is that, except for commercial thinning, commercial timber benefits accrue only at the end of the rotation, when trees are harvested. Non-timber benefits, on the other hand, accrue continuously or annually in the discrete sense.

Suppose that non-timber values of a tract of forest are given by an accumulative function (up to age t), $N(t)$:

$$N(t) = \int_0^t n(t)$$

where $n(t)$ is the (not discounted) non-timber value in a given age, t. Typically, the first derivative of $N(t)$ with respect to t is greater than zero $[N'(t) = n(t) > 0]$ and the second derivative is assumed to be less than zero $[N''(t) < 0]$. In other words, the marginal non-timber value increases over time, but at a decreasing rate.

Our objective is to choose the rotation age that maximizes the discounted stream of such benefits, recognizing that benefits fall to zero each time the tract of forest is cut. Using the continuous discounting format, this problem can be written as

$$V_n = \text{Max} \frac{N(t)e^{-rt}}{1 - e^{-rt}} \tag{7.17}$$

where V_n is the capitalized or net present value of non-timber benefits of a forest stand.

Differentiating this equation with respect to t and setting the first-order condition to zero yields an expression for determining a rotation age that maximizes only non-timber benefits:

$$\frac{N_t}{N(t)} = \frac{r}{1 - e^{-rt}}$$

(7.18)

where N_t is the derivative of $N(t)$ with respect to t and is, as noted, equal to $n(t)$.

This expression has a familiar economic interpretation: as long as the rate of increase in non-timber value exceeds the rate of increase in opportunities – given by the discount rate, r, which is modified by a perpetual term, $1/(1 - e^{-rt})$ – timber harvesting should be delayed. When that relationship is reversed, it is worthwhile to cut the trees as non-timber benefits can be improved by harvesting them. As long as non-timber values increase with age at a decreased rate, finding a maximum is possible.

Now consider both commercial timber and non-timber values over an infinite planning horizon. The objective function is then to maximize the combined benefits:

$$\text{Max } V_s + V_n = \text{Max}\left[\frac{PQ(t,E)e^{-rt} - wE}{1 - e^{-rt}} + \frac{N(t)e^{-rt}}{1 - e^{-rt}}\right]$$

(7.19)

$$= \text{Max}\frac{PQ(t,E)e^{-rt} - wE + N(t)e^{-rt}}{1 - e^{-rt}}$$

where $PQ(t, E)$ is the commercial timber benefits, $N(t)$ is the non-timber benefits over rotation t, and wE is silvicultural investment, as noted earlier. The necessary condition for maximization is

$$\frac{PQ_t + N_t}{PQ + N(t) - wE} = \frac{r}{1 - e^{-rt}}$$

(7.20)

This expression is similar to the earlier ones when only timber or non-timber benefits are considered, and is to be interpreted in the same way. The optimal rotation age occurs when the value of an additional period's growth in both timber and non-timber value is just offset by the implicit loss incurred by postponing timber harvesting for an additional period. The rotation age that meets this condition is called the *Hartman rotation age*, as noted earlier.

Rearranging this expression yields the following:

$$(PQ_t + N_t)(1 - e^{-rt}) = rPQ + rN(t) - rwE$$

(7.21)

We can examine the rotation age more closely by segregating equation 7.21 into its components. The left-hand side is the marginal benefit (MB) of additional volume growth (the additional timber from an extra period multiplied by its value, PQ_t) plus the added non-timber benefits (N_t), all restricted in a single rotation in a perpetual sequence – multiplied by $(1 - e^{-rt})$. The right-hand side is the marginal cost (MC) of delaying harvest and has three components: the immediate timber benefits forgone, rPQ, plus the non-timber benefits received over the next growth period, $rN(t)$, minus the gain from delaying silvicultural investment for an additional period, rwE. To maximize society's benefit from forests, timber harvesting occurs when marginal benefit equals marginal cost.

Other Considerations

Many non-timber values vary in different ways with forest age. Indeed, some non-timber values are higher when the forest cover is young, or discontinuous, or has a high proportion of recently harvested areas. Examples are certain kinds of wildlife or grazing values. This is illustrated in line I of Figure 7.4. Other values are higher when forest stands are older, such as certain recreational, aesthetic, or biodiversity (Northern Spotted Owl, Red-cockaded Woodpecker, or trout) benefits. This is illustrated in line III. Yet other non-timber values are unrelated to forest age (generic bird species), illustrated in line II. Aggregating these non-timber values would produce line IV in Figure 7.4.

Each forest product or service is affected by the age of the forest in a different way. If they are affected at all, they will affect, in turn, the rotation age that will maximize the aggregate net value of the forest.

Again, finding the exact relationship between all non-timber values and forest age is a challenge. Even if we found a reliable relationship for all non-timber values, we would need to add up all the non-timber values and combine them with the commercial timber value. Finally, solving for a maximum when both commercial timber and non-timber values are considered is more complicated because a single and unique maximum cannot be found.

It is clear, however, that if non-timber values increase with forest age, then the Hartman rotation age will be longer than the Faustmann rotation age. Conversely, if the non-timber value is greater in younger forests, its incremental growth with stand age will be negative, shortening the optimal rotation.

Further, a logical extension to the above discussion is that for some standing forests, the non-timber benefits might be so great that it would not be economically feasible to harvest the forests at any time in the

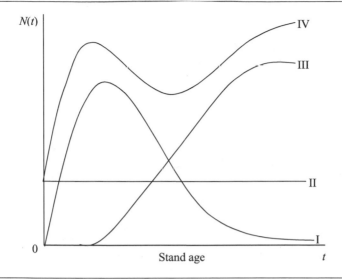

Figure 7.4   Relationship between stand age and various amenity values. |
Source: Adapted from van Kooten and Folmer 2004.

future. In this case, the environmental benefits become the primary
benefit of the forests while commercial timber value becomes a secondary benefit. Often, when "ancient" forests are used for environmental
services or for sacred, religious purposes, or when newly established
"protective" forests are used for water and soil erosion control, wind-
break, and so on, it may be worthwhile to delay harvests, perhaps indefinitely, or as long as the trees are alive.

The next chapter explains the important implications of the choice of
rotation age for the scheduling of harvests in regulated forests.

NOTES

1 The expression

$$\frac{S(t)}{(1+i)^t - 1} = \frac{S(t+1)}{(1+i)^{t+1} - 1}$$

can be solved by subtracting $S(t)$ from both sides to give

$$\Delta S = S(t)\left[\frac{(1+i)^{t+1} - 1}{(1+i)^t - 1} - 1\right]$$

and the term in large brackets can be simplified to obtain equation 7.5.

2 From equation 7.5:

$$\Delta S = \frac{iS(t)}{1 - (1+i)^{-t}}$$

$$= \frac{iS(t)(1+i)^t}{(1+i)^t - 1}$$

then subtracting and adding $iS(t)$ from the numerator:

$$= i\left[S(t) + \frac{S(t)}{(1+i)^t - 1}\right]$$

$$= i[S(t) + V_s]$$

which is the same as equation 7.6.

3 As noted, the site value, V_s, of bare land that must be planted at a cost, C, at the beginning of each rotation becomes

$$V_s = \frac{S(t) - C(1+i)^t}{(1+i)^t - 1}$$

If an annual management cost, m, must be accounted for as well, m/i must be subtracted from this value. A convenient general formulation for calculating the present value of a perpetual forest where a variety of revenues, $R_1, R_2 ...$, and costs, $C_1, C_2 ...$, are expected at different times during the rotation, and management costs, m, are expected annually, is

$$V_0 = \frac{R_1(1+i)^{t-T_1} + R_2(1+i)^{t-T_2} + \cdots - C_1(1+i)^{t-T_1} - C_2(1+i)^{t-T_2} - \cdots}{(1+i)^t - 1}$$
$$+ \frac{m}{i}$$

where t is the rotation age and $T_1, T_2 ...$ is the number of years from the present to the time the periodic cost ($C_1, C_2 ...$) or revenue ($R_1, R_2 ...$) will occur. The numerator of the first term on the right-hand side sums the value of all revenues and costs at the end of the current rotation (t)

by compounding, whereas the denominator provides for the perpetual periodicity of this value. The second term accounts for the present value of the annual management cost. In the special case where there is only one revenue at the final harvest (so that $t - T_1 = 0$) and there are no periodic costs ($C_1, C_2 \ldots = 0$) or annual management cost ($m = 0$), V_0 here becomes the site value corresponding to V_s in equation 7.4.

4 If stumpage values are expected to rise at a certain rate, α, the stumpage value of a harvest t years hence will be $(1 + \alpha)^t S(t)$, and the present value of a future series of crops becomes

$$V_0 = \frac{(1 + \alpha)^t S(t)}{(1 + i)^t} + \frac{(1 + \alpha)^{2t} S(t)}{(1 + i)^{2t}} + \frac{(1 + \alpha)^{3t} S(t)}{(1 + i)^{3t}} + \cdots$$

and the expression for the site value, equation 7.4, becomes

$$V_0 = \frac{S(t)}{\left(\dfrac{1 + i}{1 + \alpha}\right)^t - 1}$$

Manipulating the term in parentheses in the denominator (by adding and subtracting 1) yields $[1 + (i - \alpha)/(1 + \alpha)]^t$. Thus, the effective interest rate for calculating the optimal rotation is reduced to $(i - \alpha)/(1 + \alpha)$, which for practical purposes is equal to $(i - \alpha)$ as $(1 + \alpha)$ is close to 1 when α is small. It is therefore sufficient to use a discount rate of $(i - \alpha)$ in the Faustmann formula to account for the upward trend in stumpage values.

The above derivation is much simpler if we use the continuous version of the equation:

$$V_0 = \frac{e^{\alpha t} S(t)}{e^{rt}} + \frac{e^{2\alpha t} S(t)}{e^{2rt}} + \frac{e^{3\alpha t} S(t)}{e^{3rt}} + \cdots$$

$$= \frac{S(t)}{e^{(r - \alpha)t}} + \frac{e^{2\alpha t} S(t)}{e^{2(r - \alpha)t}} + \frac{e^{3\alpha t} S(t)}{e^{3(r - \alpha)t}} + \cdots$$

$$= \frac{S(t)}{e^{(r - \alpha)t} - 1}$$

Here r is the continuous discount rate. This means that when timber volume stays the same and stumpage price increases at a constant rate, α, the discount rate will become $(r - \alpha)$.

REVIEW QUESTIONS

1 Why is the growth in stumpage value of a forest stand not precisely proportional to its growth in wood volume?

2 What are the incremental costs of carrying a forest from one year to the next? What is the incremental benefit? Explain why harvesting at the age when the incremental costs and benefit are equal will maximize the return to the forest owner.

3 The merchantable volume of timber on a hectare of forest increases with the age of the stand, as shown in the following table. Assume that all timber has a value of $5 per cubic metre, and that the applicable interest rate is 6%.

Stand age (years)	Timber volume (m³/ha)
15	0
20	50
25	100
30	240
35	400
40	530
45	640
50	730
55	760

Calculate the land rent and the optimal economic rotation under continuous forest production.

4 What is the effect on the optimal rotation age of (a) an increase in the interest rate, (b) reduced reforestation costs, and (c) higher annual costs of fire protection?

5 If the recreational value of a forest is higher the older the trees are, what effect will this have on the optimum age for harvesting the timber?

6 What are the differences among the (a) Faustmann rotation age, (b) maximum sustained yield rotation age, (c) maximum growth in stumpage value rotation age, (d) Fisher rotation age, and (e) Hartman rotation age?

FURTHER READING

Amacher, Gregory S., Markku Ollikainen, and Erkki Koskela. 2009. *Economics of Forest Resources.* Cambridge, MA: MIT Press. Chapters 2 and 3.

Anderson, F.J. 1991. *Natural Resources in Canada: Economic Theory and Policy.* 2nd ed. Toronto: Nelson Canada. Chapter 7.

Binkley, Clark S. 1987. When is the optimal rotation longer than the rotation of maximum sustained yield? *Journal of Environmental Economics and Management* 14: 142-58.

Calish, Steven, Roger D. Fight, and Dennis E. Teeguarden. 1978. How do nontimber values affect Douglas-fir rotation? *Journal of Forestry* 76 (4): 217-21.

Chang, Sun Joseph. 1981. Determination of the optimal growing stock and cutting cycle for an uneven-aged stand. *Forest Science* 27 (4): 739-44.

–. 1983. Rotation age, management intensity, and the economic factors of timber production: Do changes in stumpage price, interest rate, regeneration cost, and forest taxation matter? *Forest Science* 29 (2): 267-77.

Clark, Colin W. 1976. *Mathematical Bioeconomics: The Optimal Management of Renewable Resources.* New York: John Wiley and Sons. Chapter 8.

Faustmann, Martin. 1849. Calculation of the value which forest land and immature stands possess for forestry. In *Martin Faustmann and the Evolution of Discounted Cash Flow: Two Articles from the Original German of 1849,* edited by M. Gane; translated by W. Linnard. Institute Paper No. 42. Oxford: Commonwealth Forestry Institute, Oxford University, 1968.

Hartman, Richard. 1976. The harvesting decision when a standing forest has value. *Economic Inquiry* 14: 52-68.

Heaps, Terry. 1981. The qualitative theory of optimal rotations. *Canadian Journal of Economics* 14 (4): 686-99.

Hyde, William F. 1980. *Timber Supply, Land Allocation, and Economic Efficiency.* Baltimore: The Johns Hopkins University Press, for Resources for the Future. Chapter 3.

Newman, David N., Charles B. Gilbert, and William F. Hyde. 1985. The optimal forest rotation with evolving prices. *Land Economics* 61 (4): 347-53.

Pearse, Peter H. 1967. The optimum forest rotation. *Forestry Chronicle* 43 (2): 178-95.

Samuelson, Paul A. 1976. Economics of forestry in an evolving society. *Economic Inquiry* 14 (4): 466-92.

van Kooten, G. Cornelis, and Henk Folmer. 2004. *Land and Forest Economics.* Northampton, MA: Edward Elgar. Chapter 11.

Chapter 8   Regulating Harvests over Time

Chapter 7 dealt with the age at which a "stand" of timber should be harvested for maximum economic efficiency. This chapter is concerned with the "forest," which usually consists of many stands, differing in species composition, age, site productivity, and other characteristics. Whereas Chapter 7 considered the time between harvests on a particular site, this chapter deals with the rate of harvesting within the whole forest.

How a forest should be harvested over time is one of the most fundamental issues of forest management. Decisions about the rate of harvesting and how it is related to rates of growth are the primary means of managing the structure and composition of a forest. Moreover, because harvesting is the activity that generates revenues and reduces the capital tied up in timber, its timing is critical to the economic performance of a forest enterprise, large and small.

Beyond the scale of the individual forest or firm, decisions about harvest rates govern, in large part, the economic, social, and ecological impacts of forestry. By determining the supply of timber, they dictate the size of the forest industry, the number of people it employs, and its stability over time, which are often critical factors in regional economic activity. Harvest rates also impact the spatial management of recreational and environmental forest values. For all these reasons, decisions about the level of harvesting and how it is adjusted over time are major preoccupations of both private forest owners and government forest agencies.

In this chapter, we explain the difference between a stand and a forest, discuss various timber-harvesting regulations, and examine the results in the absence of regulations. In the last section, we note some new techniques that are now widely used by forest firms whether or not they are subject to government timber-harvesting regulations.

THE STAND AND THE FOREST

As noted above, a forest stand is a group of trees sufficiently uniform in species composition, age distribution, growth rate, and other characteristics to form a distinguishable management unit. Stands vary widely in size, from a small cluster of trees to thousands of acres, but their essential characteristic is that each comprises a forested area sufficiently consistent in its composition to enable it to be managed under a single regime. A forest, in contrast, usually comprises a forested area delineated by ownership or regulatory boundaries, and usually consists of many stands. It may be as small as a woodlot or as large as a national forest of millions of acres in the United States, but the forest is the unit for management decisions relating to the forest enterprise as a whole. Although each stand must be managed with attention to its particular capabilities and circumstances, major decisions about access development, investment in forest enhancement and protection, rates of harvesting, and the measurement of economic performance normally apply to the whole forest. Accordingly, individual stands must be managed in ways that contribute to the objectives for the whole forest.

The benefits derived from a forest are not simply the sum of the benefits derived from all the component stands managed independently, because some management decisions affect many stands. For example, the location of roads and processing plants influences the cost of transportation and hence the value of timber in different areas. In addition, forestry activities such as fire protection, pest control, and silviculture need to be coordinated over many stands in order to be effective and to achieve economies of scale. Protection of watersheds and of wildlife and recreational values and production of other non-timber goods and services also require coordinated management of many stands.

Among the most important questions that must be dealt with in the context of the whole forest is how harvesting and the production of timber should be scheduled over time, which is the subject of this chapter.

The relationship between this issue and the market supply of timber should be noted. In Chapter 4, the market supply was depicted as the sum of the quantities of timber that all producers offer for sale in a particular market. Here we are concerned with how the manager of a particular forest determines the forest's production. Such production, combined with that of all other producers supplying the market, determines the market supply.

MARKET SOLUTIONS AND LIMITATIONS

To a large extent, the challenge of regulating the harvest rate is akin to the problem most firms face in managing an inventory. The stock of timber in the forest is costly to carry over time, and its opportunity cost is the potential rate of return that could otherwise be earned on its capital value. Like other inventories, the forest can be diminished through utilization and sale, and augmented through investment in more production.

In a perfectly competitive market economy, such as that described in Chapter 2, private forest owners could be expected to manage their forest inventories and timber supplies independently and efficiently in response to market costs and prices. The key to obtaining the greatest possible value from the forests has been explained in earlier chapters; they must be managed and harvested so that the net present value of the stream of goods and services produced over time is maximized. For any particular stand, this means that harvesting should occur when its growth in value, from one year to the next, no longer exceeds the incremental cost of carrying it for another year, as explained in Chapter 7, particularly Figure 7.3. If forest owners' only interest were to maximize the economic return from the forest, and if they could alter their harvesting by any amount without affecting their per-unit costs and prices, they could simply apply this rule for the optimal rotation individually to every stand in their forests. In doing so, they would generate the maximum possible value from each stand and from their forests as a whole.

Our particular interest in this chapter is the yield of timber over time that would flow from a forest under this regime. Obviously, the outcome will depend on the existing age distribution of the component stands. To illustrate the range of possibilities, consider a forest consisting of stands of uniform size and productivity, with the number of stands being equal to the number of years in the optimal rotation period. If at one extreme there was one stand of each age class from zero to the rotation age, the rule for maximizing returns described in Chapter 7 would result in the harvest of one stand each year forever. The yield would be constant over time. At the other extreme, if all the stands were the same age, the whole forest would be harvested in a single year, and no more would be harvested for one full rotation period. In practice, however, the composition of forests is never so simple. Forests rarely consist of stands of equal size and productivity, evenly distributed among age classes, so the yield of timber, harvested according to the efficiency rule, will fluctuate.

The most advantageous rate of harvesting will change over time for other reasons as well. Fluctuating markets for timber make it advantageous for producers to expand harvests when prices are high and to reduce them when prices are low. And over long periods, changes in technology, costs, and prices can be expected to alter yields, silvicultural opportunities, optimal rotation periods, and other variables, which will alter the efficient harvest rate and change the forest's efficient level of production over time.

Forest owners often find it difficult to fully adjust to market changes and other external conditions, however. Especially in the short term, significant increases or reductions in production will drive up costs, making it more efficient to smooth out sharp fluctuations in production. Owners may be obliged to maintain a steady supply to a manufacturing enterprise, or they may be hampered in changing production levels for financial or other internal reasons. Such constraints have the effect of slowing and smoothing their response to short-term changes in external conditions, but this does not imply an unchanging harvest rate; the most efficient rate for any forest enterprise can be expected to change continuously. Nonetheless, the optimal pattern of harvesting from the forest as a whole will almost certainly be smoother over time than the pattern that would result from harvesting each stand at its independently optimal rotation age.

Forest owners and governmental forest agencies vary widely in their approach to scheduling harvests, depending on the size and complexity of their forests and on their capabilities, motivation, and breadth of interests. In deciding on their harvest plans, owners of small woodlots often simply depend on their own judgments about the growth of their forests, their expectations about timber markets, and their own financial needs. Large forest enterprises increasingly rely on computer-based profit-maximization models, using linear programming or dynamic programming to plan their silvicultural activities as well as harvesting over time and space. These complex models typically incorporate large quantities of data about the forest and its inventory, growth, and responses to silviculture, as well as economic information about costs and values. They may also provide for regulatory and other constraints to protect other values, such as water supplies and environmental values.

A variety of market imperfections can, of course, distort forest owners' decisions, leading to inefficient rates of forest utilization; these include externalities, imperfect competition, divergence between private and social time preference, and other imperfections described in Chapter 2. These distortions often lead governments to intervene in various ways to offset their adverse effects; for example, regulations to mitigate externalities, such as impairment of environmental values by logging, are common. The dominant concern of governments in regulating forest harvest rates is industrial stability, arising from fears that uncontrolled producers, reacting freely to swings in forest products markets, will cause unstable employment and incomes, and that in the long term unregulated exploitation may lead to resource depletion, eroding employment opportunities and the economic base of regional economies.

With few exceptions, owners of private forestland in North America are left to schedule their harvesting as they see fit. Some government policies, such as regulations about forest practices, may constrain their decisions, but the rate of harvesting is not directly regulated and remains a function of their own needs and responses to market signals. In the later part of this chapter, we provide some empirical evidence of the effectiveness of the market as a regulator of private timber harvest over time at the regional and national level in the United States.

Governments in the United States and Canada, however, frequently regulate harvesting on public forestlands to achieve specific objectives relating to stability and sustainability. The significant economic implications of such regulation are examined in subsequent sections.

THE REGULATED FOREST

Harvest regulation or yield planning has a long history in forestry theory and practice, extending from the 14th century in Europe. The focus has always been on reconstructing the forest inventory so that a yield of timber can be sustained over long periods or forever. Underlying many of the classical approaches to forest regulation is the concept of the *normal forest*. This is a forest with an even distribution of age classes, capable of yielding the same volume of timber every year in perpetuity.

The simplest possible case has already been described: a forest of uniform land productivity consisting of stands of equal areas, with one stand of each age from zero to the rotation age. Each year, the crop that reaches rotation age can be harvested and, provided it is always restocked, the harvest can be maintained indefinitely at a rate equal to the growth of the whole forest.

This model of a normal forest, with evenly graduated age classes and steady harvests, is the foundation for the *sustained yield* forest regulation of governments in North America, Europe, and elsewhere. In some cases, the rotation age chosen for each stand is the age at which its mean annual increment is greatest, so that the growth, in terms of the volume of wood, of the whole forest is maintained at the highest rate possible. The model is referred to as a *maximum sustained yield* regime.

Although simple in concept, a normal forest presents a complicated problem in practice. Because a large tract of forestland is never uniform in productivity, a constant yield can be achieved only if the areas assigned to different age classes vary according to the varying productivity of the land. Differences in site productivity imply that the rotation period must differ as well, complicating considerably the problem of scheduling and balancing the sequence of harvests.

Once in place, a fully regulated normal forest implies a dynamic, steady-state cycle of harvests, continuously balanced by growth, which depends on specific and unchanging conditions. If a constant yield is to be maintained, no changes can be made in the land base, in rotation ages, in silvicultural effort, or in standards of utilization.

Because management planning for a normal forest inherently involves projections over long periods of time, these rigid requirements for maintaining the constancy of growth and yield are unlikely to be met. Advances in forest science and silvicultural practice can be expected to improve production over time; technological change is likely to alter the proportion of the crop that can be recovered and utilized; and changing economic circumstances will affect the costs and prices that determine the most advantageous forest management regimes. As a result, the fully regulated normal forest is never achieved in practice. Nevertheless, it is a theoretical model with considerable intuitive appeal to foresters and policy makers, and it is the goal towards which a great deal of forest regulation and management is directed. Students of economics will note that, notwithstanding its orderly conception of a managed forest, it takes no account of the costs or values produced.

TRANSITION TO A NORMAL FOREST

In many forest regions, the main challenge in implementing a sustained yield regime is not how to manage normal forests but how to create them from forests that consist of unbalanced distributions of age classes. For example, in parts of Canada, the United States, Russia, and Brazil, forests consist of original old-growth timber, well beyond the harvest age for managed forests. This stock must be removed and reforested in a carefully controlled pattern over many years to create the graduated age structure required for a sustained yield forest. In China, the United Kingdom, New Zealand, and Portugal, the problem is the opposite: with new and expanding plantation forests, the task is to build up the older age classes. For these and other circumstances, much effort has gone into methods for transforming irregular forests into a fully regulated state.

From an economic standpoint, the planned harvest schedule during the transition period is the issue of primary importance. This is because the transition phase usually extends over many decades, and the force of discounting typically reduces the contribution of harvests beyond this period to a small proportion of the total present value.

The two general approaches in transforming an irregular forest into one with an even age class distribution are referred to as *area control* and *volume control.* In the hypothetical case of a forest on land of uniform productivity, area control involves harvesting and reforesting each year a portion of the forest equal to the total area divided by the rotation period chosen for subsequent crops. Then, in the year that the last of the original forest is harvested, there will be an even gradation of age classes from zero to the rotation age, and the forest can be managed as a normal forest thereafter.

Area control will not produce a steady yield during the transition period, of course. The harvest level at any time will depend on the structure of the original forest and the sequence in which it is harvested. For example, if the forest contained old growth and this was harvested first, the yield may decline when harvesting moves on to equal areas of younger stands. In contrast, a forest of young plantations might initially require cutting stands before their planned rotation age, yielding low volumes initially and higher volumes later.

Volume control, on the other hand, aims to stabilize harvest volumes over the transition period while allowing the areas harvested to vary. It involves fixing the volume to be harvested each year, referred to as the *allowable annual cut* (AAC), using some formula that takes account of the volume and age distribution of the original forest inventory and its rate of growth.

One such formula that has been widely used is the so-called *Hanzlik formula,* originally developed in Germany and used to guide the conversion of virgin forests in the US Pacific Northwest and parts of Canada into normal forests in one rotation period. It specifies the allowable annual cut for one rotation:

$$\text{AAC} = \frac{Q_M}{t} + \text{MAI} \tag{8.1}$$

where Q_M is the existing volume of timber in the forest that is beyond the rotation age, t, planned for subsequent crops, and MAI is the mean annual increment, or average annual growth, expected on the areas occupied by immature stands over their planned rotation. Note that if the initial forest consisted entirely of old timber that had ceased to grow, MAI would be zero, and the formula would specify cutting the original stock in equal increments over t years. Conversely, if the forest consisted entirely of stands of less than the rotation age (as it would in any event at the end of the transition period), the first term on the right-hand side of equation 8.1 would drop out, indicating an allowable annual cut equal to the forest's growth rate.

In contrast to the simple form of area control described above, volume control of this kind can generate steady harvests from an irregular forest during its conversion to a normal forest. At the end of the conversion period of one rotation, however, both methods are likely to call for some adjustment to the long-term sustainable yield of the regulated forest. As suggested earlier, if the original forest contains a predominance of old-growth stands, yielding larger volumes of timber per hectare than can be expected from subsequent crops at their rotation age, the Hanzlik formula would call for a downward adjustment, reflecting the so-called *falldown phenomenon* associated with the conversion of natural forests to managed forests. If the initial forest contains mainly immature stands, however, the allowable annual cut during the transition phase must remain below the ultimate sustainable yield and growth rate.

A variety of formulas or rules can be used to smooth out adjustments to the harvest rate over time. For example, if the level of harvesting must be reduced because of the falldown when an old-growth forest is converted to a managed forest, periodic recalculation of the Hanzlik formula as the overmature inventory is depleted will provide an allowable annual cut that declines more gradually to the forest's sustainable yield. Alternatively, the derivation of the allowable annual cut can be constrained to ensure that it does not change by more than a specified percentage over any decade or other period. Some government forest management agencies in the United States and Canada have pursued an objective of "non-declining even flow," which seeks to avoid any downward adjustment. This presents complex problems for harvest scheduling in converting old-growth forests, and usually extends the conversion over a much longer period than one rotation.

Economic Implications

None of the formulas and decision rules for scheduling harvests described above takes into account the economic implications of their prescriptions although their impact on the economic efficiency of forest enterprises is likely to be substantial. The cost of this reduced efficiency is the difference between the present worth of planned harvests using the rule and the present worth of harvests under the most efficient schedule of harvests.

The loss in efficiency will depend on the characteristics of each forest, but a simple example will serve to illustrate the potential effects. Consider a uniform forest of 1,000 hectares, originally consisting entirely of overmature old-growth timber that has ceased to grow, worth $20,000 per hectare. Let us assume that subsequent stands will grow in volume and value as indicated in Table 7.1, and will be harvested at the optimal rotation age of 58 years. For simplicity, let us also assume that all costs and prices are constant over time and are unaffected by the rate of production from this forest, and that the applicable interest rate is 5%.

Under these conditions the most efficient regime would involve harvesting the entire forest in the first year, which would yield $20,000 × 1,000 = $20 million. To obtain the total value of the forest enterprise, we must add to this the present value of the future crops, which was

TABLE 8.1 Values generated under alternative harvest schedules

	Harvest schedule	
	Unregulated[a]	Regulated[b]
Harvest of original timber	$20,000,000[c]	$6,489,505[d]
Value of subsequent crops	$827,000[e]	$268,341[f]
Total value	$20,827,000	$6,757,846

Notes:

a   Involves harvesting all the timber in the initial forest in the first year.

b   Involves harvesting 1/58 of the initial inventory of timber in each of the 58 years of the forest conversion period.

c   The value of 1,000 hectares of timber worth $20,000 per hectare.

d   The present value of 1/58 of the value of the initial timber (1/58 × 20,000,000 = $344,826) accruing each year for 58 years, using equation 3.7 (page 59).

e   The site value of 1,000 hectares at $827 per hectare.

f   The present worth of the site value of 1/58 of the forest (1/58 × 1,000 × 827 = $14,259) accruing each year for 58 years, using equation 3.7 (page 59).

calculated in Chapter 7 as the site value of the bare land, namely, $827 per hectare. Thus, the total present value of the forest becomes $20,827,000, as indicated in Table 8.1.

Because we have assumed a uniform forest, the area control and volume control methods of converting such a forest to a normal forest involve the same regime, namely, harvesting 1/58 of the initial timber each year for 58 years, which is $344,828. The present value of these harvests, calculated using equation 3.7, is

$$V_0 = \frac{344{,}828(1.05^{58} - 1)}{0.05 \times 1.05^{58}} = \$6{,}489{,}505$$

Using the same formula, the present value of crops after the first rotation can be calculated as the present value of 1/58 of the total site value accruing each year over the same period, or $268,341. The total present value of the regulated harvest regime is therefore $6,757,846, as shown in Table 8.1.

In this example, spreading the harvest evenly over the first rotation period reduces the potential value of the forest enterprise by nearly two-thirds of its potential value. It should be re-emphasized, however, that this result follows from the particular assumptions of this example; the impact of harvest regulation on any forest depends on the particular conditions relating to its initial inventory, the rotation period and interest rate chosen, and other conditions.

Nevertheless, it should be clear that regulations that spread harvests further into the future than is economically efficient substantially reduce the value of forest enterprises. Our simple example illustrates two ways in which this impact is felt. One is the reduction in value of the initial forest that results from postponing its harvest. The other is the reduction in the value of subsequent crops that results from delaying the replacement of the initial forest with new stands. A third way, not illustrated in this example, is the loss in value that may result from a regulatory regime that prevents producers from responding to market swings, constraining them to produce less than would be most profitable when prices are high and more when prices are low.

Finally, note that the extensive margin of operations for harvesting original or virgin timber is unlikely to be the same as the extensive margin for continuous forestry. Land of relatively low productivity may support old-growth timber that is profitable to harvest, yet cannot generate a positive net return in artificially growing subsequent crops. Alternatively, highly productive land may support valuable timber but, once it is harvested, find its highest use in agriculture, urban development, or recreation. The history of land development and use in North America offers many examples of both these circumstances. Thus, the geographical base of forest operations will usually become smaller during the transition from original to managed stands. In the terms used in Chapter 6, this can be described as contraction of the extensive margin of forestry use of land.

Allowable Cut Effect

Another economic effect of regulations that spread harvests evenly over time is their distortion of incentives to invest in silviculture and forest improvements. To illustrate this important effect, consider the possibility of improving a stand established at the beginning of its crop cycle at a cost of $1,000 per hectare, which would double the growth in the example depicted in Table 7.1. At a harvest age of 58 years, the stand would contain 995 cubic metres more than without the treatment.

A decision maker would normally weigh the cost of the treatment against the benefits of the increased yield to be realized 58 years hence. If, however, the forest is regulated according to a rule, such as the Hanzlik formula, that prescribes a constant harvest level over the whole rotation, the enhanced yield would be spread evenly over each year of the growing cycle, through an increase in the allowable annual cut of $(1/58 \times 995)$ = 17.16 cubic metres per year. This is the so-called *allowable cut effect;* any improvement in growth, regardless of when it actually occurs, will increase the prescribed harvest rate immediately and by the same amount each year for the number of years in the forest rotation. Conversely, any loss of inventory or growth due to fire or pests will be spread over the whole rotation, regardless of whether the lost timber is close to harvest age.

The allowable cut effect, in turn, distorts the apparent economic returns from silvicultural investments. In our example, the initial cost of $1,000 would normally be compared with the present value of the resulting increased yield of $13,187 that would be realized in 58 years. Using a discount rate of 5% and equation 3.2, the benefit ($13,187/$1.05^{58}$ = $778) is so low that the project is unattractive. Using the even-flow formula (which means that this increase of 995 cubic metres in timber volume is equally allocated to each of the 58 years), however, 1/58 of this improvement would be taken each year beginning immediately, which, using equation 3.7, has a present value of benefits that exceeds its costs by more than fourfold.

The allowable cut effect thus exaggerates the returns to investments in forestry and, conversely, reduces the economic costs of inventory losses. The magnitude of the distortion depends heavily on the structure of the forest inventory and the rotation age. For forests with a preponderance of virgin timber, the apparent returns to silvicultural investments that enhance growth in new stands can be astronomical, because the allowable cut effect translates the improved growth into an immediate increase in the harvest of mature timber. This result has less to do with the real impact of the silviculture than with the artificial effect of the yield regulation.

Harvest regulation thus can have a major impact on incentives to invest in forestry. An important question, therefore, is whether the allowable cut effect should be incorporated into economic evaluations of investment opportunities in silviculture, protection, or additions and subtractions of land in the regulated forest. The answer must be that, notwithstanding the distortions, it should guide the evaluation as long as the decision maker is constrained to manage the forest within such a regulatory regime. If the regulatory policy will govern the actual outcome of the investment, it must be recognized in the evaluation. The allowable cut effect, however, is an unwarranted distortion of the returns on a forestry investment when the decision maker can use normal investment criteria and behave accordingly. And, of course, any evaluation of the regulatory policy itself must take full account of the extent to which it influences behaviour and reduces the efficiency of forest investments.

SUSTAINED YIELD RATIONALE AND CRITIQUE

To evaluate harvest regulation policies, we must weigh the costs against the benefits. The extensive literature on harvest regulation suggests a considerable variety of benefits and costs that may flow from sustained yield policies. In this section, we present the rationale for sustained yield, followed by a critique of its premises.

Economic Stability

By far the dominant rationale for sustained yield policies is regional economic and industrial stability. The argument is made that a steady harvest of timber will stabilize harvesting and wood manufacturing industries and hence also regional employment and income. This takes us into macroeconomics, the branch of economics that deals with the general level of economic activity, prices and employment, the causes of instability and growth, and the means of controlling these phenomena.

The effect of sustained yield policies depends on the linkages between the level of timber harvesting in a region and the level of regional economic activity.

First, insofar as sustained yield policies are regarded as a governmental means of promoting stability, it is important to clarify the economic variables that they are intended to stabilize. In particular, the concern for long-term economic stability must be distinguished from the problem of short-term instability. Forest products markets suffer from short-term cyclical fluctuations mainly because of shifts in demand (in contrast to markets for agricultural products, for example, which fluctuate mainly because of shifts in supply). These demand shifts are usually due to forces that have little to do with forest policy and management; often, the cause may be as remote as interest rate changes, non-forest–specific government policy, or swings in international markets. Significantly, however, sustained yield forest regulation deals only with the supply side of timber markets.

Moreover, stabilization of supply in the face of fluctuating demand cannot stabilize prices. In the absence of controls, producers reduce production when prices fall and expand when prices rise, cushioning the pressure on prices both ways. A requirement to maintain production at a constant level in the face of shifting demand has a contrary effect, aggravating swings in prices. Pressures on inventories and financial performance are increased as well.

In recognition of these effects, sustained yield regulations typically allow for some adjustments in harvest levels in response to short-term market swings. To the extent that the regulatory regime has any effect in smoothing fluctuations in harvests, however, it will tend to exacerbate market and price instability.

Second, the impact of the forest sector on the level of regional economic activity depends on its importance in the regional economy. Except in rare circumstances, the forest industry is only one among many comprising a regional economy. Each sector follows a somewhat different long-term trend in growth and employment, and macroeconomic policy is usually concerned with stability of incomes and employment in the aggregate. The impact on the stability of regional economies of regulating the forestry sector but not other sectors in the region will vary widely.

Third, harvest regulation does not stabilize the forest industry but rather the flow of raw material it uses. This distinction is important because stability in regional economies is usually sought in terms of employment and income. Over the long periods of time considered in forest regulation, the relationship between the quantity of timber harvested and processed and the amount of labour employed is likely to change significantly. Technological advances and the substitution of capital for labour tend to increase labour productivity by a few percentage points per year, so that over a period as long as a forest rotation, much less labour will be employed per unit of timber produced. Accordingly, stabilizing the production of timber over long periods will not ensure a stable employment level.

With respect to regional income, payments to labour are usually most relevant, because payments to capital are typically smaller and often accrue to investors outside the region. Labour income is largely a product of the number of persons employed and the average wage rate, which often follow opposing trends. Advancing productivity is associated with declining employment per unit of output but increasing wages. Both technological change and trends in real wage levels are governed by forces in the broader economy; how they balance out over long periods and how they affect the level of regional incomes in the forest sector will vary among regions and over time, to a large extent unpredictably.

Other changes also alter the long-term relationship between the volume of timber harvested and regional employment and income. Changes in products and manufacturing processes are likely to occur. New transportation systems are likely to alter the links between particular forests and the manufacturing centres they supply. And the organization and structure of the industry are likely to shift the geographic distribution of activity.

For all these reasons it is difficult to draw many general conclusions about the relationship between the supply of timber in a region, or from a particular forest, and the long-term stability of the regional economy. Some economists argue that there is little evidence to support the proposition that an even flow of timber over long periods will promote regional stability, and that it may, instead, retard growth, adaptation to change, and efficient reallocation of resources.

In any event, it is clear that a policy aimed at stabilizing long-term regional economic activity must look beyond the raw material supply for a particular industry and take account of the rest of the regional economy and the particular circumstances and trends in other sectors. It follows that traditional concepts of the normal forest and sustained yield are not likely to be sufficient to ensure that the forest sector will make its most effective contribution to the stability of regional economies.

National Security

Historically, an important motive of governments in promoting stable forestry was strategic. In pre-industrial Europe, wood was the primary source of energy and fuel for both domestic and commercial needs. It was also the primary raw material in construction and shipbuilding. Indeed, a good deal of early forestry was motivated by naval requirements. Because wood was such a critical raw material, dependence on foreign supplies was risky, and because it was so bulky and costly to transport, regional economies needed local supplies. Sustained yield policies were therefore part of national strategies for defence and economic security.

Economic and technical change has made these considerations much less relevant today. Wood is no longer the critical source of energy or of raw material for military and industrial purposes, and developments in transportation have further reduced dependence on local timber supplies. In any event, strategic considerations suggest a need to maintain stocks of timber that can be drawn on in case of emergency, but not necessarily sustained yields either before or during such events.

Moral Obligation

There is a widely held view, rooted in the conservation ethic and in the current search for "sustainable development," that we are temporary stewards of renewable resources. There is also a widespread perception, based on "cut-and-run" forest operations in the 19th and 20th centuries, that the forest industry, left to itself, will not adequately consider the interests of future generations. Underlying much of this discussion is a conviction that today's owners and managers have a moral obligation to future generations to ensure that forests are passed on in an unimpaired state, and that this duty can best be fulfilled through regulated sustained yield.

The matter of moral obligation raises both a technical issue and a political one. The technical issue is the effect of regulating yields on a forest's productivity. We have already noted that yield regulation is likely to diminish the productivity of a forest in economic terms. Moreover, the future productive capacity of a forest can be increased in terms of value or quantity of timber by building up the stock in the present, which implies abstaining from harvesting rather than maintaining a steady harvest. Or if productivity is measured by the growth rate, it can be increased by means of protection, silviculture, and other management measures. In short, it does not follow from a commitment to protect the future productivity of forests that they should be converted to an even gradation of age classes and steadily harvested.

The political issue is whether we have an obligation to leave our forests to our successors in as productive a state as we found them. To the extent that this proposition is a purely ethical or political view, it is not susceptible to economic analysis. A concern for the economic welfare of future generations, however, does not lead to the conclusion that forests and other resources should be maintained unchanged. Future economic productivity can be increased by shifting forestland to higher uses in some cases, and in others by investing in forest enhancement, substituting human capital for natural capital and vice versa. Indeed, changing use of resources is inherent in economic development.

Another, more recent argument for advancing conservation through sustained yield or other means is that substitutability between natural capital and human-made capital may be low. Here natural capital is defined broadly and includes the resources of a natural ecosystem that yields a flow of valuable ecosystem goods and services. It is an extension of the term "economic capital" (buildings, machinery) to environmental goods and services. When natural capital embodied in ecosystems cannot be easily substituted with other economic and human-made capital, it is argued that there will always be a minimum critical amount of natural capital needed to maintain economic production. Therefore, some natural capital, including forests, should be conserved and protected through sustained yield or other means.

Whether the historical rate of timber harvesting was too rapid is debatable, but for today's policy makers the task is to identify how unregulated resource use threatens the interests of future generations (if at all), and then to evaluate possible corrective measures. The problem and the most effective solution can be expected to vary at different times and places. In particular, they are not likely to lead to a single form of harvest regulation.

Multiple Use or Non-Timber Benefits

It is sometimes suggested that non-timber benefits of forests, such as environmental and recreational values, livestock forage, water, and aesthetic benefits, can be protected or enhanced through sustained yield management. On careful examination, this too is a tenuous notion. As noted in Chapter 5, the most favourable management regime for any particular benefit depends on the ecological and other circumstances of each forest, and different forest products and services call for different management regimes. Aesthetic values may be improved by maintaining a heavier cover of timber, wildlife and livestock might benefit from less cover, and so on.

In general, it is unlikely that an even gradation of age classes, and a constant harvest rate will maximize non-timber benefits of any particular kind, or any particular combination of them. Most of these values are affected much more by such things as how much natural forest is preserved, how roads systems are designed, and how logging is conducted than by the regularity of harvests over time.

Other Arguments

Other arguments that have been advanced in support of sustained yield policies include silvicultural improvement, risk reduction, and protection of forest owners against their own ignorance or shortsightedness. Although one or another of these may apply in particular circumstances, none offers a logical rationale for a general policy of regulated sustained yield.

TIMBER HARVESTS OVER TIME IN THE ABSENCE OF SUSTAINED YIELD POLICY: MARKET AS A REGULATOR

Earlier, we noted that with few exceptions the owners of private forests in North America schedule their harvests as they see fit. This provides a natural laboratory that enables us to see what has happened without government sustained yield regulations. More specifically, the experience in the United States, where some 70% of productive forestland (referred to as "timberland") is privately owned, sheds some light on the necessity and value of sustained yield policies in regulating timber harvests over time.

As private forest landowners respond to market forces and their own circumstances and needs, they may "overharvest" or "under-harvest" in any given period of time from a third-party or public perspective. Most relevant to public policy is the regional and long-term timber-harvesting pattern in these private forests.

Significantly, the absence of regulatory control has not, in general, resulted in "overharvesting" in the sense of harvesting more timber than has grown on private timberland. Harvests on private timberland in the United States as a whole have not exceeded forest growth since at least 1953. This is evident in Figure 8.1, which shows the per-acre growth, removal, and inventory on private timberland in the US. Since timber growth has exceeded harvests (and mortality, which is not shown in Figure 8.1 and which is about 0.8% per year on private timberland) over time, the average per-acre inventory in 2007 was much higher than in 1953.

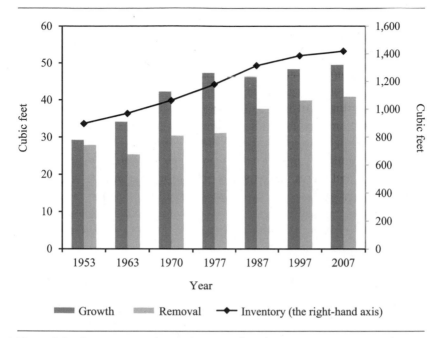

Figure 8.1   Per-acre annual growth, removal, and inventory on private timber-land in the US, 1953-2007. | *Sources:* Smith et al. 2009 and various similar reports.

At the regional level, it is interesting to note that the age class distribution of inventory on private forestland in western Oregon, where even-age forest management is popular among landowners, shows that the forest is fairly well "regulated" by the market. The acreage supporting each of the first five age classes, to age 50, is relatively consistent, at about 800,000 acres, as shown in Figure 8.2. Not coincidentally, private forests in western Oregon are typically managed on rotations within 50 years.

What does the US experience tell us? Certainly it raises questions about the need for sustained yield regulations as a means of preventing overharvesting of private forests. Market forces, supported by secure property rights in forests and forestland, appear to be sufficient to generate the investment needed to sustain and even expand the forest inventory, at least in the United States. This is not to suggest, however, that sustained yield policies are not needed in other jurisdictions where markets and economic conditions, property rights, and other institutional arrangements are quite different.

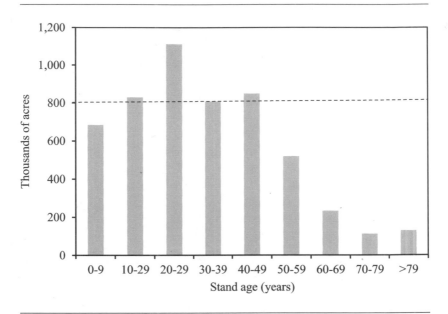

Figure 8.2   Age class distribution of inventory in private timberland in western Oregon, 1997. | *Source:* Azuma et al. 2004.

NEW APPROACHES TO FOREST REGULATION

Chapter 1 emphasized the importance of social objectives in designing public policies and assessing their performance. The objectives of forest owners and public policy makers vary and change over time, as do the circumstances in which they operate, so today they are undoubtedly different and more complicated than those that led to the development of the traditional concepts of sustained yield forestry many decades ago.

In the early years of the 20th century, the conservation movement became a powerful political force in the United States, advocating government intervention to control what was seen to be the rapacious exploitation of forests and other natural resources that threatened to deplete them, leaving American industry without its needed raw materials. The public appeal of these ideas led to lasting policy measures, particularly in the American West and Canada, to retain public ownership of forestlands, create huge public forests, and adopt sustained yield regulations to ensure timber supplies for the future. Since then, although forest production has expanded, fears of running out of timber have abated, forest practices have vastly improved, and public anxieties have shifted to the non-timber values of forests – to the protection of wildlife, water, and other environmental benefits, wilderness, and carbon sequestration. In the present environment, rigid sustained yield regulation of timber harvesting seems anachronistic.

Increasing concern for non-timber values has forced the abandonment of long-term harvest plans in many cases, perhaps most dramatically when efforts to save the endangered Northern Spotted Owl led to steep reductions in the regulated harvests in US national forests, but also as a result of increasing protection of fish and wildlife habitats, environmentally sensitive areas, and recreational and aesthetic resources. In addition, the economic shortcomings of regulating harvests on the basis of timber volumes, without regard to values and costs involved, have been increasingly recognized.

In recent years, much more sophisticated methods have been developed for identifying the most beneficial rate of harvesting a forest over time. New mathematical techniques such as optimal control theory and advances in computer technology and modelling techniques have provided practical means of analyzing complicated systems such as a forest that has hundreds of stands and whose harvest is subject not only to market forces but also to governmental (and nongovernmental) regulations that protect non-timber values.

Thus, a growing variety of computer models have become available to assist forest managers in analyzing and designing forest management plans, including harvest schedules, directed towards specific objectives and based on sound economic analysis. These new techniques are rapidly replacing traditional concepts of sustained yield regulation, especially in areas where private forests are dominant.

REVIEW QUESTIONS

1 How does the supply of timber from a particular forest affect the market supply of timber as described in Chapter 4 when (a) the forest is only one of many sources of supply for the market, and (b) when the forest is the only source of supply for the market?

2 Why is it often necessary to modify the harvesting regime that maximizes the return from an individual hectare of forest in order to maximize the return from the forest as a whole?

3 What is a "normal forest"? What silvicultural conditions must be met for a normal forest to yield a constant harvest over time?

4 Use the Hanzlik formula to calculate the allowable annual cut for a forest of 5,000 hectares of uniform productivity, half of which is occupied by old-growth timber of 2,000 cubic metres per hectare, the

other half by 20-year-old second growth having a mean annual increment of 15 cubic metres per hectare over the planned rotation period of 50 years. How does your answer compare with the allowable cut after the old growth is depleted?

5 Calculate the present value of the old growth in question 4, assuming that it will be harvested in equal annual amounts over the next 50 years. Use a discount rate of 4% and assume that all timber is valued at $20 per cubic metre. How would this compare with the value of the timber if it were all harvested immediately?

6 Why will a stable level of forest harvesting sometimes fail to ensure the economic stability of nearby communities?

FURTHER READING

Azuma, David L., Larry F. Bednar, Bruce A. Hiserote, and Charles F. Veneklase. 2004. *Timber Resource Statistics for Western Oregon, 1997.* Revised Resource Bulletin PNW-RB-237. Portland, OR: US Department of Agriculture, Forest Service, Pacific Research Station.

Bell, Enoch, Roger Fight, and Robert Randall. 1975. ACE: the two-edged sword. *Journal of Forestry* 73 (10): 642-43.

Berck, Peter. 1979. The economics of timber: A renewable resource in the long run. *Bell Journal of Economics* 10 (2): 447-62.

Bowes, Michael D., and John V. Krutilla. 1989. *Multiple-Use Management: The Economics of Public Forestlands.* Washington, DC: Resources for the Future. Chapter 4.

Buongiorno, Joseph, and J. Keith Gilless. 2003. *Decision Methods for Forest Resource Management.* New York: Academic Press. Chapters 5-7.

Davis, Lawrence S., K. Norman Johnson, Pete Bettinger, and Theodore E. Howard. 2005. *Forest Management: To Sustain Ecological, Economic, and Social Values.* 4th ed. New York: Waveland Press. Chapters 10, 11, 13, 14.

Heaps, Terry, and Philip A. Neher. 1979. The economics of forestry when the rate of harvest is constrained. *Journal of Environmental Economics and Management* 6: 297-319.

Klemperer, W. David, John F. Thurmes, and Richard G. Oderwald. 1987. Simulating economically optimal timber-management regimes: Identifying effects of cultural practices on loblolly pine. *Journal of Forestry* 85 (3): 20-23.

Luckert, M.K. (Marty), and T. Williamson. 2005. Should sustained yield be part of sustainable forest management? *Canadian Journal of Forest Research* 35: 356-64.

Nautiyal, Jagdish C., and P.H. Pearse. 1967. Optimizing the conversion to sustained yield: A programming solution. *Forest Science* 13 (2): 131-39.

Smith, W. Brad, Patrick D. Miles, Charles H. Perry, and Scott A. Pugh. 2009. *Forest Resources of the United States, 2007.* General Technical Report WO-78. Washington, DC: US Department of Agriculture, Forest Service, Washington Office.

Chapter 9   Long-Term Trends in the Forest Sector and Silvicultural Investment

In previous chapters, we discussed timber harvests at the stand level and harvesting regulations that were sometimes applied to whole forests. In Chapter 6, we noted that the forest land base may expand or contract and that forest management intensity often varies because of changes in economic conditions, development in other economic sectors, and government policies, and because of differences in the natural productivity of land. Collectively, these changes in forestland use, management intensity, harvesting, and forest growth over time reflect the responses of the forest sector to general economic conditions and changes in human demand on forests and forestland.

This chapter takes a broad look at long-term trends in the forest sector and explains how investment in silviculture can most efficiently respond to these changes. This helps in understanding the long-term transition from exploitive timber extraction to managed forests, and how forest owners and governments can determine appropriate levels of investment in silviculture. In the last part of the chapter, we connect these two themes in examining the evolution of the forest sector and the development of plantation forestry in the southern United States.

LONG-TERM TRENDS IN THE FOREST SECTOR

An economist's view of a trend often boils down to changes in prices and quantities of specific goods and services. As forests produce many goods and services, trends in the forest sector can be revealed by looking into changes in their respective prices and quantities. As we noted in Chapter 5, some forest goods and services, often labelled as environmental services, are not priced. Nonetheless, the quantities of all forest goods and services, including these unpriced environmental services, change over time. Thus, we can roughly group all forest-related goods and services into two primary goods – timber, which is usually priced, and environmental services, which are usually not – and examine the long-term trends of each. We begin with timber, illustrating the long-term trends in the timber-producing industry with reference to its development in the United States.

Timber Depletion and Transition in the Forest Products Sector

The timber side of the forest sector often involves long production cycles for forests, with standing timber inventories that are large compared with annual harvests. In these respects, the industry differs from other industries, and differs conspicuously from agriculture. This is true in all of the world's forest regions.

Early on, when Europeans began to settle in North America, timber prices were low, even negative to early settlers who wanted to clear the forests. Some forestland was more valuable for other uses, especially the production of food. Trees were removed and not replaced, and forestland was converted to other uses.

Figure 9.1 shows that the area of forestland in the US declined over 27%, from 1.05 billion acres in 1630 to 750 million acres in 1907. This steady decline was accompanied by settlement and agricultural expansion in the North and South regions of the eastern United States. Forest cover in the US (not shown in Figure 9.1), declined from 46% to 33% of the landscape over the same period. In the North, forests occupied an estimated 72% of the land in 1630 but covered only 34% by 1907. Similarly, significant reductions in forest area occurred in the South before the 20th century, with the forest proportion dropping from 66% in 1630 to 44% by 1907.

Over time, if nothing else changes, timber harvesting reduces the inventory of economically recoverable timber, and timber becomes increasingly scarce. This became evident when the demand for timber, fuelled by population growth and industrialization, began to rise. At the early stage, the primary means of meeting this increasing demand was through expansion of the extensive margin for harvesting of the natural forests. This spreading out from areas of settlement meant transporting logs over greater distances, and sometimes expansion to sites with more difficult logging conditions.

Increasing scarcity – that is, when the market supply of recoverable timber falls more than the demand for it, or the demand increases more

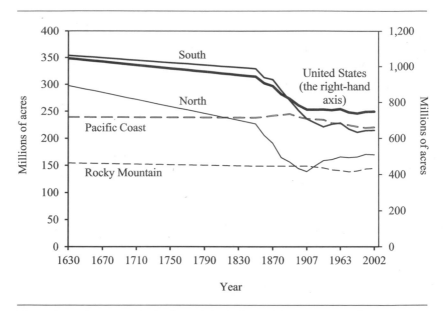

Figure 9.1   Forest area in the United States by region, 1630-2002. | *Data source:* Smith et al. 2004.

than the supply – means rising prices. In the absence of significant forest growth, which is characteristic of old-growth forests, scarcity and prices of timber continue to increase. Moreover, as logging operations extend into more difficult terrain, costs of production also rise.

The rate of price rise is not limitless, however, even when forests are not renewed initially after cutting. Because the market for natural resources is connected with the market for capital, in the absence of biological growth and access to more natural timber, market efficiency requires that timber prices rise at the rate of interest. This is *Hotelling's rule:* for non-renewable resources, net price (market price minus marginal cost for extraction) must rise at the rate of interest in a competitive market equilibrium.[1] The net prices of the initially non-renewable old-growth forests are stumpage prices, referred to as timber prices in this chapter.

Four factors dampen the upward pressure on timber prices. One is the aforementioned shift in the extensive margin. As prices increase, it becomes economical to operate in other regions that harbour primary forests. The movement of the US timber industry from the Northeast to the Great Lakes states, to the South, and to the Pacific Northwest is an example of shifts in the extensive margin. Another example is the development of the forest industry in British Columbia and northern Canada after the 1960s, which was a direct result of high timber prices in the US. Not surprisingly, US imports of Canadian forest products made from primary forests increased in subsequent decades. Such inter-regional shifts in the extensive margin constitute an immediate but limited response to increasing scarcity of timber in the US market.

Second, as timber prices in primary forests rise, more efforts are devoted to protecting these forests from fire, insects and diseases, and animal encroachment. Fire prevention policy, first introduced in the US in the 1920s, greatly reduced forest damage from wildfires. Similarly, fencing was introduced in the southern US to minimize open grazing, which was harmful to both forests and agricultural crops. By protecting the stock of natural forests, these efforts provided greater scope for continued expansion of the extensive forests to become economical at the margin of production.

Third, as timber prices rise, growing new forests becomes economically rewarding. In the US South, as in many other timber-producing regions of the world, plantation forestry grew as natural forests were depleted (this process is discussed later in this chapter). As plantation forests started to develop and natural forests were better protected, the forest land base stabilized and ceased to decline. Thus the forestland area in the United States has stabilized at around 750 million acres since the 1920s. The rise of plantation forests represented the beginning of the transition from forest exploitation to forest cultivation and forest sustainability.

Finally, higher timber prices drive technological innovation and induce increases in management intensity through investment in silviculture. On the supply side, higher timber prices and higher land costs encourage innovations in timber-growing technology, silviculture, genetics, and forest protection sciences. For example, the average productivity for eucalyptus in Brazil increased from 18 cubic metres per hectare per year in the early 1980s to some 30 cubic metres per hectare per year in the mid 2000s. The growth rate of planted pine forests in Alabama increased from 8.2% per year in the 1980s to 10.2% in the 1990s. These dramatic increases in productivity can be attributed to technological innovation and increases in management intensity. On the demand side, higher timber prices encourage economies in timber utilization in manufacturing and the development of new wood-saving products. For example, the lumber recovery factor in sawmills – the amount of lumber produced per unit of log input – increased significantly over the last half of the 20th century in the US South and British Columbia, despite

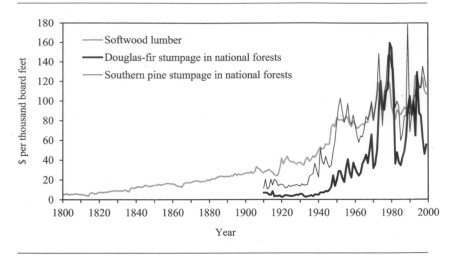

Figure 9.2   Real-price indices for softwood lumber and stumpage in terms of 1992 prices (1992 = 100). | *Source:* Zhang 2004.

progressive decline in the quality of logs. The development of new engineered wood products encourages more efficient use of wood and enables previously unusable species and timber grades to be utilized. Further, wood substitutes have been increasingly used as prices of forest products rise.

The US began keeping records of timber prices (stumpage) in national forests in 1913 and in private forests since 1955 (see Figure 4.4 in Chapter 4). As Figure 9.2 shows, the overall long-term trend of real lumber prices in the United States has been rising since 1800, despite wide fluctuations. Thus, the initial stage of the long-term trends for timber in the US in the last few centuries was characterized by intensive harvesting of natural timber stock, which led to a declining forest land base and increasing timber prices.

Gradually, timber growth will increase, initially from naturally regenerated forests and later from forest plantations as well as naturally regenerated forests. Eventually, timber growth rate will catch up, surpassing the rate of increase in timber price, and play a more important role in the dynamic relationship between the timber growth rate, the rate of timber price appreciation (or depreciation), and the interest rate, discussed below. In fact, the forest land base begins to stabilize only when annual timber growth is equal to or greater than annual timber harvests on existing forestland. Obviously, short-term movements in timber demand and supply do not always take place in an orderly fashion, but in the United States the long-term path of equilibrium for timber over the last couple of centuries has evidently been leading towards a steward-ship phase of forestry, fuelled by the economic scarcity of timber.

The long-term relationship among timber price appreciation, timber growth rates, and interest rates can be seen in the traditional Faustmann formula, but here the price variable is also a function of time, $P = P(t)$, and the long-term perspective is at the "global" scale (which encom-passes all possible shifts in extensive margins domestically and inter-nationally), while non-timber benefits and environmental services produced from forests are left out of the equation for the time being.

For simplicity, we start with old-growth forests and assume that they regenerate naturally. If we change this assumption and make silvicul-tural investment a necessity, we can treat the silvicultural investment as a reduction in the stumpage revenue. In either case, the conclusions below remain unchanged.

The land expectation value, or site value, of a typical old-growth for-est, V_s, is then

$$V_s = \frac{P(t)Q(t)e^{-rt}}{1 - e^{-rt}} \tag{9.1}$$

where $P(t)$ is the stumpage price at age t, $Q(t)$ is the standing timber volume at age t, and r is a continuous discount rate. Equation 9.1 is iden-tical to equation 7.10, except that $P(t)$ is now a function of time.

Differentiating equation 9.1 with respect to t and setting the resulting equation equal to zero, we find the necessary conditions for maximiza-tion of site value or land expectation value:

$$\frac{P_t}{P} + \frac{Q_t}{Q} = \frac{r}{1 - e^{-rt}} \tag{9.2}$$

where P_t and Q_t are, respectively, the derivatives of price and timber vol-ume with respect to age t. Thus, P_t/P is the rate of change in price and Q_t/Q is the timber growth rate.

Equation 9.2 implies that, in the long term, the rate of change in price plus the timber growth rate will equal the interest rate, r, modified by a perpetual term, $1/(1 - e^{-rt})$.

The initial rise in timber prices is readily seen from equation 9.2. As the growth rate of old-growth forests is or approaches zero, timber prices will rise initially as long as the interest rate is positive. Holding the interest rate constant, timber prices will appreciate at a slower rate as timber growth rate picks up. Eventually, timber growth rate could surpass the rate of timber price appreciation, with the latter approaching zero or even becoming negative.

Again, this long-term equilibrium does not hold in the short term, where uncertainty influences the market. In addition, if a region has trade or business relationships with other regions and obtains access to forests in other regions, its extensive margin will shift because of the new "forest growth." Finally, the interest rate in a region or country varies over time, causing timber prices and the rate of timber price change to fluctuate.

This long-term relationship shows the dynamics among three key economic and biological variables in forestry: rate of timber price change, timber growth rate, and interest rate. In the long term, timber prices change in response to various supply and demand factors. On the supply side, an increase in the timber growth rate through silvicultural investment, which is an expansion in the intensive margin of timber production, and shifts in the extensive margin of timber harvesting have a lot to do with the fact that timber price appreciation in recent decades is below historical levels and even becomes negative in some areas of North America. On the demand side, changes in demand for timber due to technological innovation in forest products manufacturing, the use of wood substitutes, or changes in what humans want from forests influence the long-term price trends of timber. One such change in what humans want is the increasing demand for environmental services from forests.

Environmental Services from Forests

Although many non-timber benefits and environmental services produced from forests are not priced, the supply of these is affected by the development of the timber side of the forest sector and timber depletion. On the other hand, human demand for these benefits and services influences the availability of harvestable timber and thus timber price. Through market processes and government regulations, society decides how much forests are used to provide these benefits and services, either exclusively or jointly with the production of timber.

As we shall see in Chapter 13, the demand for most environmental services is highly correlated with personal income, although not in a linear fashion. The evidence for such a correlation is well established in some aspects of the environment – clean air, clean water, and outdoor recreation – but the relationship is generally true for all environmental services.

Because some environmental services are provided outside of formal markets and are unpriced, there is little incentive to provide them and their supply may be inadequate. On the other hand, a growing population and rising personal incomes increase the demand for these services. Thus, their supply is low while demand is high and increasing, but there are no prices to signal relative scarcity and to induce socially appropriate changes in production and consumption.

This imbalance in supply and demand suggests a need for government intervention. Indeed, when such a disparity grows over time, governments often step in and intervene through forest practice regulations, zoning, and land set-asides, which reduces the land base for timber production. All else being equal, reductions in land used for timber production will lead to a rise in timber prices, which of course can be alleviated by an increase in timber supply through such means as forest plantation development, increased use of wood-saving technologies, and the use of other materials for traditional wood products, as noted earlier.

Because competing demands for timber and environmental services contribute to scarcity of land, it is logical to substitute other factors, especially capital and labour, for land that is fixed in supply. Using more capital and labour in tree growing is characteristic of traditional silviculture, the purpose of which is to grow trees more intensively and to enhance economic returns on a given land base. If increasing silvicultural investment generates sufficient timber to meet society's demand for it, more land can be used to produce environmental services, as noted in Chapter 6. In other cases, silvicultural investment is aimed primarily at providing environmental services.

In any event, silvicultural investment is important for both timber production and environmental services. Before turning to the optimal level of silvicultural investment for timber production and environmental services, however, we conclude this discussion of long-term trends in the forest sector with an example that is significant at the local, regional, and national levels in the US. This example demonstrates that general economic and social development, advancement in other economic sectors, and changes in human demand have had a large impact on forests and the forest sector in the northeastern United States and the rest of the country.

An Illustration: The View from John Sanderson's Farm
Our example is based on Hugh Raup's 1966 essay "The View from John Sanderson's Farm: A Perspective for the Use of Land." We use it to demonstrate the changes in land use and forestry in a small New England

community, from 18th-century wilderness to mid-20th-century third-growth pine forests.

In the essay, Raup traces his family's early New England history back through events that forever altered the course of forestry in the US Northeast and beyond, and presents his thesis: evolving human attitudes and values lead to land-use and forest change. This true story is relevant to the long-term development of the forest sector: land-use and forest composition changes and the depletion and renewal of forests are associated with changing human demands for forest goods and services. These changing demands often result from events that take place far beyond the forest sector itself.

John Sanderson's farm was located at Petersham in central Massachusetts. Petersham saw its first settlers in 1733, and forests there consisted mostly of oaks and chestnuts with some hemlock, beech, sugar maple, and white pine. In 1771, when the first inventory of land was done, some 10% of the forest in Petersham was cleared. By 1791, this figure had risen to 15%.

The first Sanderson came to Petersham in 1763 and began farming with 50 acres of land. By 1830, Sanderson's farm had expanded to 550 acres; at the same time, some 77% of the forest in Petersham had been cleared. Agriculture was prosperous, and the Sandersons intended to stay. By 1850, some 90% of the forest within Petersham's boundaries had been cleared.

The Industrial Revolution was running its course by then, but southern New England's farm economy began falling to pieces. The Sandersons sold their farm in 1845 and opened a bank. By 1870, at least half of the agricultural land had been taken out of farming.

Meanwhile, in Ohio, Raup's grandfather had given up his saw and grist mill in 1854 and had begun working for a manufacturer of farming implements such as racks and small tools. Evidently, at that time, there was a strong demand for farming implements in the Ohio Valley and the whole US Midwest.

Between 1830 and 1870, something had happened that led, on the one hand, to the demise of Sanderson's and others' farming operations in New England, and on the other hand, to the strong demand for agricultural tools in the Midwest. Most likely, it was the completion of the Erie Canal in 1825 and its expansion in 1860. The canal, shown in Figure 9.3, crossed New York state from Albany, the head of navigation on the Hudson River, to Buffalo on Lake Erie, opening the fertile farmlands of the Midwest to equally fertile consumer markets in the East. In the ensuing market competition, New England's farmers found that their small

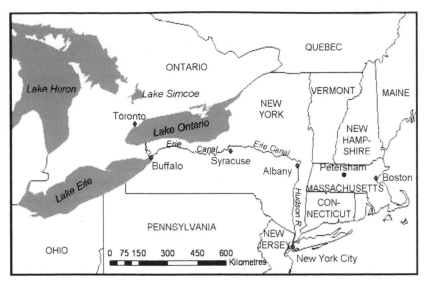

Figure 9.3   The Erie Canal.

farming operations were no match for the production from vast and fertile farms in the Midwest. As Raup observes, the sandy loam soils in Petersham were droughty, stony, and essentially infertile, and fields were always small. Thus, Midwest farmers could produce more for less, which created both opportunity and capital for fledgling farm implement manufacturers, whose early reapers and rakes made a monumental contribution to the efficiency and profitability of Midwest farming.

"It was this that destroyed the agricultural prosperity of the Sanderson family in Petersham and of similar families in southern New England," Raup writes. "The farms here could remain prosperous only so long as they had no serious competition. Once the latter appeared, their economy collapsed. And it did so rather suddenly and on a large scale over wide expanses of the landscape. Agricultural use of the land was simply abandoned as the Sandersons and others like them sought prosperity elsewhere."

Trees began to return on abandoned farmland, only this time mainly with a pioneer species – white pine. This laid the cornerstone for the Northeast's vast lumber and paper industries a few decades later. By 1900, most of the New England farms that had been carved from forests in the early 1800s were forests again. Around 1900-10, portable sawmills were brought there and lumber was made of timber harvested from these white pine forests. The lumber was used to make boxes and containers to transport goods back to the Midwest. Raup remembers that everything his family bought in the grocery store came out of a wooden box or barrel when he was a boy in Ohio in the years around 1908 and 1910. He had not known then what he knew when he wrote the essay – that all of these boxes, barrels, and pails came from one or another of the hundreds of factories that were turning them out in quantity in southern New England at the time, a large percentage of them within a few miles of Petersham. Further, they were being made from white pine lumber that was coming from vast, pure stands. This pine, mature for cutting, was coming from land that barely 60 years earlier had supported the prosperous farm communities of which the Sandersons were a part.

As white pine became a major crop of industrial timber, it encouraged early 20th-century foresters to invest heavily in replanting white pine, which created the third forests in the region. By the time the crop matured in the 1960s, however, its primary value was not for wooden box–making because cardboard and plastic packaging had replaced wooden containers for agricultural products.

Sanderson's farm would eventually become part of Harvard University's Harvard Forest, where Raup served as director in the 1960s. Today, the tumble-down remnants of stone walls that divided pastures and ownerships are all that remains of 19th-century farming in New England.

Raup notes that by the mid-1800s, "summer people" began to move into the old farm towns. They were mainly affluent city dwellers in search of vacation hideaways. They bought up old farmhouses and renovated them in order to have a quiet place for the summer. Before long, they were remodelling main streets and imposing their own ideas about what a rural town ought to look like. The primary demand for land in Petersham and other New England communities changed from agriculture to recreation in the last half of the 1800s. It changed again in the early 1900s, when the Northeast's forests caught up with the Industrial Revolution and created an entirely new set of demands that favoured a new industry: lumbering. The demand for boxes, barrels, and pails to carry the products of a burgeoning Midwest farm economy over a rapidly expanding railroad system created a market for the pine that had grown back on abandoned farmland in the Northeast. Nowadays, the primary demand for Northeast forests arguably involves non-timber benefits and environmental services, while timber production has become secondary and less important than in the early 1990s.

We see in this story that the initial depletion of natural forests, self-renewal of the second forests with pioneer species, and cultivation of the third forests in New England were associated with the expansion and subsequent demise of agricultural operations, industrialization, and changing human demands. Changing human demands led to changes in use of the land – from farming to production of industrial timber to environmental services. Today, forests in New England and elsewhere are increasingly valued for recreation and environmental services.

A CONCEPUTAL MODEL FOR SILVICULTURAL INVESTMENT

This section and the next present an overview of silvicultural investment from a public policy perspective. Thus, they are most useful for analyzing the effects of government policies that encourage or hinder private landowners with respect to investment in silviculture, and for examining the responses of landowners to market signals. All individual and practical silvicultural investment decisions – for example, how a forester would go about evaluating stands for precommercial thinning, how a landowner would select the optimal thinning regime, and whether a landowner should choose artificial reforestation or natural regeneration – are simple extensions and applications of the tools of investment analysis with regard to the stand treatment problems discussed in Chapter 3. We note that some knowledge of calculus will help readers go through this section.

Silviculture, defined as the art and science of controlling the establishment, growth, composition, health, and quality of forests to meet the needs and values of landowners and society, can take place throughout the life cycle of forests. Traditional silvicultural activities often include site preparation, selection of the right species and seedlings with desired genetic traits and quality, tree planting, herbaceous weed control, thinning, fertilization, and final timber harvesting. Harvests of the timber produced provide financial benefits to landowners. Thus, with the exception of timber harvesting, we consider silvicultural activities, and the total amount spent on them, as *silvicultural investment.*

Although silvicultural investment takes place at various stages of forest growth, we can discount the cost of all these activities, using the appropriate interest rates, to their equivalent value in the year when the stand was at age 0, as explained in Chapter 7. This yields a total silvicultural investment or silvicultural cost, wE, where w is the unit factor cost of silviculture and E is the total silvicultural effort.

The Faustmann model provides the conceptual basis for silvicultural investment. Forest landowners' afforestation and reforestation decisions are assumed to be based on maximizing the net present value of timber production. Present value in this case is the expected total timber revenue minus the costs of silvicultural effort.

Timber production function, $Q(t, E)$, is influenced by age, t, and silvicultural investment effort, E. It is often assumed that the first derivatives of this production function with respect to t and E are positive, that is

$$Q_t = \frac{dQ}{dt} > 0$$

and

$$Q_E = \frac{dQ}{dE} > 0$$

This is because, all else being equal, an increase in time or silvicultural effort will result in more timber growth.

The second derivatives, denoted as Q_{tt} and Q_{EE}, are negative ($Q_{tt} < 0$, $Q_{EE} < 0$) because of diminishing marginal rates of return. Again, the silvicultural investment is the product of the silvicultural effort, E, and its unit cost, w, where w, $E \geq 0$.

The objective of a landowner is to maximize

$$V_s = \frac{PQ(t,E)e^{-rt} - wE}{1 - e^{-rt}} \tag{9.3}$$

where V_s is the expected land value or site value, r is the continuous discount rate or opportunity cost of capital for the landowner, and P is the expected stumpage price. Equation 9.3 is identical to equation 7.9.

The first-order condition of equation 9.3 with respect to E is

$$V_E = PQ_E e^{-rt} - w = 0$$

or

$$PQ_E e^{-rt} = w \tag{9.4}$$

In economic terms, equation 9.4 means that the optimal level of silvicultural investment is reached when the marginal revenue product of effort is equal to the unit cost of the effort, w.

Next, we would like to know the effect of stumpage price on silvicultural effort. Intuitively, an increase in stumpage price would be expected to lead to an increase in silvicultural effort. To demonstrate such an effect, we need to find out whether the derivative of silvicultural effort with respect to stumpage price, dE/dP, is indeed positive.

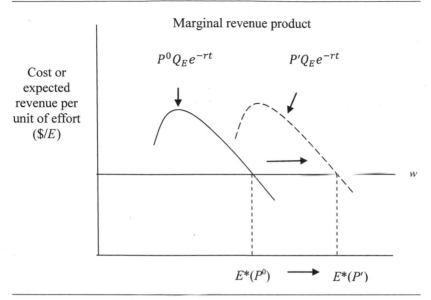

Figure 9.4   Optimal reforestation effort, E^*, changes when stumpage price increases. | *Source:* Adapted from Hyde 1980.

In order to get dE/dP, which is the marginal effect of stumpage price, P, on optimal silvicultural effort, E, we totally differentiate equation 9.4 with respect to P and E while holding other variables constant. This yields

$$\frac{dE}{dP} = \frac{-Q_E}{PQ_{EE}} \qquad (9.5)$$

Since $Q_E > 0$, $Q_{EE} < 0$, equation 9.5 is positive, implying that the landowner will intensify silvicultural effort when the expected stumpage price goes up.

Figure 9.4 demonstrates the change in the optimal reforestation effort, E^*, when the stumpage price increases from P^0 to P'. The optimal reforestation effort is determined by the marginal revenue product of effort and the unit planting cost, w. When the stumpage price is P^0, the optimal effort level is $E^*(P^0)$. When the price increases to P', the marginal revenue product of effort curve shifts to a higher level, and the optimal level of effort rises to $E^*(P')$.

Similarly, we totally differentiate equation 9.4 with respect to E and w while holding other variables constant to get

$$\frac{dE}{dw} = \frac{1}{Pe^{-rt}Q_{EE}} \tag{9.6}$$

Finally, totally differentiating equation 9.4 with respect to E and r while holding other variables constant yields

$$\frac{dE}{dr} = \frac{rQ_E}{Q_{EE}} \tag{9.7}$$

Equation 9.6 is negative because $Q_{EE} < 0$ and $P > 0$. This implies that the landowner will exert less silvicultural effort when the unit silvicultural cost increases. The horizontal unit silvicultural cost curve in Figure 9.4 will shift upward when w increases. Because the marginal revenue product curve does not change, the optimal effort will decrease accordingly. Government assistance usually reduces the unit silvicultural cost burden for private landowners. As a result, landowners are expected to intensify their silvicultural efforts when government subsidies are available.

Equation 9.7 is also negative, as $Q_E > 0$, $Q_{EE} < 0$, and $r > 0$. When the interest rate increases, the marginal revenue product of unit effort will decrease, as can be seen in Figure 9.4. This leads to a lower level of silvicultural effort from the landowner. In other words, when landowners' discount rate increases, they will invest less in silviculture.

FACTORS INFLUENCING SILVICULTURAL INVESTMENT ON PRIVATE LANDS

Based on the foregoing discussion, we conclude that timber prices (P), unit silvicultural cost (w), the availability and amount of government subsidy, and interest rates (r) are the major factors influencing silvicultural investment on private forestlands. The magnitude of the effect of these factors is related to the responsiveness of timber growth to silvicultural effort (Q_E and Q_{EE}). Anything else that influences these factors will, in turn, affect private silvicultural investment.

Let's look at the timber price variable first. Because current silvicultural investment brings in future timber revenue, timber price in the Faustmann formula is actually the expected price in the future. Because expected prices are currently unknown and can only be estimated, current prices are often used as an approximation of the expected prices in empirical studies and as a guide for landowners in making silvicultural investment decisions.

Moreover, the relevant timber price for landowners is the net price received, after all expenses, such as taxes and consulting forester fees, are deducted. Thus, a high yield tax or high income tax on timber income will have a negative impact on silvicultural investment. There is evidence that, in some developing countries, when government taxes on timber income are high, forest farmers do not invest much in afforestation and reforestation activities or make much effort to protect their forests from human and animal encroachment. Consequently, some forestlands are left idle without trees, while other forests remain poorly stocked for some time.

Government regulations and forest certification could increase the expenses of landowners, reducing the timber prices they receive from the harvest. Thus, forest landowners often prefer to have little or less onerous government regulation, and have resisted expensive forest certification programs.

The unit cost of silvicultural activities is often determined by the market, and is something that individual landowners rarely have any control over. If government subsidies are available, however, landowners may take advantage of these subsidies to lower their unit silvicultural cost.

In the United States, government subsidies often pay landowners a certain percentage of their expenses for afforestation, reforestation, and stand improvement activities. Sometimes, an additional land rental is also paid to landowners if they agree to provide some environmental goods and services from their forests by holding their land in forest cover for more than 10 years. Examples of the former include the Stewardship Incentives Program and the discontinued Forest Incentives Program. Examples of the latter include the Soil Bank Program from 1956 to 1962, and the Conservation Reserve Program since the later 1980s. There is evidence that all government subsidy programs in the US have motivated forest landowners to invest in silviculture.

From equations 9.5, 9.6, and 9.7, we can see that the magnitude of these marginal impacts (dE/dP, dE/dw, and dE/dr) have a lot to do with responsiveness of tree growth to silvicultural effort (Q_E and Q_{EE}). Such responsiveness depends on the natural productivity of the land, which is influenced by soil quality, slope, and weather conditions, as well as the species and the genetic traits of seedlings. Technical assistance is sometimes provided to help landowners select the most appropriate species, seedlings, and silvicultural treatments.

Finally, interest rates vary among forest landowners, who have differing opportunity costs of capital and tolerance for risk. Nonetheless, market interest rates can often serve as a rough guide of interest rate movement for a "typical" forest landowner over time. When market interest rates increase, landowners are likely to invest less in silviculture.

SILVICULTURAL INVESTMENT ON PUBLIC LANDS

When forestlands are publicly owned, government officials make decisions on silvicultural investment. Depending on how the public forestlands are used and managed, silvicultural investment is made in one of the following ways.

First, when public forestland is directly managed by a government forest agency, the agency directly conducts silvicultural activities. The agency may have the equipment and labour to do all of the jobs or it may contract private firms to carry out the work.

The amount of silvicultural investment when forestland is directly managed by the government forest agency is allocated through a government appropriation process: silvicultural projects have to compete with all other public projects for funding. In theory, this can ensure that all public expenditures – including spending on education, health care, and transportation infrastructure – are assessed based on economic and social criteria. In practice, economically beneficial silvicultural projects may be sacrificed for non-economic reasons, and investment may be made in land that cannot yield a net return. Governments have often established silvicultural programs funded by revenues from timber sales to support silvicultural investments, but few of these arrangements have been long-lasting and silvicultural funding provided in this way has proven to be unstable.

Second, when public forests are used and managed by firms through various tenure arrangements, the government landowner can require these firms, under the terms of their tenure contracts, to undertake certain silvicultural measures, the cost of which is sometimes reimbursed. These arrangements require a process for planning, approving, and inspecting silvicultural activities.

Finally, governments can require tenure holders in public forests to make the necessary investment with their own money after timber harvesting. Often the tenure holders are required to make sure that trees reach certain standards (such as "free to grow," referring to a seedling or small tree free from direct competition from other vegetation) within a certain period of time. Because tenure holders are required to make the silvicultural investment on public lands, they often lower their bids for the right to harvest public forests. Silvicultural investment is thus treated by these tenure holders as a cost of obtaining public timber.

Although this approach could eliminate the need for a large public forest bureaucracy and ensure that all cutover lands are reforested, silvicultural investment is based not on the present value of the future benefits it may bring but as a necessity or cost to the current stands, and there is some inefficiency when investment is made in places where it should not be made based on economic principles. Furthermore, since the tenure holders often do not expect to benefit much from the next forest crop, they have little incentive to invest anything more than the minimum that is required by the government or that can pass future government inspection, unless they can benefit from the allowable cut effect noted in Chapter 8.

FOREST PLANTATION DEVELOPMENT IN THE SOUTHERN UNITED STATES

In this section, we use the southern United States as a case study to illustrate the roles of economic forces, government policy, and natural factors in the development of the forest sector in a region over the long term. In the US South, the transition from forest exploitation to forest cultivation and sustainable forestry is well advanced. The forest industry is increasingly dependent on forest plantations, and as of 2007 some 24% of all forests and 56% of all pine forests were planted forests. As in many other timber-producing regions of the world, the development of plantation forestry in US South has been shaped by the interactions of nature, markets, and government policies.

The US South consists of 13 states of the continental United States, including Virginia and Kentucky in the north, Texas and Oklahoma in the west, and Florida in the south. It is an important forest region: with some 2% of the world's forestland, it produces some 16% of global industrial timber. Its forests also contribute significantly to environmental services such as water supply, biodiversity, and carbon sequestration. Nearly 90% of the forestland in the region is privately owned.

As noted earlier, southern forests declined significantly until the early 1900s, mostly because of agricultural expansion. Today, forestland is under heavy pressure from population growth and urbanization, but the rate of decline is much slower. Between the 1970s and the 2000s, forestland in the region stabilized at about 215 million acres.

Initiation

As in other parts of the United States, the forest industry in the South prior to the 1930s depended on harvesting the abundant natural timber. Even the powerful conservation movement of the early 1900s had little impact on this removal of natural forests and the continued disregard of reforestation.

US prices of lumber and other wood products followed an upward trend from 1800, indicating that they were becoming increasingly scarce relative to demand. This, in turn, increased the stumpage value of standing trees. As natural forests diminished and the cost of harvesting them increased, reforestation and production of timber in plantations became economically attractive. In the early 1910s, a few private firms and landowners began to experiment with growing forest plantations in the region.

In 1924, the first government program for promoting private plantations was established by the Clarke-McNary Reforestation Act. This act authorized a federal government agency to cooperate with universities and state agencies to assist forest landowners in the production and distribution of seeds and seedlings for the creation and utilization of forest growth. It also initiated government efforts to control wildfires. Federal forest research stations were set up after 1928.

Reforestation continued to be of little importance, however, until the Great Depression of the 1930s, when the federal government provided incentives such as the Civilian Conservation Programs and Agricultural Conservation Programs to encourage tree planting. In 1928, a few thousand acres were planted. Private tree planting in the South reached 49,572 acres in 1945.

Acceleration

Tree planting in the South accelerated in the 45 years after the Second World War, stimulated by both rising timber prices and favourable government policies. Planting on non-industrial private forestlands in the South increased from 44,461 acres in 1946 to 887,000 acres in 1990, an annual growth rate of over 7%. Tree planting on industrial lands increased from 4,579 to 929,000 acres in the same period, an annual growth rate of nearly 13%.

Significant tree-planting efforts were associated with federal incentive programs, including the Soil Bank Program from 1956 to 1963, the Agricultural Conservation Program from 1936 to 1997, and the Forestry Incentives Program from 1974 to 1997. As shown in Figure 9.5, a tree-planting spike in the late 1950s was associated with the Soil Bank Program, designed as a tool for reducing agricultural surplus. Another tree-planting spike occurred around 1988, associated with the Conservation Reserve Program that was initiated in 1986.

A second, perhaps more significant government program was the capital gains tax treatment for timber income, implemented in 1944.

Under this tax policy, only 40% of timber income for qualified forest landowners is taxed, compared with ordinary income. With a much lower tax rate for ordinary income, this favourable treatment for timber income was eliminated in 1986 and then partially restored in 1991. Since 1997, the capital gains tax rate for qualified timber income has been much lower than that for ordinary income. Finally, environmental policies such as the Endangered Species Act and Clean Water Act gradually reduced timber harvesting on public lands, causing further increases in timber prices. The massive (nearly 80%) reduction in federal timber harvesting in the US Pacific Northwest resulting from the listing of the Northern Spotted Owl as a threatened species in 1990 (which afforded it the same level of protection as endangered species under the Endangered Species Act of 1973) was a major contributor to the sharp rise in stumpage prices in the early 1990s, which continued in the ensuing decade.

Meanwhile, the wide application of technological advances in genetics, nursery management, planting techniques, herbaceous weed control, pest and disease control, fertilization, and plantation management increased land productivity and economic returns to landowners. A single negative factor during the 45 years since the Second World War was an increase in the cost of tree planting.

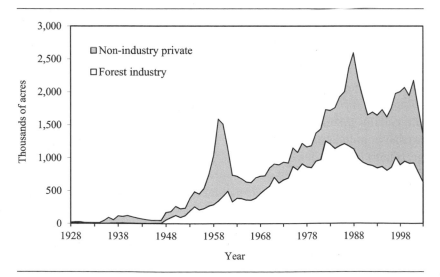

Figure 9.5   Private tree planting in the US South by ownership, 1928-2003. |
Source: Li and Zhang 2007.

Steady Period

Since the mid-1990s, annual tree-planting activities in the South have held roughly steady, although tree planting by both industrial and non-industrial forest landowners declined after 2001. As of 2003, non-industrial private and forest industry tree planting in the region covered 716,000 and 629,000 acres, respectively.

Four factors have limited tree-planting activities during this period. First, timber prices did not rise as much as in the acceleration period, due partly to increases in forest products imports and partly to two economic recessions, in 2001-02 and 2007-09. Timber harvest did not increase much, and even declined from the mid-1990s to 2010, resulting in less land being available for reforestation. Second, some forestland in the region was converted to higher and better uses, further reducing the land available for reforestation. Third, as noted in previous sections, landowners have less incentive to reforest or afforest when stumpage prices are low. The collapse of the housing market in 2007-08 and the resulting financial crisis have further reduced stumpage prices and limited the financial resources available for landowners to carry out reforestation and afforestation. Finally, government subsidies for silvicultural activities declined during this period.

Nonetheless, limited federal incentives have encouraged tree planting. The reforestation tax credit incentive, enacted in 1980 and lasting until 2003, allowed landowners to receive a tax credit up to $1,000 against their income tax for reforestation investment. Also, private landowners can deduct $10,000 over an eight-year period on their income tax for reforestation and afforestation purposes. Since 2004, forest landowners have been eligible to deduct from federal income taxes $10,000 per qualified stand in reforestation costs each year. More information on reforestation tax incentives in the US is presented in Chapter 11.

Summary

As we shall see in the next chapter, well-defined and enforced private property rights are a significant factor in determining private investment and production activities. The US Constitution protects private property rights and thus encourages landowners and other investors to invest in long-term projects such as tree planting when financially feasible.

Prior to 1900, forest resources were abundant and timber derived from natural forests could easily meet the demand for forest products in the United States. After the First World War, tree planting became financially attractive because of three main factors: (1) timber became scarce and timber prices rose substantially; (2) wild forest fire was under control; and (3) extensive fencing had minimized open grazing. With a favourable climate, the South soon became an attractive region for a forest industry that relies on the third and fourth growth of naturally regenerated forests and forest plantations.

In addition to rising timber prices, various government policies enhanced the development of forest plantations in the South in the 20th century. In fact, empirical evidence shows that, between 1977 and 2000, some 40% of all NIPF tree planting in this region was covered by various cost-sharing programs, and that an even greater proportion of NIPF landowners (59%) used the reforestation tax incentive. The application of research findings from public institutions and private firms also enhanced economic returns to landowners.

Forest plantations in the South are expected to continue expanding. The expansion could accelerate if the combination of timber price appreciation and timber growth is greater than the establishment and management costs of plantations, and if public policies enhance the economic returns from such plantations. We turn to two of the most important subjects of forest policies, property rights and taxes, in the next two chapters.

NOTE

1 The intuition behind Hotelling's rule is straightforward. Net price has to rise at the rate of interest as a condition of equilibrium, otherwise the present value of the net price that could be received from selling in some periods would be higher than in other periods. Another way to understand Hotelling's rule is as a condition of inter-temporal arbitrage that ensures that the last unit extracted in any time period earns the same returns (in present-value terms). In the forestry context, the net prices for old-growth forests is their stumpage prices, which equal the market prices of delivered logs, minus marginal costs of extraction (logging and transportation).

REVIEW QUESTIONS

1 Describe the long-term dynamics of forest land use and timber production in the United States.

2 How is demand for environmental services incorporated in the dynamics of the timber side of the forest sector?

3 What factors influence the long-term dynamics of timber production?

4 What have you learned from the story of John Sanderson's farm? What factors have led to the changes in land use in New England in the last 200-plus years?

5   It is argued that increase in agricultural land productivity is one of the main reasons that declines in forestland were halted in the United States in the first half of the 20th century. Identify the main factors that may lead to loss of forestland in the US and Canada in the 21st century.

6   What factors directly influence private landowners' silvicultural investment decisions? What are the indirect factors? Is forest certification a factor? If so, is it direct or indirect?

7   Why is the US South able to produce 16% of global industrial timber production with only 2% of global forestland?

FURTHER READING

Binkley, Clark S. 2003. Forestry in the long sweep of American history. In *Public Policy for Private Forestry: Global Initiatives and Domestic Challenges,* edited by L. Teeter, B. Cashore, and D. Zhang. Chapter 1. Wallingford, Oxon, UK: CABI International.

Clawson, Marion 1979. Forestry in the long sweep of American history. *Science* 204: 1168-74.

Hyde, William F. 1980. *Timber Supply, Land Allocation, and Economic Efficiency.* Baltimore: The Johns Hopkins University Press, for Resources for the Future. Chapter 3.

Li, Yanshu, and Daowei Zhang. 2007. A spatial panel data analysis of tree planting in the US South. *Southern Journal of Applied Forestry* 31 (4): 175-82.

Raup, Hugh M. 1966. The view from John Sanderson's farm: A perspective for the use of land. *Forest History* 10: 2-11.

Smith, W. Brad, Patrick D. Miles, John S. Vissage, and Scott A. Pugh. 2004. *Forest Resources of the United States, 2002.* General Technical Report NC-241. St. Paul, MN: US Department of Agriculture, Forest Service, Northern Research Station.

Zhang, Daowei. 2004. Market, policy incentives, and development of forest plantation resources in the United States. In *What Does It Take? The Role of Incentives in Forest Plantation Development in Asia and the Pacific,* edited by T. Enters, and P.B. Durst, 237-61. RAP Publication 2004/27. Bangkok: Food and Agriculture Organization of the United Nations, Asia-Pacific Forestry Commission.

PART 4

Economics of Forest Policy

Chapter 10   Property Rights

Chapter 1 noted the basic policy questions about forest resources. Who will own them, who will utilize them, who will manage them, and who will get the economic benefits from them are among the fundamental issues of forest policy. In large part, these questions are resolved through property rights arrangements or forest tenure systems, which govern the rights of owners, users, and others over forestland, timber, and other forest resources.

People commonly think of property in terms of tangible goods, such as land and buildings that they may own. Lawyers, however, are taught to think of property as a bundle of rights, like a bundle of sticks, each stick representing a right, such as the right to use a particular thing, to enjoy the economic benefits it provides, to sell it to someone else, to exclude other people from using it, and so on. The bundle may be big or small, reflecting the range of rights that the holder has over particular goods or assets. For our purposes here, we will adopt the lawyers' conception of property, emphasizing rights rather than objects themselves. Thus, property rights mean rights to property; they define the extent to which the holder of the rights can enjoy the benefits of particular goods or assets.

Property rights comprise a major branch of legal and economic studies, supported by extensive literature and long tradition. Here, we need consider only property rights in forestland, timber, and other forest-dependent resources, and their economic implications. In western countries, these rights take a variety of forms. Some forests are owned as private property, subject to varying degrees of government regulation. Some are held by governments, referred to in Canada as federal or

273

provincial Crown land and in the United States as federal, state, or county public lands. Even where forests are owned by governments, however, companies and individuals often hold rights over them through various forms of leases, licences, and permits. The simple distinction commonly made between private and public property is inadequate and potentially misleading in this area. In fact, property rights in forestland and timber take many forms, and because they establish the rights, responsibilities, and economic benefits that accrue to tenants, landlords, and governments, they are important subjects of forest policy.

Most instruments of forest policy are about, or work through, or at least affect property rights. Obviously, private ownership and direct public ownership are fundamentally different property rights arrangements; so are tenure arrangements that enable private persons and firms to use and manage public forests. Laws and regulations governing private forestland or public forests under forest tenures often try to align private interests with public interests, even though some regulations, such as log export restrictions, are intended to benefit certain interest groups. By enhancing the property rights of some and restricting certain rights of others, these laws and regulations work through adjustments in property rights. Finally, government incentive programs such as tax credits, subsidies, and technical assistance increase the value of forest owners' property rights, whereas taxes and fees have the opposite effect. It is therefore logical to start with property rights when analyzing the economics of forest policy.

In our discussion of the conditions for efficient use of resources in a market economy in Chapter 2, we noted the importance of producers' control over their inputs. We also noted that producers of forest products rarely enjoy complete control over the forestland and timber they use. The link between the rights held by users of forest resources and the efficiency with which the resources are developed and used is the subject of this chapter.

PROPERTY RIGHTS, PROPERTY VALUE, AND ECONOMIC EFFICIENCY

In this section, we begin with the concepts related to the value of property rights and its relation to economic efficiency, or the use of resources in such a way as to maximize the production of goods and services. We also introduce transaction costs and the Coase Theorem, which highlights the relationship among property rights, transaction costs, and economic efficiency.

Property Rights and Value

The value of a property right depends upon two things. One is the inherent physical and economic properties of the resources or goods over which the property rights extend; these govern the economic benefits they can yield. The other is the extent to which the property right enables its holder to enjoy these benefits. Property rights over even very valuable resources will not be worth much if the rights themselves are highly restricted or truncated. If the rights extend for only a short period, if the holder is restricted from selling the rights, or if his rights allow others to share the benefits, the value of the property rights will be correspondingly lower.

Thus the value of property rights over a forest depends, first, on the inherent productivity of the forest in terms of the value of forest products and services it can yield and, second, on how much of these benefits the rights allow their holder to enjoy. The rights will be worth more if their holder is permitted to use the forest forever than if he or she is allowed to use it for only a year or two. If they include the right to use the water, minerals, and agricultural values, they will be worth more than if they include only the right to use the timber. If the benefits from the forest are not taxed, the rights will be worth more than if the benefits must be shared with the government, and so on.

Property rights include formal and informal rights. The former have government authorization and protection while the latter often do not. Often, informal property rights are recognized or accepted on a first-come, first-served basis, by tradition, or through the use of force. For example, tailgating before football or basketball games in some US colleges creates a kind of informal property rights arrangement over parking and partying space: some families and groups tailgate (party) at certain spots before and after games without a licence or official approval from the colleges or universities, whereas others, often latecomers, accept and respect the former's "rights" to use the spots.

Economists put great importance on property rights because they govern the efficiency of resource use throughout an economy, as well as the distribution of income. In the theory of the pure, perfectly competitive market system discussed in Chapter 2, all factors of production are owned privately and property rights are "complete" in the sense that they are not restricted or qualified. The bundle of sticks is big. With unrestricted rights to use their assets to their best advantage, owners are consistently driven by self-interest to put such assets to their highest and most rewarding use, thereby contributing to total economic production.

Through this same process, ownership of productive factors determines the distribution of income. Economists of the property rights school have argued that comprehensive and "complete" property rights, precisely defined, in all factors of production would ensure maximum possible efficiency in economic production through market mechanisms.

Transaction Costs and the Coase Theorem

An important concept in studying property rights is *transaction costs.* Unlike concepts such as opportunity costs, however, there is no consensus on the definition of transaction costs, even though most economists understand and agree on its general meaning. From a macroeconomic perspective, transaction costs can be seen as the costs of running economic systems. From a microeconomic perspective, they can be seen as the costs of establishing and maintaining property rights. In other words, they are the costs of capturing and protecting property rights and transferring them from one party to another. These costs include the costs of discovering exchange opportunities, negotiating contracts, monitoring and enforcing implementation, and maintaining and protecting the property rights from others. For example, the commission that all consulting foresters in the United States charge on timber sales is a transaction cost to owners of the timber.

Transaction costs can be categorized according to the sources from which they arise. Information costs about market opportunities can be part of transaction costs. Measurement costs in timber sales and negotiation costs of contracts are all transaction costs. Once an exchange is made, monitoring, protection, and enforcement costs are all transaction costs. In cases such as buying groceries, the transaction costs, as a percentage of the exchange value, is usually small. In other cases, such as timber sales, transaction costs can be a significant part of the property value. The costs of protecting one's forests, when trees grow for many decades, can be high where timber theft or illegal logging may occur. In jurisdictions where a forest can be taken by the government without fair compensation, the cost of protecting it may be prohibitively high to the forest owner.

Transactions costs are an impediment to freely competitive markets, and the theoretical perfectly competitive market economy would achieve maximum economic efficiency only if there were no transaction costs. Indeed, if there were no transaction costs, maximum economic efficiency would be achieved regardless of how property rights were distributed to begin with. This is the *Coase Theorem*, attributed to economist Ronald Coase, which we can illustrate with a simple example.

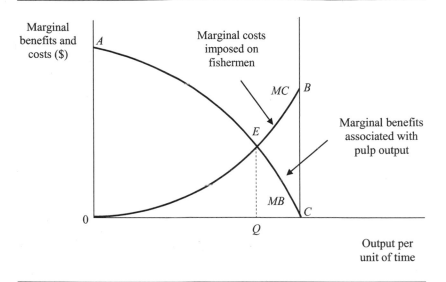

Figure 10.1   The Coase Theorem.

Consider the case of an upstream pulp mill that pollutes a river, re-ducing the downstream fish stocks upon which a group of fishermen depend. Let us consider the benefit/cost calculus of the pulp mill owner and the fishermen.

In Figure 10.1, the horizontal axis measures the pulp mill's output per unit of time, increasing to the right. Curve *AC* represents the marginal benefit of the pulp mill in producing this output. The marginal benefit curve declines with increases in output because the rate of return on additional production in a given unit of time generally declines. The mar-ginal cost curve, *0B*, represents the externality to the fishermen caused by the pulp mill's production. The marginal cost measures the additional cost created at each output level by additional effluent, which is directly related to the level of output. The marginal cost curve rises as output increases.

Suppose that the fishermen hold the right to a clean river so that the owner of the pulp mill will have to consider the costs to the fishermen. Assuming that the cost of bargaining is zero, we can predict what will happen with the help of Figure 10.1. For levels of output up to *Q*, we ob-serve that *MB* (marginal benefits for the pulp mill associated with its pulp output) is greater than *MC* (marginal costs imposed on fishermen). The pulp mill's profits on additional units are greater than the additional pollution costs borne by the fishermen. This means that the owner of the pulp mill would be willing to buy, and the fishermen would be willing to sell, the right of using the river to produce at these levels. Since the total cost of the externality caused by this output (Q) is represented by the area under the marginal cost curve, $0EQ$, a bargain can be struck between the two sides to produce Q through an agreement on how to divide the surplus marginal benefit, $AE0$. Beyond Q, MC is greater than MB. A bargain to produce these levels other than Q cannot be reached. So Q, where $MB = MC$, is the equilibrium outcome when the fishermen hold the right to a clean river. This is the *damager liability rule.*

Suppose instead that the pulp mill holds the right to pollute the river. We then have the *victim liability rule.* Under this arrangement, the pulp mill has the right to dump effluents in the river and the fishermen must pay the pulp mill to produce less and not pollute the river. Up to Q, MB is greater than MC, which means that the fishermen cannot offer a large enough sum of money to induce the pulp mill to reduce its output and its effluent. Beyond Q, the situation is changed: the pulp mill is willing to accept an offer of money to reduce its output. The fishermen are therefore willing to make it worthwhile to the mill to reduce its effluent to the point where $MB = MC$ at Q.

Therefore, the assignment of property rights (*liability rule*) of the river to either party determines only the distribution of benefits and costs; the efficient production outcome remains at Q.

This example shows that the benefit/cost calculus for both parties will include the costs of polluting the river, and that the river will find the same level of uses (or pollution), provided that the transaction costs are zero and there are no legal restrictions on transfer of these rights to pollute or to have a clean river. It also shows that assignment of property rights at a given point in time has wealth effects or distributive effects but has no impact on economic efficiency: the allocation of resource (the level of output of the pulp mill and the resulting level of pollution in the river) is unchanged regardless of which side has the property right.

The Coase Theorem highlights the importance of transaction costs. Transaction costs are ubiquitous; at minimum, the owner of a property and potential buyers incur the cost of obtaining information about the use and potential value of the property. They arise whenever there is exchange among economic agents and whenever there is a need to protect the property. In some cases, they are relatively low; in other cases, they can be prohibitively high and exchange will not occur.

We can learn at least three things from our example of pulp mill versus fishermen. First, if a competing form of production (fishing versus pulping) is not marketed, this will lead to misallocation (an externality), as noted in Chapter 2. Further, if both forms of production are owned, divisible, and marketed and transaction costs are nil or very small, bargaining will lead to an optimal result; in such a case, the initial allocation of property rights does not matter. Finally, high transaction costs may, and often does, prevent an optimal result, and government intervention to improve resource allocation may be needed.

EVOLUTION OF FOREST PROPERTY RIGHTS

Property rights have evolved slowly through centuries of tradition, conflict, and court decisions. Theories about the origins of property suggest that private rights emerged from an original regime with no property, or with common property, when the available supply of a resource was more than sufficient to satisfy all the demands on it. When demand began to exceed the available supply, users, lacking a means of defining and establishing their claims, began to interfere with each other's production and caused other costly inefficiencies. Sooner or later, the potential gain from eliminating this interference exceeded the cost of organizing exclusive private property rights to eliminate it. In short, as long as resource values are low, the benefits from organizing property do not justify the associated transaction costs, and the system of users' rights remains crude. As resource values rise, however, raising as well the potential gain from improved allocation arrangements, more sophisticated systems of property rights can be expected to emerge.

The evolution of property rights over natural resources in North America is consistent with this model. Early settlers helped themselves to the abundance of fish, timber, water, and other resources of the land. There was no scarcity in the economic sense, and no allocation problem, so there was no need to worry about the complications and cost of organizing property rights.

Gradually, with settlement, demands on land and resources increased, and it became necessary to allocate them among users. There arose the need to parcel out land in settlements for living space and around them for agriculture, but not beyond the "frontier," which, by definition, was unappropriated. In pioneer days, there was less urgency to allocate rights over water, fish, and timber, but as the frontier receded and development proceeded, natural resources had to be allocated among competing uses and users. Thus, we find in North America well-developed systems of property rights over resources that became valuable long ago, such as urban and agricultural land and minerals, but rights to fish, wildlife, and water remain relatively primitive in many jurisdictions.

In the early British colonies, the forms of property rights over land were derived from English common law. Frequently, the land was acquired from Aboriginal occupants by the Crown through treaty or conquest, and the government then granted title to settlers, land development companies, railroads, and other private interests. Early grants usually carried with them the full range of rights under traditional common law, including unrestricted rights to use the surface of the land, whatever lay beneath the surface, the water and timber on it, the wildlife, and so on. In the early days, grants of title were the simplest and least expensive, and were therefore the normal way of obtaining rights to resources.

Later, mainly in the decades around 1900s, governments throughout North America turned away from the policy of granting complete or outright title to land and resources except for land needed for urban development and agricultural purposes. For other resources, notably timber, governments designed rights in the form of leases, licences, and permits, which they could issue to private parties to allow them to use specified resources while the government continued to hold the title to the land. Some of the early leases involving timberland were much like outright private ownership insofar as they had long terms and provided their holders with the exclusive use and benefit of the resources, with few restrictions and controls. Others, such as timber sales, authorized the use of a specific amount of timber, for a specific purpose, within a short period. The result is a spectrum of property rights in forestland and timber, ranging in duration, comprehensiveness, exclusiveness, and other characteristics that have important economic implications, as discussed below.

Governments have also developed regulatory means of reconciling competing uses and users of public resources, to some extent as alternative and implicit systems of property rights. The cost of organizing and establishing property rights over some resources, such as those that yield general environmental benefits, can be exceedingly high, so governments have developed ways of regulating their use instead. Examples include pollution controls and recreational fishing and hunting regulations. Thus, today we find a complicated mix of public, private, and common property, and policies that rely on a combination of market processes and government regulations.

The United States and Canada are common-law countries. Historically, the courts have created some property rights. A good example is the change in the liability rule. Tort law – the law of private injury – which helped stimulate industrialization in the 18th and 19th centuries, evolved in part from trespass laws and originally dealt with liability regardless of fault. If, for example, a worker was injured by a falling log, the employer would be liable. Such *strict liability,* of course, was a serious obstacle to the Industrial Revolution. Judges generally sided with new and growing industries by making fault a requirement for liability: a person was liable for injury only if he or she acted negligently, that is, in a way not in conformity with what a reasonable person would have done in similar circumstances. Nowadays liability is *limited* only to such a person or firm that acted unreasonably and was thus at fault.

Another example is the recognition of regulatory takings by US courts. The Fifth Amendment of the US Constitution stipulates, "nor shall private property be taken for public use without just compensation." Prior to 1922, taking of private property in the US had been limited to physical taking. In 1922, however, the Supreme Court ruled in one case that government exercise of regulatory power was so excessive that it resulted in a taking of private property and that the owner of the private property was entitled to fair compensation. Thus, the court created a precedent and a new right for property owners: they can demand fair compensation if there is a regulatory taking of their property.

It is worth noting that differentiating between regulations (no compensation needed) and regulatory taking (compensation required) has become controversial in the United States. The questions concern what constitute public uses of private property and how excessive the regulation must be before compensation to the property owners is required. Recently, the US Supreme Court ruled that economic development or the need for tax revenue is a public use. Thus, private property owners have been forced to sell their property to local governments that want to use it to build a Walmart or a shopping mall. In light of this ruling, some states have enacted laws that prohibit state and local governments from using economic development as a public use to take private property.

The excessiveness of the regulation is more difficult to determine. One of the criteria for determining excessiveness is how much the property value is lowered because of a regulation. The closer the devaluation is to 100%, the more likely the regulation becomes a regulatory taking requiring compensation. In the 1990s, some state governments defined a threshold for regulatory taking through legislation. Similar legislation failed in the US Congress, however.

In summary, property rights often evolve in response to economic needs. They may be allocated officially by laws and regulations, and through the common law developed from court cases.

DIMENSIONS OF PROPERTY AND THEIR ECONOMIC IMPLICATIONS

Dimensions of Property

The dimensions of property that are of economic importance include comprehensiveness, duration, benefits conferred, transferability, security, and exclusiveness.

Comprehensiveness

Comprehensiveness refers to the extent to which the holder of the property has rights to the full range of benefits from an asset. For example, those who hold a tract of forestland under a private freehold can usually claim the full range of values generated from timber, agriculture, recreation, water, and so on. If their rights are in the form of a timber sale licence, however, they are usually restricted to the benefits of timber production alone.

In this and other dimensions of property, there is a spectrum of possibilities. Each form of property right occupies a particular place on the spectrum.

The degree of comprehensiveness of owners' property rights has important implications for the economic efficiency with which they will manage and use a forest. Those who have full, comprehensive property rights over a forest can be expected to maximize the aggregate value generated by all its attributes and possible uses, sacrificing one in favour of another whenever it is advantageous to do so. In contrast, those who hold the rights to the timber in a forest but not to the water, wildlife, or other benefits that are affected by their timber operations will tend to ignore any adverse effects of their actions on these other values, if they are not liable to pay anyone any compensation. Such holders will seek to maximize the benefits that they can claim, disregarding those that they cannot claim and any external costs or benefits that they may inflict on others. When users thus fail to take account of all the effects of their decisions, the aggregate benefit from all the values will fall short of its potential. Government regulation is the only means of overcoming these impediments to socially desirable patterns of resource use, and regulation may be required on a global scale to provide for some environmental benefits, such as clean air and a stable climate.

Such problems are common in forestry. Firms exercising their rights to cut timber impinge on the benefits of those who have rights to recreation, aesthetic enjoyment, or wildlife in the same forest. In these circumstances, no single decision maker – whether companies, recreationists, or government regulators – has an economic incentive to search for the optimal combination of uses, as described in Chapter 6. Thus, the optimal combination of uses often requires a mix of market and regulatory tools.

Private forest owners as well as governments often regulate the activities of those who contract to cut timber on their lands. In this way, they protect their interest in values other than the timber, such as aesthetics and the continuing productivity of their land.

Sometimes rights are fragmented among private parties, who separately hold rights to water, timber, minerals, and so on. In addition, they are often fragmented among governments, as when a provincial, state, or local government claims rights over the timber while a federal government has authority over fish or wildlife habitats. This often results in serious conflicts among governments and various private holders of resource rights.

It is important to note that these conflicting interests are created by artificially separating rights to the same resources. If they were all held by one party, the conflict – ownership externality – would not arise; that party would determine the most advantageous combination of uses, as noted above. Even if the rights to the different resources or attributes of a forest were held by different people, market processes could still produce an optimal result as long as the rights were freely tradable and transaction costs were small. This sort of beneficial transaction is commonplace where rights are private property. Such market processes cannot be relied upon, however, when property rights are not well defined and are freely purchased and sold. Misallocations are then likely to occur, and conflicts often remain unsettled unless government regulation is invoked to resolve them.

Duration
Duration refers to the length of time over which the property rights extend. Private freehold grants rights in perpetuity, whereas leases and licences normally have finite terms.

The duration of property rights is important because it determines the extent to which holders will take account of the future impact of their actions. If the rights over a forest extend for a long period, holders can be expected to carefully consider the relative economic advantage of harvesting now or in the future and the returns to investments in silviculture; if the rights terminate shortly, however, they will disregard such long-term considerations. In the framework of Chapter 3, the effect is tantamount to holders' discount rate rising to an infinitely high value beyond the date at which their rights expire.

Because of the extraordinarily long term considerations involved in forest development and silviculture, the duration of rights over forestland is a special problem. Unless their rights extend over the several decades it takes to grow forest crops, those who harvest timber will lack incentives to provide for reforestation. Thus, subsidies or regulations might have to be devised to ensure appropriate investments in resource management.

Moreover, the duration of property rights over forests is often a primary determinant of the holders' security of supply of timber. Security of raw material supply, in turn, has a major influence on decisions about investment in manufacturing and other facilities, and hence on the efficiency of resource use.

Benefits Conferred

"Benefits conferred" refers to the extent to which the property provides its holder with a right to enjoy the potential economic benefits from an asset such as a forest. This is often constrained by government restrictions on how the forest can be harvested, managed, or utilized. Regulations that restrict the rate at which timber may be harvested require loggers to recover uneconomic timber, impose measures to protect the environment, prohibit exports of unmanufactured timber, and direct some of the return into the public purse through taxes, royalties, and other charges, all of which impinge on the benefits that flow to the holders of forest property.

Restrictions on the extent to which holders of rights over a forest can enjoy its potential benefits obviously affect the value of their forest property and hence the distribution of income. Moreover, they usually create incentives to alter the way resources are used, and thereby also affect efficiency. In extreme cases, government impositions may eliminate all the benefits to private forest owners, as was the case in Canada and the US during the Great Depression, when the burden of taxes caused many owners to abandon their forestlands, which then reverted to the government.

Transferability

The transferability of property is its capability of being bought and sold or assigned to someone else. Transfers of forest property are sometimes restricted. For example, the terms of temporary licences and leases often restrict the licensees from transferring their rights to someone else, or require them to obtain the consent of the government or private licensor to do so.

If a property is absolutely non-transferable, it has no market value. The only way its holder can benefit from it is by exercising the rights himself or herself. Non-transferability obviously affects the distribution of income and wealth, and it also impedes efficient allocation. Economic efficiency depends upon the acquisition of resources by those who can generate the most value from them, who are thereby able to offer more and bid them away from less efficient users in a competitive market economy. Impediments to the marketability of assets prevent them from being transferred to those who can use them most productively.

A related issue is the divisibility of property. To take full advantage of economies of scale and changing economic opportunities, entrepreneurs must be free to divide and combine rights to resources. This is sometimes restricted in forest property by government prohibitions on the subdivision of rights such as leases and licences on public land, or regulations about maximum and minimum allocations under particular forms of rights.

Security

Security refers to the extent to which property rights are recognized and enforceable. It depends heavily on the political, legal, and cultural environment in which the property is held. Where property rights are respected, there is little risk of losing them without due process and fair compensation, so owners are encouraged to make efficient use of their resources. Where property rights are insecure, perhaps because of the risk of confiscation by government, of excessive regulation, or of private encroachment, the incentives to invest and use resources efficiently are weakened and the value of property rights is correspondingly reduced.

Exclusiveness

Exclusiveness refers to the extent to which the holder of the property can claim sole rights to exclude others. Rights to timber are typically (but not always) defined spatially, so that the owner has exclusive rights within a clearly defined area and can carry out operations without interference from others. In the case of other resources, such as fish or water, there are often many users of the same resources, so that none of them has sole control or the incentive to use the resources most efficiently.

When rights are not exclusive, and their holders compete with others for the same benefits, resources are likely to be too quickly and inefficiently exploited. Moreover, incentives for users to conserve for the future and invest in future yields will be weak because the users cannot expect to capture the full benefits of their individual actions. Thus, the ability to exclude "third parties" is a fundamental element of property rights and has important economic implications. We shall return to this important dimension of property below.

Combining Various Dimensions in Forest Tenures

The six dimensions of property rights discussed above are the most important economic forces that influence forest property rights and tenure systems. Together they govern the extent to which producers can obtain control over forest assets, identified in Chapter 2 as a primary condition for economic efficiency in a market system.

These dimensions or characteristics of property rights are interdependent. Security, for example, is influenced by duration, transferability, and exclusiveness. Comprehensiveness and benefits conferred are obviously linked.

Each of these characteristics varies across a spectrum. Consider the dimension of exclusiveness. At one extreme, the holder of rights over the forest has absolute *exclusive* rights; that is, the holder can exclude all other users. This is exemplified by the traditional freehold ownership, where private landowners are given exclusive rights to use the land and its attributes, such as timber.

At the other extreme, the case of *no property* rights allows nobody to hold any special rights and allows no one to exclude anyone else. The best example of this is the high seas, where no one, nor any nation, can assert legal rights over another. There is no forestry counterpart to the high seas because all forests are claimed by national governments at least. There are, however, many examples of public forests to which all citizens have free and equal access for certain purposes such as recreation. Between the extremes of no property and completely exclusive property rights, there is a wide range of possibilities, as illustrated in Figure 10.2.

Several forms of *common property* exist. In frontier times, it was not uncommon for forests to be accessible to anyone who wanted to cut timber. Such unregulated exploitation of timber is now rare, but other forest products, such as game, are commonly available to an unlimited number of users. An interesting example of timber managed in this way is the ancient anomaly of the coastal strip of Newfoundland, which resulted from the early influence of fisheries traditions. Centuries ago, it became internationally accepted that coastal states had the exclusive right to fish within three miles of their coast, and in western countries the fisheries

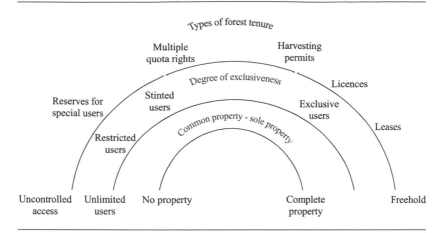

Figure 10.2   Degrees of exclusiveness of forest tenure.

within this coastal sea were open to all citizens. In Newfoundland, a corresponding strip of forest extending three miles inland was kept available for all residents to exploit as they wanted. The effect on forestry was similar to that in open-access fisheries everywhere: overexploitation of the resources, absence of incentives for users to conserve or invest in them, overcapacity in utilization, and dissipation of economic rents. This phenomenon has been labelled the *tragedy of the commons.* Another example is the traditional open range in western North America, which allowed ranchers to graze unlimited numbers of cattle, leading to overgrazing. Each rancher found it advantageous to add more cattle, and none had incentive to restrain their use to the sustainable capacity of the grasslands.

Somewhat less vulnerable to this sort of economic waste are common property resources accessible only to a restricted number of users. In North America, this has become the most common regime for commercial fisheries; fishermen are required to hold licences and the number of licences is limited, but they compete with each other for undefined shares of the available resources. Forestry examples of such *limited common property* are sometimes found in the rights of users of minor products such as fuel wood and Christmas trees.

In other cases, common property resources are exploited by holders of licences that authorize them not only to use the resources but also to take a specified quantity. The available harvest is thus *stinted,* giving the users more well defined rights. Water rights, grazing rights, and fishing

rights often entitle their holders to take a specific quantity of the resource, used in common with others. Timber-harvesting rights sometimes take this form, where several users are authorized to harvest
quotas of timber in a public forest without holding exclusive rights to
any defined tract. Most common in forestry are various forms of *sole
property*, where specific resources are reserved to a single user.

There is thus a spectrum of possibilities with respect to this critical
dimension of exclusiveness. They are held in the form of permits, licences,
leases, and titles, which overlap and merge in varying ways.

A corresponding spectrum of possibilities exists for each of the other
dimensions of property noted earlier. Comprehensiveness can range
from narrowly defined rights to rights that include all attributes and
uses of the land and resources; duration from a brief period to perpetuity; benefits conferred from zero to all of the potential resource rents;
and transferability from completely non-transferable to complete freedom to transfer, divide, and combine rights.

Figure 10.3 illustrates how any property right is a combination of
these characteristics, each defined in terms of degree. Each brick in the
diagram refers to a particular characteristic, such as comprehensiveness
or duration, and the degree of the characteristic increases from zero at
the left end to the maximum possible, at the right end. A property em-
bodying the maximum possible characteristics illustrates "complete"
property. The dotted lines denote a combination of characteristics that
are truncated in varying degrees. An infinite variety of such combinations is possible, producing a correspondingly infinite variety of possible
forms of property. As a result of centuries of evolution and development
of the law of property, however, most property rights over forestland
and timber now fall into a few broad categories, discussed under
"Common Forms of Forest Tenure" below.

Dimensions of Property Rights and Transaction Costs
All of the six dimensions bestow benefits on the holder of a property
right and enhance the value of the property right. But how do they relate to the transaction costs of the property right itself, and to the transaction costs of the property right holder in obtaining other factors of
production?

An enhancement in some characteristics (comprehensiveness, exclusiveness, duration, benefits conferred) may decrease or increase the
transaction cost of the property right itself. Take benefits conferred as an
example. An increase in benefits conferred may entice holders to spend

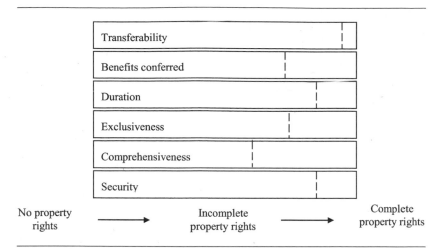

Figure 10.3   Combinations of attributes in forest property.

more resources in protecting their right. An enhancement in security and transferability will, by definition, reduce the transaction costs of the property right.

On the other hand, when the value of the property right increases, holders can use it to their advantage, by reducing the costs of other inputs. Often, holders have an easier and lower-cost access to credit (capital) and labour when their property is worth more. Thus, the characteristics of property rights and the transaction costs of factor inputs are always inversely related. Notably, recent forest tenure reforms in some countries were to provide secure property rights to farmers and lower the transaction costs of inputs so that they could use more labour and capital in long-term forest production.

COMMON FORMS OF FOREST TENURE

In North America, property rights in forestland and timber are commonly divided into two general categories: *freehold,* or what is commonly referred to as private title, fee-simple, or deeded land; and *usufructuary rights,* or rights to use resources owned by others. In Canada and the United States, the latter most commonly take the form of a licence or lease over public and private lands, in contrast to Britain and other parts of Europe, where it is often over lands of private landowners.

The typical forms of property rights to timber, forestlands, and other forest resources that we find in North America today, used by both governments and private parties, are listed in Table 10.1, along with their characteristics in terms of the six dimensions of property discussed earlier. The forms listed are only representative; there are many other varieties of forest tenure that do not fit easily into these common types, and there are considerable variations within them. For example, the traditional *freehold* under English common law was comprehensive, carrying with it the rights to all the attributes of the land, including the minerals and other resources under it, the water flowing over it, the fish and wildlife on it, and so on. Over the last century, however, most governments granting title to public lands in North America passed statutes that progressively stripped away many of these rights, reserving for governments themselves the rights to such things as subsurface minerals, water, and wildlife when they made new grants to private owners. As a result, the comprehensiveness of freehold tenure over lands granted long ago is typically more complete than that over lands granted more recently. The title to lands granted in the more recently settled western states and provinces often excludes everything but the right to use the surface of the land.

Where other attributes of the land, such as minerals and water, are excluded from the landowner's property rights, governments often allocate the right to use them to other private parties under some form of usufructuary right. With few exceptions, however, trees retain their traditional status under common law as part of the land, so title to land carries with it the title to any tree on the land. In some African countries such as Ghana, trees belong to the state while land belongs to communities.

In granting title to lands, some governments have kept a financial interest in the forest by requiring the owner to pay a royalty, or fee, when timber is harvested. To this extent, the benefits conferred have been narrowed. Property taxes, zoning, and land-use regulations, all of which have burgeoned in recent years, similarly impinge on the owner's right to enjoy the potential economic benefits from the land and forest.

An important variant of freehold tenure in some jurisdictions is held by owners who contract with the government to dedicate their lands to continuous forest production. Typically, in return for an undertaking to manage the lands according to a sustained yield plan and other provisions approved by the government, owners receive a property tax concession. This arrangement reduces the degree to which taxes encroach on the benefits enjoyed by holders, but their economic benefits are reduced to the extent that they must forgo more advantageous ways of using or developing their forestland.

TABLE 10.1 Characteristics of the typical forms of forest tenure

Tenure	Comprehensiveness	Duration	Security	Benefit conferred	Transferability	Exclusiveness
Freehold	Complete	Perpetual	Very secure	All, except for taxes	Unrestricted	Exclusive
Timber lease	Most attributes	Long-term	Secure	Most timber benefits	Few restrictions	Exclusive
Hunting lease	Limited to wildlife	Short-term, often renewable	Secure	Wildlife benefits only	Few restrictions	Exclusive
Forest management agreement	Limited to timber harvest and forest management	Long-term, renewable	Varies	Most timber benefits	Some restrictions	Exclusive
Woodlot licence	Timber harvest and forest management	Long-term, renewable	Varies	Most timber benefits	Usually restricted	Exclusive
Timber licence	Timber harvest only	Short-term	Varies	Share in timber benefits	Usually restricted	Exclusive
Cutting permit	Timber harvest only	Short-term	Usually secure	Most timber benefits	Usually restricted	Exclusive or non-exclusive

Early usufructuary rights on public land were typically in the form of *timber leases*. These often carried very long terms and some persist today. In some cases, they are almost as comprehensive as freehold, carrying rights to other resources as well as timber. Typically, however, they require holders to pay an annual fee or rent, and a royalty or similar charge on timber harvested, reducing the benefits that accrue to them. Often, the terms of a lease restrict its transferability to the extent of requiring its holder to obtain the approval of the government or private landlord before selling the lease to someone else.

The modern version of timber lease on private land has some similarity to timber lease on public land – that is, the duration is long enough so that the leaseholders can grow timber for one or two rotations. Thus, in the 1950s and 1960s, when forest products firms needed to secure certain amount of timber supply to support their newly built forest products (mostly pulp and paper) manufacturing plants in the southern US, they leased some timberland from non-industrial private forest landowners for 60 years, long enough to support two rotations. *Hunting leases* on private land in the United States, on the other hand, are typically short-term (one to several years, renewable) and limited to specific wildlife benefits.

The term "forest management agreement" is used here to describe a long-term licence granted over an extensive tract of public forestland to be managed by its holder as a coherent sustained yield forest. This form of licence is now common in Canada's major timber-producing provinces, but it is rare in the United States. In some cases, these licences integrate freehold lands held by the licensee with public lands under a single management plan. Under detailed provisions, they assign to the licensee not only the right to cut timber but also extensive responsibilities for silviculture, forest development, protection, and management under approved plans.

Forest management agreements, sometimes called tree farm licences, forest management licences, or concessions, generally carry long terms of 20 years or more and are renewable. As Table 10.1 suggests, the rights of holders are confined to timber, their benefits are constrained by annual fees and levies on timber harvested, and the transferability of their rights is usually qualified or restricted to some degree.

The *woodlot licence* has a well-established but relatively modest place in the forest tenure policies of many jurisdictions in North America, conveying rights to very small tracts of public forests. They are usually not intended to support industrial operations but are designed to complement farms by providing sources of fuel, fencing, and other building materials or seasonal employment in timber production. They usually convey long-term rights to the timber only, in return for modest charges and general management obligations.

The term "timber licence" refers to a large category of usufructuary rights to timber, especially on public lands. These are short-term licences to harvest defined, relatively small tracts of timber. The most well known form is the traditional "timber sale," but there are many variations. Licensees' rights are typically limited to harvesting the timber in an approved manner, in return for which they usually must pay a licence fee, an annual rent, and a charge against timber harvested. The licence often imposes a variety of other obligations as well, relating to such things as road building and reforestation. This is the most common form of tenure for users of timber in US National Forests, and is used in some form and degree by most forest management agencies in North America. Rights to harvest timber on private forestlands are also in the form of timber licences or contracts, which are often granted for a period of less than one year.

The term "cutting permit" describes minor forms of licences to cut timber that are issued for short terms and for special purposes. Authorizations of this kind are employed to clear public timber from rights-of-way and reservoirs, to salvage timber in fire-damaged forests, to supply materials for mining, to cut fuel wood, and for a variety of other minor purposes.

These common forms of property rights in timber reveal considerable diversity in their basic dimensions, but they refer primarily to interests in timber and some wildlife species. Rights to other forest values, such as livestock forage, water, and recreation, also take a variety of forms. Rights to timber and other benefits in the same forest area may be allocated in quite different forms to different parties, so that the patterns of overlapping rights are sometimes varied and complicated.

ECONOMIC ISSUES OF TENURE SYSTEMS

Bearing in mind both the multidimensional character of property rights in forests and the wide range of property forms, we can now review briefly the main features of forest tenure systems that affect the efficiency of resource use and the distribution of benefits.

Methods of Allocation

Rights to timber and forestland can be obtained in various ways. In Canada and the United States, timber staking traditionally provided a first-come, first-served method of establishing private property rights in the public domain. Preemption and improvement was another way of obtaining grants of public lands in the frontier era, which allowed early homesteaders to acquire title to the lands they used and improved. Railroads were granted extensive tracts of forest in the west by the federal government.

Nowadays, it is common in Canada's major timber-producing provinces for corporations to secure rights over resources through bilateral negotiations with governments. In the US, rights to harvest timber on public lands are usually allocated by competitive bidding. And, of course, rights of most kinds can be acquired by purchasing them from others. Recently, some countries formerly organized under central planning systems have been allocating their collective forestland to members of the collectives on an equal basis. On the other hand, Brazil has been auctioning some national forests in the Amazon region to communities, companies, and nongovernmental organizations since 2006.

From an economic viewpoint, the important issue is whether the method of allocation enables the resources to find their way into the hands of those who can make the most productive use of them. Competitive allocations encourage this process, because the most efficient users can outbid the less efficient ones. Non-competitive allocations may not result in the most productive use but, as noted earlier, much depends on whether the rights are transferable once allocated. As long as they are freely transferable, they will tend to be reallocated through subsequent purchase and sale to the highest users. In this case, the only long-term consequence of the original allocation method involves the distribution of income; if the subsequent transactions are competitive, all the benefits not captured in the initial allocation will be captured by the first recipient of the property rights.

Scope of Rights
This relates to the comprehensiveness of property rights, as discussed above. Generally, the more restricted the rights held by users, the more externalities can be expected. In contrast, if a single user's rights include a full range of values, she will internalize her costs and benefits and attempt to maximize the aggregate net return. Where the incentive to take account of some values is eliminated by the narrow scope of the users' rights, they can be protected only through government regulation.

Even the most comprehensive forms of property rights exclude some values. Certain externalities cannot be easily internalized, for technical reasons. Examples include the amenity of forest landscapes and values associated with fish and wildlife that migrate in and out of the relevant forestland. These are public good externalities noted in Chapter 2.

Private property rights everywhere are subject to certain rights or powers held by governments. In the United States, the states' rights over private property include property taxation, eminent domain (taking private property for public use with fair compensation), regulation, escheat (where a property goes to the state when its owner dies with no heir and no will), and overflight. In addition, the US government has wide powers to regulate private landowners under the Commerce Clause, Tax and Spending Clause, and Treaty Clause of the US Constitution.

Security and Economic Efficiency
The security of rights to resources is a major influence on entrepreneurial investment and operational planning. Insecurity invokes risk and, as noted in Chapter 3, investors are generally averse to risk. Hence, insecure rights impede investment in resource development and management, and encourage overexploitation. Empirical studies show that insecure rights, resulting from weaknesses in tenure arrangements on public lands in Canada, regulatory restrictions on private forests in the US, uncertainties about rights in collective forests in China, and state ownership of mature forests in Ghana, discourage silvicultural investment and encourage earlier, wasteful timber exploitation.

As noted above, a primary determinant of security is the duration of the rights. Private title carries an infinite term, whereas licences and permits have terms of only a year or two. Where the terms are finite, however, the security of holders depends not only on the duration of their rights but also on the provisions for renewal. If there are no rights of renewal, the term is critical; if renewal is automatic, the term is unimportant. Between these extremes, there may be variety of provisions for qualified rights of renewal.

Against the licensee's customary desire for security lies the licensor's interest in retaining the ability to alter allocation arrangements. Landlords, both private and public, have an interest in preserving their flexibility to reallocate resources in the face of changed circumstances or to alter the terms and conditions under which rights to their resources are assigned to others.

The licensee's interest in security on the one hand, and the licensor's interest in flexibility on the other, conflict on the issue of the term of rights. A recent innovation in forest tenure arrangements reconciles these interests by providing for "evergreen renewal" – that is, a provision in the licence to enable the licensee and licensor to negotiate the terms of a new licence to replace the existing one when its term has only partly expired. This technique provides a regular opportunity to change the terms and conditions while ensuring that licensees never have to face the imminent expiry of their resource rights.

Finding a balance between property rights holders' desire for security and the government or private licensor's interest in flexibility is critical. The production function of firms or individual producers is constrained by the structure of property rights, just as it is constrained by the state of technology. Thus, property rights can be seen as a factor of production. Because secure rights to resources reduce the uncertainty and risk faced by forest enterprises, and enhance their opportunities for long-term investments, holders of rights over forestland always seek more security for their property rights.

At the broader level of society, however, greater security might not be beneficial. The benefits that would flow to the holders of rights must be weighed against any losses of non-market values, external benefits, and flexibility for public works that might result. Obviously, the most beneficial balance between these private and social interests will vary in differing circumstances, and over time. This balancing act is often carried out through government intervention.

Scope for Intervention

An important consideration in forest tenure arrangements is the scope for governments or others to interfere with the activities of those who hold rights to use the forest, and any ambiguity about how holders may exercise their rights. Governments can employ legislation to intervene on both private and public lands, and where they are the licensors of users of public resources, the terms and conditions of the licences provide them with an additional means of regulating forest operations. Having broad scope for regulatory control enhances the government's power to protect all social values and interests, but it restricts the ability of property holders to use the resources to their best advantage Susceptibility to government regulation thus diminishes the quality of property rights and their value.

The optimal amount and form of government regulation is difficult to identify, not least because both the costs and benefits of intervention are exceptionally difficult to estimate. In theory, a country can have too many regulations, such that if and when regulations themselves become a tragedy of the commons as different regulatory agencies engage in a turf war and compete for more regulations, society will bear the economic consequences in terms of decreased economic activity and lower efficiency.

Allocation of Management Responsibilities

Many forms of usufructuary rights over forests not only provide their holders with rights to use resources but also assign them management responsibilities. In Canada especially, elaborate forest management agreements require licensees to manage public forests according to plans approved by the responsible forest agencies.

It is important to distinguish who is responsible for resource management from who is to pay for it. This is because licensees who assume such contractual responsibilities are often reimbursed, directly or indirectly (for example, through lower stumpage payment than otherwise), for the costs. The ultimate impact on the distribution of resource rents thus depends on these financing arrangements as well as on more direct fiscal measures.

Distribution of Resource Rents

Taxes, royalties, stumpage fees, and other levies determine how much of the potential resource rents will accrue to the property holder. Most fiscal devices also affect incentives to conserve forests and invest in future yields. These issues are considered in Chapter 11.

PRIVATE AND PUBLIC OWNERSHIP

The preceding sections suggest that the customary distinction between private and public ownership of forests is inadequate. The property rights of private owners are almost always qualified by restrictions in their deeds and titles, by laws and regulations, and by various taxes and charges. Those who utilize public forests are usually private corporations or individuals who hold some form of usufructuary rights. In most jurisdictions where forestry is important, those who utilize forests do so within a tenure system that consists of a range of public and private property rights, which overlap and blend in various ways.

Public ownership of forests is an important form of property right and is common throughout western, eastern, and Third World countries. National, regional, and local governments, as well as a variety of public boards and agencies, hold title to forestland. Much is managed for timber production, but special public needs, such as those associated with parks, greenbelts, and watersheds, often provide the main purpose of government control. In most western countries, public ownership of forests is far more prevalent than public ownership of agricultural and other types of land.

As noted early in this chapter, the widespread public ownership of forestland in the United States and Canada is the result of early political decisions to abandon the traditional procedure of granting title to land and resources sought by private users. The shift in policy towards maintaining public ownership of forestland appears to have been motivated by three economic considerations. One was a perceived threat, vigorously asserted by the powerful conservation movement early in the 20th century, when eastern forests were being depleted and western forests were opening up, that the remaining forests would fall into the hands of land barons and developers who would exploit them too fast, leaving inadequate resources to support the industrial needs of future generations.

Another concern of the US conservation movement, which also influenced Canadian policy makers, was that speculators would lay claim to resources and withhold them from development in the expectation that their value would rise, to the detriment of the present generation. This fear obviously conflicted with contemporary concerns about too rapid resource depletion, but both propositions concerning undesirable rates of exploitation by private owners gave support to advocates of public ownership of land and resources.

A second motive was to protect the public financial interest in natural resources. Timber provided the foundation for early regional economies, and many people saw the expansive natural forests as a bank of public assets that could be realized fully only by retaining them for sale as demand for them, as well as their value, rose over time. It is worth noting that this view presumes that potential private purchasers have expectations about future values that are more pessimistic and less accurate than those of governments, otherwise the prices they would be prepared to pay for the resources would match the present value of the returns that governments could expect from later sales. At any rate, shortsightedness on the part of private corporations was a popular assumption among reformists a century ago.

The third motive, of more recent vintage, was to protect the public interest in forest values other than timber. This motive rests on the twin assumptions that private owners will pay insufficient attention to these other values and that they can be better protected through public ownership than by creating more comprehensive private property rights or by regulating private owners. This motive has led to the creation of multiple-use and ecosystem service paradigms in recent decades.

All these arguments in support of public ownership of forests are subjects of continuing academic and political debate. As noted in Chapter 3, economic theory does not demonstrate conclusively that the social interest lies in exploiting resources either faster or slower than would result from market forces, and experience does not reveal that governments consistently manage forests better or worse than private owners. In any event, in the US and Canada today there appears to be a firm political commitment to maintaining at least the present degree of public ownership of forests. Privatizing public forests is sometimes proposed in both countries but has yet to attract wide support.

This commitment to public ownership is paralleled by an equally entrenched commitment to have the private sector utilize the resources. In western countries, virtually all timber, as well as most other forest products and services, are recovered and used by private corporations and individuals. These circumstances place a heavy onus on the form of property rights used to provide private users with access to public resources.

But the task of designing usufructuary rights that will encourage their holders to utilize resources in the interests of society as a whole is more difficult for forests than for most other natural resources. As we noted earlier, the unusually long planning and investment periods in forestry and the time it takes for the full impact of activities in a forest to be felt, mean that users will have incentives to respond to all the future costs and benefits of their actions only with rights of very long duration. The prevalence of multiple uses and products, public goods, and external effects similarly complicate the design of property rights.

For all these reasons, the property rights of users of both public and private forests are rarely adequate to ensure that their incentives always converge with the interests of society at large. The primary alternative to improving property rights systems in order to promote this convergence is government regulation of behaviour.

REVIEW QUESTIONS

1 If property can be described as a "bundle of sticks," with each stick representing a right, what are some of the rights that can comprise someone's property interest in a forest?

2 Explain the externality that may result if someone has a well-defined interest in the industrial timber produced in a forest but no one has a

comparable economic interest in the wildlife. Why must government take responsibility for protecting the wildlife values?

3 What are transaction costs? Why are transaction costs important?

4 Describe and explain the Coase Theorem.

5 If someone holds only short-term harvesting rights over a forest, how are his or her profit-maximizing decisions likely to be less efficient from the viewpoint of society as a whole than if those rights were perpetual?

6 Why is the transferability of property important in promoting the efficient use of resources?

7 How would you explain the greater political support for public ownership of forestland than of farmland?

FURTHER READING

Allen, Douglas. 1992. What are transaction costs? *Research in Law and Economics* 14: 1-18.

Barzel, Yoram. 1997. *Economic Analysis of Property Rights.* 2nd ed. Cambridge, UK: Cambridge University Press.

Coase, Ronald H. 1960. The problem of social cost. *Journal of Law and Economics* 3: 1-44.

Demsetz, Harold. 1967. Toward a theory of property rights. *American Economic Review* 57 (2): 347-59.

Fortmann, Louise, and John W. Bruce, eds. 1988. *Whose Trees? Proprietary Dimensions of Forestry.* Boulder, CO: Westview Press. Chapter 8.

Furubotn, E., and S. Pejovich, eds. 1974. *The Economics of Property Rights.* Cambridge, MA: Ballinger Publishing.

Libecap, Gary D., and Ronald N. Johnson. 1978. Property rights, nineteenth century federal timber policy, and the conservation movement. *Journal of Economic History* 39 (1): 129-42.

North, Douglas C. 1981. *Structure and Change in Economic History.* New York: W.W. Norton.

Pearse, Peter H. 1980. Property rights and the regulation of commercial fisheries. *Journal of Business Administration* 11 (1 & 2): 185-209.

–. 1988. Property rights and the development of natural resource policies in Canada. *Canadian Public Policy / Analyse de Politiques* 14 (3): 307-20.

Posner, Richard A. 1991. *Economic Analysis of Law.* 4th ed. Boston: Little, Brown. Chapter 3.

Randall, Alan. 1987. *Resource Economics: An Economic Approach to Natural Resource and Environmental Policy.* 2nd ed. New York: John Wiley and Sons. Chapter 8.

Scott, Anthony D. 1985. Property rights and property wrongs. *Canadian Journal of Economics* 14 (4): 556-73.

–. 2008. *The Evolution of Property Rights.* New York: Oxford University Press.

Zhang, D., and E. Oweridu. 2007. Land tenure, market and the establishment of forest plantations in Ghana. *Forest Policy and Economics* 9: 602-10.

Zhang, D., and P.H. Pearse. 1996. Differences in silvicultural investment under various types of forest tenure in British Columbia. *Forest Science* 44 (4): 442-49.

Chapter 11   Forest Taxes and Other Charges

Governments tax private forests, and both public and private forest owners levy various charges on forestland, timber, and other forest products and services when forests are used by others. These charges take the form of royalties, rentals, lease fees, stumpage charges, taxes, and a variety of other levies. They all influence the way forests are managed and used, and the distribution of the benefits they produce.

This chapter considers taxes and other charges applied specifically to forest resources. It does not deal with other direct and indirect taxes on businesses, individuals, goods, and services. These general fiscal devices bear on forestry in the same way that they bear on other economic activity, and are better dealt with in texts on public finance. Our concern here is with the kinds of levies that are applied directly or indirectly to forestland and timber.

In previous chapters, we mentioned two specific tax issues that have broad implications for forestry in the United States. One is that differential corporate income taxation contributed to the demise of industrial timberland ownership and the rise of more tax-advantaged institutional timberland ownership in the 1990s and 2000s. The other is capital gains tax treatment for timber income. The capital gains tax, applied to gains from capital investment such as stock held for longer than one year, is lower than the tax on ordinary income because of larger exemptions and/or lower tax rates. When forest landowners hold their timberland longer than one year, they are eligible for capital gains treatment for their timber (stumpage) income. Capital gains tax treatment for timber income reduces the tax burden on landowners and has a positive impact on the development of forest plantations in the United States.

CHARACTERISTICS OF FOREST CHARGES

Traditionally, four criteria have been used to judge the quality of taxes: their *neutrality,* their *equity,* their *simplicity*, and the *stability of the revenue* they generate.

Neutrality

A neutral tax is one that does not create incentives to change behaviour. It will not distort the allocation of resources or the efficiency of resource use.

Few taxes and charges are entirely neutral. Most induce those who must pay them to lighten the burden by altering their economic activity in some way. For example, forest owners faced with a tax on their harvests will find it advantageous to postpone their harvests to a longer rotation period and invest less in silviculture, among other things. Further, differential property taxes applied to rural land in different uses can cause land-use changes at the margin, yet property tax laws in many countries apply differently to land used for different purposes. It is therefore important to identify the incentives created by particular kinds of levies, the distortions in behaviour they cause, and the economic losses or costs that result.

Equity

Equity, or the distribution of income, is important because levies on forest resources shift income and wealth from the payers to the receivers. Indeed, this is usually their primary purpose. Because most charges are not neutral, and therefore cause changes in economic behaviour, they have subtle and complicated effects on the distribution of income, well beyond the obvious transfer of income from payers to receivers. For example, the tax on forest harvests referred to above, because it will induce changes in forest harvesting and silviculture, will also affect the incomes of those employed in these activities. As a result, the ultimate *incidence* of a levy, which refers to the final distribution of its burden, requires careful analysis.

As noted in Chapter 1, equity or fairness is a subjective concept, but it is particularly important in taxation. Two well-established principles of equity in taxation are the *ability to pay principle* and the *benefit principle.* The first of these implies that people should be taxed in accordance with their wealth and income, and those in similar circumstances should be taxed similarly. Income taxes reflect this approach. The benefit principle implies that people should be taxed according to the benefits they receive from public expenditures. Property taxes that pay for local services to property owners are consistent with this idea. These two principles of equity in taxation are not mutually consistent, however, and modern tax systems, including taxes and charges on forests, embody a mixture of them.

A tax with a tax rate that increases as the tax base increases is a *progressive* tax, whereas a tax with a tax rate that decreases as the tax base increases is said to be *regressive*. Often, the tax base is the income or expenditure of a person or firm. A property tax, however, is based on the value of the property regardless of the owner's income. Thus, a progressive (regressive) property tax is one in which the tax rate increases as the value of the property increases (decreases). If a property tax imposes a heavier burden on a poor and less valuable site than on a good site, the tax is regressive.

The burden of a tax on land can be measured in two ways. The first measure is the *site burden* (SB), which measures the reduction of land value as a result of the imposition of a property tax:

$$SB = \frac{V_s \text{ without tax} - V_s \text{ with tax}}{V_s \text{ without tax}} \tag{11.1}$$

where V_s is the site value or land expectation value. This measure can be used to compare the tax burden on different land uses and to compare the tax burden on the land portion of different sites with the same forest use. If the site burden for agricultural land is less than that for forestland, forest landowners will bear a heavier burden of property tax. The second measure is *forest burden* (FB), which measures the reduction in the value of the forest (trees and land) as a result of the imposition of a tax:

$$FB = \frac{FV \text{ without tax} - FV \text{ with tax}}{FV \text{ without tax}} \tag{11.2}$$

where *FV* is the forest value, which includes the value of both bare land and standing timber. The forest burden provides a measure of a forest landowner's loss in wealth of forest holdings. The forest burden can be used to compare the tax burden of different sites with the same forestry use.

Simplicity

Simplicity refers to the complexity of fiscal arrangements for administration and enforcement, their understandability, and the costs of compliance. For administrative reasons, one would like to have a tax that is easy to collect. Some taxes and charges, however, require a lot in the way of data, of econometric calculations, or of resources required to administer and police them. These costs are often significant and have efficiency and distributional consequences of their own.

Stability of Income Generated

Finally, forest levies have an effect on stability of incomes, both of the payer and of the receiver. Some levies, such as stumpage charges on standing timber or timber harvested, increase and decrease with the income generated by the payer. Others, such as land rentals and taxes, tend to be more constant over time, regardless of the payer's level of production and earnings.

The desirability of stable payments over time depends on particular circumstances. Other things being equal, payers will usually prefer payments that have a stable relationship to their earnings and ability to pay. Receivers generally favour steady payments, but for governments with a multitude of revenue sources this is less important. Moreover, revenues that fluctuate with the level of economic activity have an automatic stabilizing effect, which is usually considered desirable.

Thus, taxes and other levies have important implications for efficiency in the way resources are allocated and used, the distribution of income, administrative complexity, and stability of revenues. This chapter examines forest taxes and charges in terms of these effects.

TYPES OF FOREST LEVIES

Levies on forests take various forms, and each form has particular economic impacts. The most common types are classified in Table 11.1 according to their tax or assessment base. As shown in the leftmost column of the table, the most common bases are the land, the standing timber, the harvest, and the property right. The last of these refers to charges for the acquisition or renewal of property rights over the forest, such as fees for licences and leases.

The form of these levies, the way their amounts are determined, and how they are assessed vary widely among jurisdictions. Nonetheless, the most common government taxes often include property taxes, income taxes, sales taxes, capital gains taxes, inheritance and estate taxes, and export taxes and import duties. Property taxes are levied on real property, whereas sales taxes are levied on personal property. Income taxes, including payroll taxes, apply to individual and corporate incomes, and capital gains taxes apply to gains from long-term capital investment. Private forest landowners often charge stumpage when they sell their

TABLE 11.1 Common forms of levies on forest resources

Base	Common form	Usual method of determination	Usual method of assessment
Land	Land tax	Percentage rate on land or site value or forest productivity, or a fixed tax per acre	Annual levy
	Land rent	Usually arbitrary	Annual levy
Standing timber	Timber tax	Percentage rate on value of timber	Annual levy
Land and timber	Common property tax	Combined ad valorem tax on land and standing timber	Annual levy
Harvest	Royalty, severance tax, and cutting fees	Specific fees for volumes harvested, usually arbitrary	According to harvested volumes and product, or value
	Yield tax	Percentage rate on value harvested (or net income generated from timber sales)	According to harvest value
	Stumpage fees	Competitive bidding or appraisal	According to harvest or lump sum
Rights	Land rentals, licence fees, and administrative charges	More or less arbitrary	Lump sum or annual charge

timber, and fees such as hunting lease fees when they lease their land to others.

Since forestland and timber are considered to be real property, we begin with property taxes. Forest property taxes are often levied on land and timber simultaneously, but to understand the difference between taxes on land and on timber, we shall consider them separately first. When we refer to taxes on land, we need to be careful about whether we mean a tax on the value of bare land (no timber) or a tax on capitalized land value, for which timber is included. When harvesting removes timber from land, severance taxes are often levied.

Taxes on Land

Where forests are privately owned, the most important of these levies are property taxes on forestland and timber, commonly based on estimates of the market value of the land and timber. The land tax itself is usually determined annually by applying a percentage tax rate to the base value. These are therefore *ad valorem taxes,* meaning that they are determined according to the value of the asset taxed.

Property taxes on forestland are often assessed in three ways in North America. First, they may be based on estimates of the value of the land's forest productivity (*land productivity taxes*), which in fact includes the value of both bare land and trees. Second, they may be based on the value of the bare land, ignoring the value of trees, and thus are referred to as *site value taxes.* Because such taxes do not fall on trees growing on the land, they are sometimes combined with a *yield tax* on timber harvested. Finally, when the same amount of tax per acre is collected on any acre of forestland regardless of its value, we have a *flat property tax.*

Thus, although this subsection is entitled "Taxes on Land," we must recognise that land productivity taxes target both land and trees, whereas site value taxes and flat taxes simply target bare land. Land productivity taxes are used by more than half of the 50 American states and are also used widely in Scandinavian countries.

The tax base for land productivity taxes is usually determined by estimating the present worth of the historical or expected mean annual revenue from the forest. When the mean annual increment of a forest is multiplied by the stumpage price, the gross mean annual revenue (GMAR) is obtained:

$$\text{GMAR} = \frac{PQ(t)}{t} \tag{11.3}$$

where P is the stumpage price, $Q(t)$ is the stand volume of timber in a stand t years old, and $Q(t)/t$ is the mean annual increment (MAI). Using equation 3.6 in Chapter 3, the capitalized value of GMAR would be GMAR/i, where i is the annual interest rate. The capitalized value would form the basis or assessed value of property taxation.

To arrive at the *net mean annual revenue* (NMAR), mean annual management expense is subtracted from gross mean annual revenue:

$$\text{NMAR} = \frac{PQ(t)}{t} - C \tag{11.4}$$

where C is the mean annual management expense. Mathematically, the tax base under net mean annual revenue is $(GMAR - C)/i$. Where forests are managed for sustained annual yields and future stumpage values are assumed to be constant, the tax base under both types of forest productivity tax will be stable over time, regardless of tree stand ages.

To simplify administration, forestland is sometimes classified into a few broad categories of productivity and each is assigned to an assessed value (tax base), to which the fixed tax rate is applied. In the US, the tax base for different categories of productivity is often determined at the state level, while tax rates vary by county and municipality.

All three types of land taxes – forest productivity taxes, site value taxes, and flat property tax – have the rare quality of *neutrality.* Because the amount payable does not depend on the timber inventory, the amount harvested, the choice of rotation, or other variables that can be manipulated, it does not create incentives to alter management decisions and so does not distort efficient resource use. Whatever management regime maximizes private returns without the tax will continue to maximize returns to the owner with it; the levy is neutral in this sense.

The effect of land taxes is mostly distributional, shifting resource rents from forest owners to governments or from tenants to landlords. In terms of *equity,* the tax burden based on the gross mean annual revenue method is heavier on less productive land than on more productive land, which is regressive in terms of both site burden and forest burden. So are flat taxes. On the other hand, the tax burden based on the net mean annual revenue places a lighter burden on less productive land and is thus progressive. The site value tax, applied as a percentage to site value or land expectation value, imposes the same site burden on forestland of differing productivity. Nonetheless, if the tax burden is measured in terms of forest burden – the reduction in the value of the forest (both the land and the trees), the tax burden tends to be heavier on less productive forestland than on more productive land.

The tax rate used in assessing a tax based on land value or productivity can be adjusted to capture any desired proportion of the economic rent, and if the tax is consistently administered it will capture a consistent share of the rent across forestlands of varying productivity and value.

A land tax is *neutral,* as noted previously, as long as it captures less than the entire economic rent, but if it exceeds the rent it will make the land unprofitable to use at all. Thus, a uniform levy on forestland of varying productivity will shrink the extensive margin of forestry. In terms of the discussion in Chapter 6 and Figure 6.3, a levy on marginal land will impose a loss on the owner, driving the land out of productive forestry use and, if the levy is not also on land used for other purposes, into other uses.

Taxes on land tend to be fairly *stable* because they do not depend on the level of production or earnings, although some forms vary with timber prices, as has been noted. Finally, such taxes are typically *simple* to calculate, administer, and collect as no annual timber inventory data are needed. Data requirements are demanding only in the more elaborate attempts to estimate land productivity and to calculate the land value.

Taxes on Timber
When property taxes are levied on standing timber, a percentage tax rate, or mill rate (1 mill is equal to $1 tax for each $1,000 of assessment), is applied against the appraised value of the timber inventory. Appraisals are usually based on information about recent sales of comparable timber or calculations of recoverable volumes and values.

Property taxes of this type generate assessments that are proportional to the current value of the forest crop. Other things being equal, they increase as the forest grows in volume and value. Forest owners are therefore faced with repeated annual assessments on their timber, increasing in amount, for as long as they continue to grow the crop.

Even modest tax rates applied to growing forests can generate payments that, with accumulating interest, will capture a substantial portion of the value of the crop by the time it reaches harvesting age. For example, if the stand described in Table 7.1 were subject to a tax of only 1% per year, the tax payments and interest on them would accumulate to nearly one-quarter of the value of the crop by the time it reached the harvesting age of 58 years. With less productive land or a higher tax rate, the tax could wipe out any gain to the owner and may impose a loss. As a result, forest owners sometimes find it advantageous to keep their land denuded of valuable forest. By thus reducing the profitability of forestry, and by lowering the returns to growing timber relative to the returns to other land uses, such taxes may drive land out of timber production altogether.

Thus, in terms of *fiscal neutrality*, a tax on the value of timber will induce shorter forest rotations and reduce the forest land base. Lack of neutrality gives forest landowners an incentive to cut and remove timber from their property earlier and even abandon forestry to avoid the tax.

Governments have sometimes sought to take advantage of the incentive to harvest resulting from a property tax on timber. Especially in developing countries and regions with extensive private holdings of virgin forests, such taxes have been used to encourage harvesting and expansion of the forest industry. If the tax continues to apply to subsequent managed crops, however, it will tend to discourage continuing timber production.

In terms of *equity*, such taxes on timber are regressive, as illustrated in the next subsection. Property taxes on timber are also criticized on the grounds that they fail to synchronize tax liabilities with owners' revenues from their crops. In contrast to levies on harvests, these taxes recur every year the crop is growing, imposing on owners the added burden of financing tax payments.

In terms of *simplicity*, such taxes require the maintenance of tax rolls with specific data on each forest property, and annual or at least periodic updating of the tax rolls, including forest inventories, which is costly in terms of information and administrative requirements. The revenues from individual forests can be expected to fluctuate over time as land and timber values change and forests grow and are harvested, but the aggregate revenue generated from such taxes within a taxation district is likely to be more stable, depending on the number and diversity of forests taxed.

Taxes on Land and Timber

Taxes on forestland and timber are combined in property taxes on forests; this is analogous to taxes on residential property, which normally include a levy on both residential lots and the houses on them. Such property taxes are intended to tax owners according to the value of their assets. These values change constantly, of course, in response to shifting supply and demand. This presents a particularly difficult challenge for tax assessors of forest properties, for which continuing information from market sales of comparable properties is typically much less available than it is for residential properties. Moreover, forests change in value over time through growth and harvesting, implying a need for continual reassessment. In practice, however, few government agencies update their tax rolls every year.

The effects of such ad valorem taxes on land and timber are similar to that of an ad valorem tax on standing timber – that is, they are neither fiscally neutral nor easy to administer; they are regressive with the possibility of revenue stability.

Much debate has centred on whether the ad valorem property tax on land and timber places a heavier tax burden on forest landowners than on agricultural landowners. Since the tax on agricultural land taxes only the value of the land whereas the ad valorem tax on forestland collects tax on the value of both the land and the growing crop, it is evident that an ad valorem tax of the same rate always places a heavier tax burden on forestland. Another question concerns whether it imposes a heavier burden on less productive forestland than on more productive forestland. When the ad valorem tax is administered perfectly, the tax rate has the same impact as an increase in the interest rate. Consequently, it imposes a heavier site burden on less productive forestland than on more productive forestland.

For all these reasons, ad valorem taxes on land and timber are much less used in the United States. In fact, only seven states use this tax system. Table 11.2 summarizes the effects of property taxes.

Land Rentals and Fees

Rentals are commonly paid by holders of forest leases and licences to public or private landlords. Unlike taxes on land, rentals usually take the form of fixed annual charges per hectare of land licensed or leased.

Because rental fees, particularly those levied by governments on holders of usufructuary rights in public forests, are usually assessed as fixed amounts payable per hectare, they capture an inconsistent proportion of resource rents. On the other hand, the economic effects of land rentals and fees are similar to those of a tax on land: they are fiscally neutral, simple to administer, and stable.

Levies on Harvests

A considerable variety of taxes and charges are levied on timber harvested. In contrast to property taxes, which are levied on owners of forest property, levies on harvests are usually paid to public or private landowners by those who harvest their timber. These levies go to the public treasury when forests are publicly owned or to private landowners in the case of private land.

Levies on timber harvests are assessed in three ways, with differing economic effects. One takes the form of assessments specified in dollars payable on the *quantity of timber harvested, with little or no regard to its value.* This includes royalties, severance taxes, and cutting fees or dues. Severance taxes in some jurisdictions, however, apply at different unit rates, in dollar per cubic metre or per thousand board feet, to hardwoods and softwoods, or to timber of different quality. The second form of taxes on harvests takes the form of a percentage of the *value of timber harvested,* usually called a *yield tax.* The third is a charge based on the *value of the timber to be harvested,* referred to as a stumpage price.

The terminology used for various types of taxes and charges is often confusing and inconsistent. In the following paragraphs, each type is defined in a specific way to facilitate discussion. These definitions are not universally adopted, however.

TABLE 11.2 The effects of forest property taxation systems

Characteristic	Tax on land				Tax on standing timber
	Gross productivity	Net productivity	Site value	Flat	
Fiscal neutrality	Yes	Yes	Yes	Yes	No
Tax burden					
Site burden	Regressive	Progressive	Proportional	Regressive	Regressive
Forest burden	Regressive	Progressive	Regressive	Regressive	Regressive
Revenue stability	Yes	Yes	Yes	Yes	Possible
Administrative simplicity	Yes	Yes	Possible	Yes	No

Note: The effects of tax on land and standing timber are similar to those of tax on standing timber.

A basic distinction must be made between a tax on the one hand and a royalty or stumpage price on the other. A tax is always a government levy that, in effect, expropriates for the public purse a portion of private wealth or income; a royalty or stumpage price is a purchaser's payment for a commodity – timber. As we shall see, however, the incidence or impact of taxes and other levies may be similar, depending upon how they are assessed.

The three general types of charges on timber harvested can be described as follows.

Royalties, Severance Taxes, and Cutting Fees or Dues

These are typically specific dollar charges, or schedules of charges, for various categories of timber, levied on each cubic metre or other unit of timber harvested. They are modest charges in most cases, and because they are specified in monetary terms, they tend to be eroded by inflation over time.

Traditionally, a royalty represented the value of a resource that the sovereign reserved to the Crown when granting rights to land and resources to a private person, which is still common in Canada, as described in Chapter 10. Severance taxes, most common in the US, were similarly conceived as payments to the government for severing personal property (logs) from real property (standing timber). Like royalties, cutting fees and dues are usually simply schedules of charges for public timber. All these levies are fixed in dollars per unit of timber harvested (sometimes differentiated by product such as sawtimber and pulpwood) and so, regardless of their original intent or justification, they have similar economic effects.

The direct distributional effect of these levies is to shift the economic returns of timber from the harvesters to the government. They also affect economic efficiency and the allocation of resources in harvesting by making it unprofitable for the owner to harvest marginal stands of timber and to recover marginal logs.

The value, per cubic metre, of logs of different species, from different stands, and even from individual trees varies widely. Figure 11.1 illustrates how the value, per cubic metre, of logs harvestable from a tract of forest typically declines with diminishing quality, which can be measured in terms of their size, species, grade, and other features. At some level of quality, g in Figure 11.1, the log value is just sufficient to cover the harvesting cost, indicating the marginal or "cut-off" grade of log, and the limit of profitable harvesting. A royalty or tax of a fixed amount per cubic metre harvested has the effect of lowering the net value to the

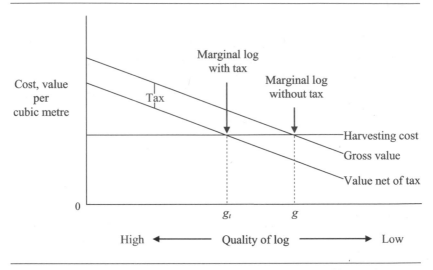

Figure 11.1   Effect of a royalty or severance tax on the range of log quality that can be profitably harvested.

owner of all logs by the amount of the assessment, reducing the range of quality that can be profitably recovered, by $g_t g$ in Figure 11.1. Such levies thus encourage "high grading" in timber harvesting.

Such levies have the effect of contracting both the intensive and extensive margins of recovery. The contraction of the intensive margin of recovery can be illustrated with reference to Figure 6.1 on page 169, where a levy on all timber harvested (or logs recovered) will lower the marginal revenue product of labour on any log, and hence constrain the amount of labour to be applied in logging and the intensive margin of recovery.

Correspondingly, this contraction of the intensive margin of recovery is paralleled by a contraction of the extensive margin of recovery as well; that is, stands of timber at the margin of economic recoverability are made submarginal by such levies. This development can be illustrated with reference to Figure 4.8 in Chapter 4 (or Figure 11.1), where a levy on all timber harvested would have the effect of lowering the value of the inventory and shrinking the extensive margin.

Further, the growing of forests on marginal sites is also rendered unprofitable. Again, this effect can be illustrated with reference to Figure 6.1 on page 169, where a levy on timber will lower the marginal revenue product of labour on any site, thus constraining the intensive margin of forestry. In all these ways, such levies impede efficient forest management and use.

Royalties and other charges of this kind are relatively simple to administer because they require information only about the timber removed from the forest, which is usually collected for other purposes as well. The payer's liability to these assessments depends on his harvests, which facilitates his ability to pay, although at the expense of stability of revenues to the recipient.

Yield Taxes

Yield taxes are government levies on timber cut on private land. They take the form of a percentage rate applied to either the selling price of the logs (the *gross value* of the harvest) or the stumpage value of the standing timber (the *net value* of the timber harvested). They are thus ad valorem taxes, being based on the value of the tax base (the harvest). Yield taxes are sometimes adopted as an alternative to property taxes on timber in North America.

For the forest owner, a yield tax levied on the sale value of logs has the same effect as a lower selling price. It constrains both the intensive and extensive margins of profitable harvesting. The effect is similar to that of a royalty or severance tax, illustrated in Figure 11.1, except that the yield tax reduces the gross value of logs by a constant proportion rather than a fixed dollar amount. If we draw a figure that is similar to Figure 11.1, the value net of tax curve will start at the same point as the gross value curve touches the horizontal axis, and the gap between these two curves will be greatest when they reach the vertical axis.

This distortion of the economic margin does not occur when the tax is levied as a percentage rate applied to the *net value* of the timber. At the margin (the point where the gross value curve and harvesting cost curve cross each other, or when quality of log is at point g in Figure 11.1), the net value of timber is zero and a tax assessed on that value is zero also, so the margin does not change. In this respect, the economic effects of the tax are identical to those of a stumpage levy, described below.

Whether a yield tax is applied to the gross value of the harvest or the net value of the harvest, it will affect the rotation age and silvicultural investment, as noted in Chapters 7 and 9, respectively.

Yield taxes applied to the *net value* of the timber are more complicated to administer because they call for information not only about the quantity of timber harvested but also about its market value and, where they are based on net value of the timber, about its harvesting cost also. Procedures for collecting and averaging price data, allowing for fluctuations and policing sale information, add considerably to the cost of assessment and collection.

Compared with annual property taxes on timber, yield taxes have the advantage of avoiding the accumulation of payments and interest on payments as the forest grows and the resulting financial disincentives to carrying crops noted earlier. They enable governments to take a consistent share of the value of forest production when it is produced, facilitating the payer's ability to pay. Both the level of harvest and the price of harvested products tend to fluctuate in response to market cycles, however, so the revenue from yield taxes tends to be less stable than that from property taxes, which is a disadvantage to many local governments that depend heavily on revenues from such taxes.

Stumpage

Stumpage charges are levied by private or public forest owners on their timber harvested by others. In this respect, they serve the same purpose as royalties, but the latter are usually relatively modest fixed rates applied uniformly, whereas stumpage charges are typically more discriminating levies that are intended to capture for the owner all or a substantial share of the value of timber, which differs on each tract harvested.

Stumpage charges are based on the *net value* of the timber as it stands "on the stump," and can be designed to capture all or part of this value. They are determined in a variety of ways: competitive auctions, appraisals, and various types of negotiated arrangements.

In the US, most public timber is sold at competitive auctions, where potential buyers are invited to bid for the timber in terms of dollars per unit of volume harvested. Bids are usually subject to a minimum acceptable price, or "upset price," which is set by public forest managers based on appraisals. The highest bidder wins the harvesting licence and thereby sets the stumpage charges. Competitive bidding is also used to allocate certain minor forms of cutting rights in public timber and in private forests in Canada. As long as competition is sufficiently vigorous, this method provides the seller maximum assurance of capturing the full net value of each tract allocated for harvesting.

In the absence of competition among buyers, the upset price becomes the stumpage price, and the share of the net value of the timber captured by the seller depends heavily on the way these administered or appraised prices are calculated. Appraisal procedures vary, but they have certain common elements. First, the volume of each species and grade of logs recoverable from the forest tract is estimated from inventory data. Second, the value or selling price per cubic metre of each category of logs is determined from market information or from calculations of the value of the products that can be manufactured from them. Third, the cost of harvesting the timber per cubic metre, including costs of building roads, logging, and transporting the logs to the market where the selling prices can be realized, are estimated and subtracted from the selling price. Finally, the remainder, referred to as the "conversion return," is reduced by an allowance for the operator's profit and risk to yield the stumpage price payable to the seller of the timber.

These calculations can be illustrated as follows:

	Value or cost per cubic metre
selling price of the logs	$100.00
minus cost of logging and delivery	−60.00
equals conversion return	40.00
minus allowance for operator's profit and risk	−15.00
equals appraised or upset price	25.00

Either the calculation is done separately for each species of timber or an average is calculated by weighting the selling price of each species by its percentage of the total volume of the timber to be harvested.

Appraisals of this kind are exceedingly complex and highly demanding of data relating to the timber and operating conditions, current costs, and market prices. They also depend on judgments about such matters as allowances for risks, depreciation, and reasonable returns to operators. As a result, the administrative costs tend to be high and the results are often contentious.

The greater the proportion of the net value of timber that the stumpage charges are intended to collect, the heavier the onus on these calculations. In some cases, owners use relatively simple techniques to set stumpage rates that will secure for them a reasonable share of the resource values, such as by extrapolating values from competitive sales of comparable timber in the region or by rules of thumb for dividing revenue from the sale of the logs between the owner and the logger.

The effects on economic efficiency of stumpage charges depend not only upon how they are determined but, more importantly, upon how they are assessed. Once the stumpage price for a tract of timber is determined, it may be applied to the estimated volumes of standing timber and assessed as a *lump sum*, or the same amount may be collected in several equal payments over the planned harvesting period, or the stumpage rate may be assessed on the timber harvested as it is cut and scaled (*per unit* or *pay as cut*), or it may be collected in a variety of other ways. If the charges are payable in a lump sum, in predetermined annual amounts, or in any other way that is unaffected by the entrepreneur's actual harvesting, his economic behaviour will not be influenced by the charges regardless of how they were determined in the first place. They become neutral fixed or sunk costs, affecting none of his production decisions at the margin. His profit-maximizing pattern of production will be the same with them as without them.

In contrast, stumpage charges assessed uniformly on the timber actually recovered from the forest, in dollars per cubic metre of logs removed, will affect the operator's decisions. In this case, his net return on each cubic metre is reduced by the amount of the stumpage levy, rendering marginal timber unprofitable to recover through the same effect as a royalty or severance tax, illustrated in Figure 11.1. Because stumpage charges are often much more substantial, the effect is correspondingly greater.

This incentive to "high-grade" timber in logging operations can have significant consequences. If the stumpage price per cubic metre is set at the average net value of the timber in the stand, loggers will earn a profit on only half the economically recoverable volume of timber – that portion worth more than the cost of logging plus the stumpage price. On the other half, they would incur a loss, giving them financial incentive to refrain from harvesting it. Some regulatory agencies impose utilization standards in order to offset such incentives, but these rules are inevitably more or less arbitrary, and rarely take account of the variations in the value of logs in different stands and under different logging conditions, or with changing costs and prices.

In short, stumpage charges levied on the *logs recovered* from the forest, in contrast to assessments on the standing *timber allocated* for harvesting, constrain the intensive and extensive margins of profitable logging operations, resulting in the waste of potentially valuable timber.

Assessments of the standing timber are thus preferable on grounds of neutrality in a logging operation. Assessments on the volume removed call for less exacting data about the forest itself, however, and remove some risk from operators. Moreover, logs recovered from the forest are usually scaled for other purposes in any event, so it is often convenient to base stumpage charges on these measurements.

Stumpage charges fixed in advance of harvesting leave payers vulnerable to any miscalculations or mistakes in determining bids or administered prices. They also impose on them all the risks associated with unforeseen changes in harvesting conditions, costs, and timber prices. These risks are greater the longer the period between the commitment to make the payments and the completion of harvesting. They are most onerous when the charges are paid as a lump sum in advance and the harvests extend over a long period.

The risk involved and the vulnerability of timber buyers to federal timber sales was well illustrated in the US Pacific Northwest in the late 1970s and 1980s. Expecting a period of timber shortage in an economy with high rates of inflation in the late 1970s, these buyers bid up the prices of federal timber sales dramatically until a severe economic recession in the early 1980s caused timber prices to plummet. As a result, in the early 1980s the buyers held timber-harvesting contracts that obliged them to pay prices up to five times the value of the timber. Some loggers went out of business. Others lobbied for, and in 1984 secured, federal assistance of billions of dollars. This example demonstrates the importance of risk in timber-related charges, and the potential cost of miscalculations.

The burden of risk borne by the buyers of timber can be alleviated in various ways. One is by designing the assessments to capture only a portion of the net value of the timber harvested, so that the risks will be shared proportionally between the owner and the producer. Another, already noted, is by basing the payments on the timber actually recovered rather than fixing them in total in advance, thus removing the financial uncertainty about recoverable volumes. And the risk of changing market conditions can be offset by tying the assessments to some index of product prices. Thus, some public agencies adjust stumpage charges, determined at the beginning of a harvesting contract, according to a sliding scale as the market price of specified forest products changes. All these measures shift risk from payers to receivers of stumpage payments.

Charges for Rights
For completeness, we must consider charges for rights to timber and forestland, as distinct from charges for the timber itself. It is usual for private landowners to charge rental fees for the use of their forestlands not only for harvesting timber but also for hunting and other purposes, and for governments to charge forest licensees and leaseholders annual fees for fire protection, rentals, and other purposes. These charges are typically justified as payments by licensees to have resources reserved for their exclusive use (or, conversely, as payments to the government for withholding the resource from use by others). In the US, some state governments also levy a fire protection fee on private forest owners.

Such fees are often payable when contracts are signed or licences are issued, or as fixed annual assessments to maintain the rights in good standing. Initial once-and-for-all licence fees have the same neutral effect on behaviour as any other lump sum levy, and annual fees are neutral in the same way as annual rentals, described earlier. As long as the payments are not affected by the licensee's behaviour and do not extract all the potential economic rent from the forest operation, they will not distort efficient operations.

DIRECT TAXES

Most jurisdictions tax income earned by forest landowners and forestry enterprises under general taxes on income and capital gains, although many provide special arrangements for forestry enterprises that have important economic implications. Three deserve attention here, and we illustrate them with examples from the US.

Capital Gains Tax for Timber Income and Depletion Allowance

As noted earlier, all income from timber sales in the United States after the owners have held the timberland for longer than one year qualify for long-term capital gains tax treatment. Many landowners have taken advantage of this tax law when they pay annual income tax to the federal and state governments. Since the capital gains tax rate is usually much lower than the tax rate on ordinary income, it benefits forest landowners and provides a strong incentive for them to invest in silviculture and other forestry activities.

Timber income taxation in the US can be complicated, however. From the beginning, the total cost of acquiring purchased forestland should be appropriately allocated among capital accounts for the land, the timber, and other capital assets acquired with them. The fair market value of gifted or inherited forestland should be allocated similarly. The cost basis for gifted land is the donor's cost plus any gift tax paid on the appreciation. The basis for inherited land is the fair market value at the time of the predecessor's death. This establishes the *timber basis.* Those who did not establish their basis when they first acquired their timber can do so retroactively if they so choose.

Of the original purchase cost, only the portion attributable to timber is recoverable through the depletion allowance (or income tax exemption). US tax law specifies a procedure for calculating a "depletion unit," usually expressed in dollars per thousand board feet or cords or tons. Included in this depletion unit are the various carrying charges and capital expenditures. Successive cutting requires successive adjustments to the depletion unit. Landowners can use depletion allowances to get part or all of their income exempted from their income tax payments.

Only the gain from appreciation of the standing timber can qualify as a capital gain. The value added by cutting and hauling the timber is ordinary income. For example, if one were to sell 200 tons of pine saw-timber out of a total of 1,000 tons on the entire tract for $8,000, and the basis for the entire tract is $10,000, and sale expenses were $900, the depletion unit would be $10 per ton ($10,000/1,000 tons), and the taxable gain would be $5,100 ($8,000 – [200 tons × $10/ton] – 900). Capital gains tax would be paid on this $5,100. In this way, she gets $2,000 of her timber income exempted from capital gains tax.

Reforestation Tax Credit

Current US tax law allows for up to $10,000 of reforestation expenditures to be deducted per qualified timber property (timber stand). Any additional reforestation expenditures can be amortized (deducted) over an 84-month period. This tax treatment allows small landowners to have significant tax savings and to recover their reforestation costs faster, and is thus a great incentive for tree planting.

Further, forest landowners who manage their forestland for profit (*investors*) or who are *actively* involved in forest management can also deduct from their income tax ordinary and necessary timber management expenses such as costs associated with brush control, fire, insect and disease protection, timber stand improvement, and fertilization. Other forest landowners cannot deduct their timber management expenses from their income tax. They can capitalize their expenses, at zero interest rate, and deduct these capitalized expenditures when they cut their timber.

Of course, landowners, whether they are investors or passive landowners as far as income tax is concerned, can choose to capitalize the expenses until timber harvest instead of depletion. In this way, the landowners do not recover their expenses as early. If there is a long lag between when expenses occur and the final timber harvest, the landowners can, in fact, cover only a portion of their capitalized costs, because statutory tax laws do not consider annual capital carrying cost in interest rates and inflation.

Because an income tax always represents a percentage reduction of net revenue, it does not distort forest management decisions. Thus, a tax on ordinary income is neutral. Any special tax treatment for timber income and reforestation tax incentive can change management intensity and rotation age, however.

The personal income tax is usually progressive, but not the corporate income tax. An income tax can be difficult to administer, especially with complicated, voluminous, and evolving tax laws and regulations.

Estate and Inheritance Taxes

Generally known as "death taxes," both estate and inheritance taxes are imposed only once, at the death of the owner. They differ from each other in that estate taxes are levied on the entire estate as a unit and paid from the deceased person's estate, whereas inheritance taxes are imposed on the share left to each heir and are paid by the beneficiary. Further, estate taxes are a federal tax while inheritance taxes are a state tax in the United States. Transfers of assets for less than fair and adequate consideration are taxed as gifts for transfers during life and as part of an estate at death. The current estate tax rate is 45%, higher than the tax rate of the highest income bracket in the United States. Some states add an inheritance tax on top of the federal estate tax.

Proponents of estate and inheritance taxes argue that they represent a policy aimed at avoiding excessive concentration of wealth. They are also regarded as a fair and easy way to raise revenue – fair because it is consistent with the liberal principle of equal opportunity for all citizens. Generally, a portion of the estate is exempted from taxation. The rest is then taxed progressively up to a percentage or at a flat rate.

Death taxes impose a decided disadvantage on individual forest ownership compared with corporate forestry, since a corporation usually has a longer life expectancy. Under current US tax law, a rural property (including forest property) can be valued at its current use rather than highest and best use and its value can be reduced by up to $1 million for 2009 as far as the estate tax is concerned. Nonetheless, heirs of an estate sometimes have to sell part of the forestland or cut some timber in order to pay the tax. The resulting fragmentation of forestland and accelerated harvesting may disrupt efficient forest operations.

TAX INCIDENCE AND DEADWEIGHT LOSS

As noted earlier, most taxes affect both economic efficiency and the distribution of income. In this section, we refer to excise and sales taxes, which are levied either as a fixed dollar amount per unit of good (an excise tax) or as a certain percentage of the value of the good (an ad valorem tax), to indicate how tax burden is distributed. Excise and sales taxes are often charged on forest products such as paper and lumber, and, in the form of tariffs and export taxes, on forest products traded internationally, as discussed in Chapter 12. More importantly, as the markets for forest products, logs, and standing timber are linked, such taxes on forest products tend to be shifted back onto the income of loggers and forest landowners. Similarly, a tax imposed on forest landowners tend to be passed on and reflected in the market prices of timber, logs, and forest products and impacts the incomes of producers and consumers of forest products.

Economic analysis of the way the burden of a tax is ultimately distributed, the *tax incidence,* leads to two important conclusions. First, the incidence levied on a good is the same whether the tax is levied on sellers (suppliers) or buyers (demanders). This principle can be seen with reference to Figure 11.2, where a $5 excise tax on a good is levied on suppliers, causing an upward shift in the supply curve (from S_1 to S_2, where $S_2 = S_1 + 5$). As the tax on suppliers makes them receive a lower after-tax price, they produce a smaller quantity (from Q_1 to Q_2). On the other hand, if the buyers have to pay a higher price (from P_1 to P_c), they, too, will buy less (again, from Q_1 to Q_2).

If the $5 excise tax is levied on buyers instead, the demand curve in Figure 11.2 will shift downward in parallel, from D_1 to D_2, by $5 and the supply curve will stay at S_1. As the tax on buyers makes buying less attractive and buyers demand a smaller quantity, the sellers will supply less than without the tax, which is a movement downward along the supply curve. With less quantity supplied, the sellers receive a lower price (at P_p), and thus they are worse off with the tax.

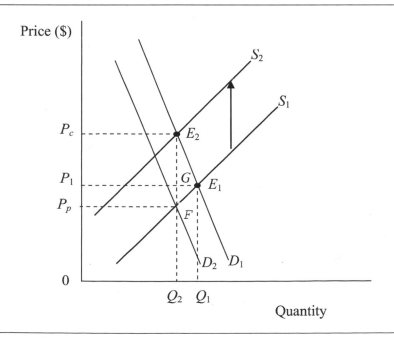

Figure 11.2   The relative burden of tax.

In either case, the new equilibrium price and quantity for buyers and sellers will be the same. The share of the full tax burden ($5 × Q_2) borne by sellers or buyers is the same regardless of which party pays the $5 excise tax.

The second conclusion, which follows from the first, is that the exact tax incidence of such an excise tax depends on the elasticities of supply and demand:

$$T_s = \frac{E_d}{E_s + E_d} \tag{11.5}$$

and

$$T_d = \frac{E_s}{E_s + E_d} \tag{11.6}$$

where T_s is the tax incidence to sellers, T_d is the tax incidence to buyers, and E_s and E_d are the absolute values of supply and demand elasticity, respectively.

Equations 11.5 and 11.6 can be derived arithmetically in three steps:

1 Calculate the changes in price on the demand and supply curves in Figure 11.2 (P_1P_c and P_1P_p, respectively) using the concept of demand and supply elasticity.
2 Compute the absolute amount of loss in consumer surplus (the area $P_1GE_2P_c$, which is equal to $P_1P_c \times Q_2$) and producer surplus (the area $P_1GFP_p = P_1P_c \times Q_2$).
3 Derive the share of tax burden (the area $P_pFE_2P_c$, which is the tax revenue) for consumers and suppliers.

Intuitively, one can see the buyers' and the sellers' actual tax share by looking at the change in their respective prices. As P_1P_c is greater than P_1P_p in Figure 11.2 – price increase for consumers is bigger than price reduction for producers – consumers share more of the tax burden in this example. Why has price increased more for consumers than it has dropped for producers? Because demand is less elastic than supply. If the relative elasticity were the other way around, producers would take on a larger share of the tax burden.

Thus, *the tax burden falls more heavily on the side of the market that is less elastic.* In extreme cases, when supply is perfectly elastic or demand is perfectly inelastic, consumers carry the full tax burden; when demand is perfectly elastic or supply is perfectly inelastic, producers assume the full tax burden.

These two principles are equally applicable under an ad valorem tax system. When an ad valorem tax is applied to suppliers, the supply curve will shift upward from its intercept to the price axis (or y-axis) and rotate by a percentage (the tax rate). If it is applied to consumers, the demand curve will rotate downward from its intercept to the quantity axis (or x-axis) by the same tax rate. In either case, it is proven that the new equilibrium point and tax incidence are the same. Again, the greatest share of the tax burden falls on the side of the market that is less elastic – the part of the transaction that is affected least by a change in price. It is the relative value of supply and demand elasticity that determines the share of the tax burden between suppliers and consumers.

Note that the area E_1FE_2 in Figure 11.2 is called *deadweight loss* or deadweight cost of taxation. It occurs because the price received by sellers is less than the cost to buyers when a tax is introduced. This means that fewer trades occur and that the individuals or businesses involved gain less from participating in the market. This deadweight loss can be eliminated only when either the supply or demand curve is perfectly inelastic. Because tax revenue represents a transfer or redistribution of income, the higher the tax revenue relative to the deadweight loss, the more efficient the tax system is.

THE COST OF RISK

Throughout this discussion of taxes and charges, we have referred to the way risk and the uncertainty of changing economic circumstances are borne by forest owners and users. Clearly, the way this burden is shared is affected by the way taxes and other charges against timber are levied. Assessments that fix the amount and timing of charges in advance minimize payers' uncertainty about their revenues but impose on them all the risks of miscalculation or unforeseen events. Conversely, to the extent that payments are based on the timber that loggers actually recover and the market conditions that they encounter, their risks are reduced and those of the receiver correspondingly increased. As we have seen, various techniques can be employed to distribute risk between payers and receivers.

The importance of this derives from the general rule that entrepreneurs are averse to risk, as discussed in Chapter 3. As a result, they demand higher expected returns from risky ventures than from secure ones. It follows that if those who use the forest are forced by fiscal means to bear more risk and uncertainty, the rents that can accrue to the forest owner will be lower. In the often unstable economic circumstances of the forest industry, this effect can be significant.

If a private entrepreneur has a choice between a tax fixed once and for all and another that is expected to yield the same amount but is based on his profits, he is likely to prefer the latter. It represents less of a threat in the event of unexpected adversity and low returns because the risk is shared with the government and so the tax is less burdensome. For this reason, it is often argued that governments can collect higher revenues from public forests in the long run if they design their taxes and other charges to shift the financial risk and uncertainty from taxpayers to governments themselves.

In this connection, it is also important to note that governments are usually considered to be less vulnerable to fluctuations in any single source of revenue, such as a tax on forests, than are enterprises that must pay it. Governments have an interest in stable revenues, but forest revenues are often a minor component in a government's large and diverse revenue system, so that fluctuations in such revenues alone are not highly disruptive. In contrast, a firm in the business of harvesting timber is likely to be much more sensitive to changes in such charges. Thus, the cost to a government of assuming the financial risks of a forestry operation is likely to be less than the gain to a forest enterprise from avoiding them.

The vulnerability of enterprises and governments to fluctuations in the tax payments on a particular forest tract is therefore dependent on their diversity and scale of operations. The same is true of the burden of financing charges, such as a yield tax or stumpage charge on a timber stand that is collected until harvesting takes place or is about to take place; if the timber stand is part of a large forest in which various timber stands mature and are harvested continuously, the financing problem is alleviated.

OTHER ECONOMIC CONSIDERATIONS

It must be emphasized that all taxes and fees are complicated and change over time, and that the impact of any particular tax or charge on forest resources is affected by the whole fiscal and institutional environment within which forest production takes place. Some levies are allowed as deductions in calculating income taxes or capital gains taxes, others plug loopholes or offset incentives that would otherwise distort activity, still others compensate for inequities that would otherwise exist, and so on. These circumstances vary, and so the effects of forest levies must be analyzed with careful attention to the circumstances of each time and place.

In the next few paragraphs, we review some of the relevant issues and economic considerations related to forest taxes and charges in North America and elsewhere.

Forest Charges in Various Institutional Settings

In some countries, forest industry firms offer assistance in the form of seedlings, management plans, and technical advice to non-industrial private forest landowners; in return, they request these landowners to notify them when they are ready to sell their timber. In some cases, these firms obtain a *first right of refusal* on the landowners' timber: that is, when the landowners offer timber for sale, they will sell it to the industrial firms as long as the industrial firms will match the highest offer the landowners have received from other buyers. Non-industrial private forest landowners in the US often sell their timber through sealed bids or negotiations.

When certain US forest industry firms sold their timberland to Timberland Investment Management Organizations (TIMOs) in recent years, they negotiated long-term timber supply contracts with the TIMOs as a condition of the sales. These contracts run from 10 years to 50 years and may contain a minimum timber supply clause, stipulating a minimum amount of timber to be harvested and sold to the forest industry firms each year. Stumpage prices are set once monthly or quarterly based on the amount, species, and product harvested, and on the market conditions.

In Scandinavian countries, non-industrial private landowners sometimes establish local forest landowner associations or cooperatives. These associations or cooperatives in turn negotiate with timber buyers on prices of stumpage or of delivered logs on behalf of the landowners.

Tax and the Intensive Margin of Forestry

As we have seen, the only neutral levies on forest resources are those that capture economic rent, such as taxes on bare land values, land value based on forest productivity, annual rentals, and lump sum stumpage assessments, which do not affect management decisions. All others generate incentives to alter production decisions in order to reduce the burden of the charges. We have noted how charges based on the volume or gross value of timber harvested narrow the range of materials that timber harvesters find profitable to recover. We have also observed how taxes on the forest inventory give owners incentives to shorten forest rotations, whereas taxes on harvests create incentives to lengthen them. In addition, any charges that reduce the economic return from growing timber constrain the extent of profitable investment in silviculture and management. These are all impacts on the intensive margins of forestry.

Tax, the Extensive Margin of Forestry, and the Current Use Rule

Most taxes and charges affect the extensive margins of forestry as well. Levies on timber harvested make marginal forests uneconomic to harvest, and taxes on forest inventories can make it unprofitable to grow timber on marginal sites.

The concept of neutrality therefore also applies to the allocation of land among uses, discussed in Chapter 6. Only a property tax that imposes the same burden on land regardless of its use can avoid distorting patterns of land use. Such a tax cannot, of course, be based on timber or any improvements to the land. It must be based on the market value of the land.

The market value of forestland can increase drastically, however, if a potential higher and better use is found. In other words, as population increases and the economy grows, forestry may no longer be the most productive use of a particular piece of land. Thus, its fair market value, assessed for tax purposes, will exceed its capitalized income-producing capability under forestry use. While such appreciation benefits the landowners and the government, it does not improve the landowners' ability to pay taxes and often forces land-use change, usually to development.

In order to restore balance between the taxable value of rural properties (including forest properties) and their income-producing potential, every state in the US has implemented programs or current use value laws that allow or require preferential property tax treatment of farmland and sometimes other rural lands. The most common policy is assessment of rural land according to its current use rather than its market value. This is the *current use rule.*

Current use value laws are seen as providing tax relief to rural landowners and enabling them to retain land in traditional uses, which are considered socially desirable. Like farms, forests are sometimes taxed at relatively low rates to encourage rural land use, stimulate employment, improve the welfare of low-income groups, or stabilize rural communities. Empirical studies have found that the effectiveness of the current use tax law (or current use value law) varies with states. In most cases, however, its impact is limited as development pressure, driven by population growth, economic advancement, and urbanization, can be overwhelming.

Interestingly, the statutory property tax laws in various states often create artificial differential tax treatment among rural uses. For example, the tax base for agricultural land may be limited to one-third of its market value while that for forestland may equal 60% of its market value. In one extreme case, the best (class one) forestland in a state pays a higher property tax than the best (class one) agricultural land. The difference can be expected to cause the extensive margin for agriculture to expand at the expense of forestry use, thus distorting the efficient pattern of land use.

A related concern that historically has influenced the design of taxes and charges on timberland is about speculative holdings of timber rights. The history of the forest industry in North America is rich in examples of investors seeking to acquire and hold extensive rights over timber in anticipation of growing scarcity and increasing resource values. Governments have tended to view this practice with disfavour and have sought to discourage it. As noted earlier, property taxes have sometimes been used to encourage owners to harvest timber. For the same reason, annual rentals have been levied on licences and leases over public timber to make it unattractive to hold rights to resources in excess of those needed for planned production. This policy calls for levies on forests that extract more than the present worth of anticipated gains that speculators can expect from holding timber.

Economics offers no support for the presumption that speculative acquisition and holding of resource rights is contrary to the public interest. Speculators can usually be assumed to assist in the efficient allocation of resources over time by constantly trying to identify the most advantageous time to utilize these resources. If governments nonetheless consider it desirable to discourage speculative activity, levies on timber rights provide an expedient means of doing so.

REVIEW QUESTIONS

1 What is a "neutral" tax? Give an example.

2 Using the data presented in review question 3 of Chapter 7 (page 224), calculate the amount to which tax payments and interest on them would accumulate if a tax of 1.5% were levied on the value of the timber every year until the stand reached the age of 50 years. (For simplicity of calculation, assume that the timber volume indicated for each fifth year remains constant at that level for the following four years.)

3 How does a yield tax affect incentives to invest in silviculture?

4 In what way will the extensive margin between forestry and agriculture be affected by a land tax law that applies a higher tax rate on forestland than on farmland?

5 How will a logger's incentive to "high-grade" a stand be affected by stumpage charges that are (a) assessed on each cubic metre of timber recovered, and (b) assessed as a lump sum for the stand, regardless of the logger's actual harvest?

6 What is the current use rule as applied to property tax?

7 Describe the taxes and charges that a typical forest landowner pays in your area, and assess the extent of these taxes and charges as a percentage of his forest income in one full rotation.

FURTHER READING

British Columbia Royal Commission on Forest Resources. 1976. *Timber Rights and Forest Policy in British Columbia.* Report of the Royal Commission on Forest Resources, Peter H. Pearse, Commissioner. Victoria: Queen's Printer. Chapter 13.

Boyd, Roy G., and William F. Hyde. 1989. *Forestry Sector Intervention: The Impacts of Public Regulation on Social Welfare.* Ames, IA: Iowa State University Press. Chapter 7.

Chang, Sun Joseph. 1996. US forestry property taxation systems and their effects. Proceedings of Symposium on Nonindustrial Private Forests: Learning from the Past, Prospects for the Future, Washington, DC, February 18-20.

Duerr, William A. 1960. *Fundamentals of Forestry Economics.* New York: McGraw-Hill. Chapters 26 and 27.

Gregory, G. Robinson. 1987. *Resource Economics for Foresters.* New York: John Wiley and Sons. Chapter 8.

Mattey, J.P. 1990. *The Timber Bubble that Burst: Government Policy and the Bailout of 1984.* New York: Oxford University Press.

Musgrave, Richard A., and Peggy B. Musgrave. 1989. *Public Finance in Theory and Practice.* 5th ed. New York: McGraw-Hill. Part 3.

Polyakov, Maksym, and Daowei Zhang. 2008. Property tax and land use change. *Land Economics* 84: 396-408.

PART 5

Forest Economics in Global Perspective

Chapter 12　Forest Products Trade

In today's globalized economy, forests in one region of the world are linked to and influenced by events associated with forests in other parts of the world. This linkage has come about through expansion of international trade, investment, movements of labour forces, enhanced communications, as well as global environmental changes. Today's foresters need to have an international perspective; it is not good enough to just be foresters of local or domestic forests.

This chapter is about the international forest products trade. As we shall see, such trade has increased in the last few decades. Not only does the increased trade have implications for forest and forest products production, employment, and wages but it also impacts global forest conservation and the environment, which is the subject of the next chapter.

In this chapter, we begin with the level, products, and patterns of global forest products trade and then move on to fundamental concepts and issues, such as comparative advantage, specialization, and exchange rates. The latter sections cover various restrictive trade measures and trade remedies, the political economy of trade measures, and trade dispute systems. We end with a look at foreign direct investment in the forest industry.

TRENDS IN INTERNATIONAL FOREST PRODUCTS TRADE
Global trade in primary forest products has been expanding over the past decades in both value and volume, and as a proportion of global production. Table 12.1 shows that the value of global forest products exports increased 15 times in nominal terms from 1970 to 2006, when it reached nearly US$200 billion. In addition, exports of secondary

TABLE 12.1 Value and share of different forest products in global exports, 1970, 1996, and 2006

	1970		1996		2006	
Product	Value	%	Value	%	Value	%
Industrial roundwood	1,790	14.1	8,053	6.0	10,926	5.4
Sawnwood	2,683	21.2	24,891	18.5	33,031	16.3
Paper and paperboard	4,414	34.8	65,699	48.9	100,006	49.2
Wood pulp	2,528	20.0	16,669	12.4	24,438	12.0
Wood-based panels	1,146	9.1	16,604	12.4	30,968	15.3
Other	106	0.8	2,462	1.8	3,752	1.9
Total	12,667	100.0	134,377	100.0	203,121	100.0

Notes: Values are in millions of US dollars. "Other" includes chips and particles, wood charcoal, treated wood fuel, and wood residues.
Source: FAO 2010.

processed products such as mouldings, doors, furniture, and non-wood forest products (NWFPs) such as rattan, rubber, nuts, oils, and medicines make a substantial contribution.

Export volume expanded for all primary forest products from 1970 to 2006, as shown in Figure 12.1. The trade volume of industrial round-wood has experienced the smallest increase, at about 1.15% annually. Sawnwood trade volume nearly tripled, as did wood pulp; each increased at 2.8% per year. Trade volume in paper and paperboard quadrupled, and that in wood-based panels increased fivefold. The average annual growth rates for these two product categories were 4.6% and 6.1%, respectively.

As of 2006, the value of trade in forest products accounted for about 3% of the value of global merchandise trade. The relative importance of countries as forest products exporters and importers changes over time, along with variations in markets, resource conditions, and domestic product production and consumption levels. Table 12.2 shows the major exporters and importers of forest products in 2006. Canada, the United States, Germany, Sweden, Finland, and Russia were the top 6 exporters. The US, China, Germany, Japan, the United Kingdom, and Italy were the top 6 importers. The top 15 importers and exporters accounted for some 73% of all forest products imports and 76% of all exports.

The list of major forest products importers in Table 12.2 gives an impression that those countries tend to have a large population base, have high per capita incomes, and/or are experiencing high rates of economic growth. The major exporters, on the other hand, are countries that are

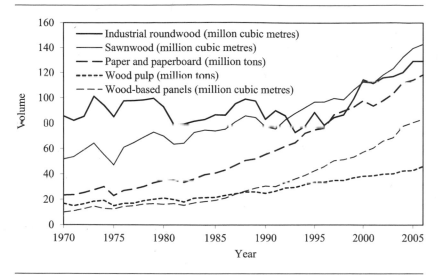

Figure 12.1 Global export volume of different forest products, 1970-2006. | *Source:* FAO 2010.

TABLE 12.2 Major global importers and exporters of forest products, 2006					
Rank	Importers	Value	Rank	Exporters	Value
1	United States	31,689	1	Canada	28,223
2	China	20,367	2	United States	18,482
3	Germany	16,012	3	Germany	18,179
4	Japan	12,778	4	Sweden	14,553
5	United Kingdom	11,343	5	Finland	14,343
6	Italy	10,456	6	Russian Federation	8,740
7	France	9,628	7	France	7,699
8	Spain	6,326	8	China	7,618
9	Netherlands	6,248	9	Austria	6,649
10	Belgium	5,858	10	Indonesia	6,170
11	Canada	5,133	11	Belgium	5,626
12	Mexico	4,356	12	Brazil	5,618
13	Korea, Republic of	4,300	13	Italy	4,785
14	Austria	3,683	14	Malaysia	4,035
15	Poland	3,268	15	Netherlands	4,030
	Sum	151,444		*Sum*	154,749
	World	207,349		World	203,121

Note: Values are in millions of US dollars.
Source: FAO 2010.

abundantly endowed with forest resources and that have adequate capability in forest products manufacturing. Some countries are both major importers and exporters. For example, the United States, although a net importer of forest products, especially wood products, is a major exporter of paper and paperboard. Similarly, although China imports a large amount of logs and pulp and paper products, it has in recent years increased its exports of wood products, especially those that are labour-intensive. Germany imports forest products mostly from the Nordic countries and sends its exports mostly to other European countries. Canada, the largest forest products exporter, also imports wood products, especially those made from hardwoods, of which it has only limited domestic supply.

COMPARATIVE ADVANTAGE AND THE PRINCIPLE OF SPECIALIZATION

Why is the US both a big importer and a big exporter of forest products? Why is Canada the largest exporter of forest products? More fundamentally, why does international trade take place, and how does it expand?

International trade takes place largely for the same reason domestic trading takes place: both parties gain. Ordinary and domestic exchanges, such as trading a bicycle for a portable computer, may occur because two persons have different preferences with respect to demand for goods and services: one prefers to have a bicycle, the other prefers to have a computer. Many exchanges are demand-driven, although there are also many supply-driven exchanges. The opportunity to gain from international trade is different insofar as countries differ in their endowment of productive resources – the size and quality of their labour force, capital, natural resources and other raw materials, and technical sophistication – which gives each a relative advantage in producing some goods and services and a disadvantage in producing others. Even if the people in two countries have identical preferences, they can gain by exporting goods and services that they can produce most advantageously, and importing those produced most advantageously by others. Thus, international trade is often purely supply-driven. In particular, international trade is driven by comparative advantage and specialization in production among countries.

The Principle of Specialization

Specialization refers to a situation in which the tasks associated with the production of a good or service are divided – and sometimes subdivided

– and performed by many different individuals to increase the total production of the good or service. Virtually all known peoples have engaged in specialization. Those who specialize in some productive effort for a while know that they eventually produce more and better because they have become used to doing it and have developed the necessary skills. When tasks are divided, permitting individuals to concentrate on a single element in the production of a good, output increases over what it would be if each individual produced the entire good or service. Specialization also occurs among various goods and services.

Thus, specialization and division of labour lead to increased output. All modern nations, states, firms, and individuals specialize in the production of economic goods and services, to varying degrees. Individuals, the most basic economic entity, specialize to the greatest extent. To decide on which task one should specialize in, however, will require a comparison and recognition of one's comparative advantage over others or of the opportunity costs of doing various tasks. The same logic applies to countries and producers in various countries, even though our illustration in the next subsection uses individuals.

The Principle of Comparative Advantage
Comparative advantage refers to the ability of an individual (person, firm, or nation) to produce a particular good or service at a lower opportunity cost than others. In contrast, absolute advantage is an individual's ability to produce a good with a smaller quantity of inputs than another individual. Opportunity cost is the measuring stick for comparative advantage, whereas productivity is the basis for determining absolute advantage.

Even when an individual has an absolute advantage in producing two goods, the opportunity cost of producing both goods could be high, implying that it would be beneficial to concentrate on producing the good that has a lower opportunity cost and to trade with another individual for the good that has a higher opportunity cost. Comparative advantage explains why trade is advantageous even if one has an absolute advantage over another in producing both goods.

For example, a lawyer may be competent at both law and typing. He or she should concentrate on the law, however, and hire a secretary for word processing because, when measured in terms of income (the opportunity cost), the lawyer's comparative advantage lies in legal affairs and not in secretarial work. This shows that comparative advantage arises because of differences in opportunity costs among producers.

TABLE 12.3 Production possibilities, opportunity costs, and total gains from specialization and trade in the example of Adam and John

	Adam	John	Total
Fish	4	3	
Deer	1	2	
Ratio (opportunity cost)	4.0	1.5	
Autarky* for two days	4 fish + 1 deer	3 fish + 2 deer	7 fish + 3 deer = 10
Specialization for two days	8 fish	4 deer	8 fish + 4 deer = 12
Gains from specialization and trade			1 fish + 1 deer

* Autarky means no trade; each is on his own.

A simple numerical example will illustrate how comparative advantage is applied to trade. Two students, Adam and John, like fishing and deer hunting. Suppose Adam can catch 4 fish or kill 1 deer in a day, and John can catch 3 fish or kill 2 deer a day. Over a two-day weekend, Adam can catch up to 8 fish (with no deer), or kill 2 deer (catching no fish), while John is expected to catch 6 fish (with no deer), or kill 4 deer (with no fish). These are their individual production possibilities. For simplicity, let us not give a monetary value to fish and deer but merely note that Adam and John like to have more deer and fish for food, and they are roughly equivalent in consumptive value. So both Adam and John are likely to devote one of the two days to catching deer and the other day to catching fish when they are alone.

Table 12.3 shows that the fish-to-deer ratio is different for Adam and John. This ratio is in fact the opportunity cost to each of deer hunting. Conversely, we can calculate the deer-to-fish ratio, and as long as these ratios are not equal, there will be an opportunity for specialization and trade.

Because Adam's fish-to-deer ratio (4.0) is higher than John's, he should concentrate on fishing for two days. John, on the other hand, should concentrate on deer hunting. At the end of the two days, they would collectively produce 12 food items, instead of the 10 that they would when each one hunts for a day and fishes for another day.

Again, acting individually, they can catch 7 fish and kill 3 deer. If they specialize according to their comparative advantage, however, Adam catches fish on both days and John kills deer on both days. Their total increases to 8 fish and 4 deer. They can then make a trade, and both would be better off.

What about the terms of trade? It will be somewhere between 4.0 and 1.5 (fish-to-deer), which are the opportunity costs for these two individuals. Beyond either point, there would be no trade, as each could do this well on his own. The final result for the terms of trade will depend on the bargaining skills and power of the two.

If Adam can catch 4 fish and 2 deer per day and John can catch 5 fish and 4 deer per day, John would have an absolute advantage over Adam in producing both goods. Even so, both would be better off specializing and trading based on their comparative advantage. Interested readers can compute the gains from specialization and trade by following the procedure presented in Table 12.3.

PRODUCTION POSSIBILITIES AND TERMS OF BILATERAL TRADE

To place this argument about specialization on a nation-to-nation basis, we note that the pattern of specialization in particular goods across countries may be observable. For example, although many goods consumed by Americans and Canadians are obtained largely from domestic sources, there is no domestic production of certain commodities in either country. The entire amount of crude rubber is imported. Exotic spices are not grown commercially in Canada, although they could be grown there with greenhouses and sufficient expenditures. Canada does produce a fair proportion of the world's wheat, softwood lumber, newsprint, and oil and gas. Likewise, no fine silk is produced on the tiny Caribbean island of Grenada, but nutmeg and other spices are. The explanation for such production patterns across countries rests on fundamental economic concepts of – comparative advantage, specialization, and trade.

A Two-Country Model

When will trade in a particular forest product occur between two countries? If there is a demand for the product in both countries, it all depends on the difference in domestic prices and associated transportation costs (including insurance and tariffs, if any). If the difference in prices between the two countries is greater than transportation and related costs, trade is likely to take place.

In Figure 12.2, the supply of and demand for plywood in two countries are shown in one graph. Each side of Figure 12.2 represents the domestic supply and demand in one country, measured in that country's currency. For simplicity, let us assume the exchange rate between these two currencies is 1. As both countries share a common price axis, their respective quantities increase as they move away from 0 in opposite directions.

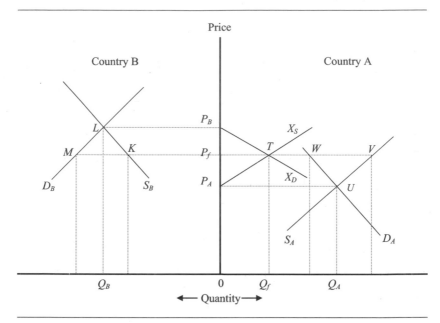

Figure 12.2 Determination of price and quantity of plywood to be imported and exported when trade is free, transportation costs are negligible, and all else remains constant.

Suppose that S_A and D_A are supply of and demand for plywood in Country A, and S_B and D_B are supply and demand in Country B. If there is no trade, the domestic prices will be set at P_A and P_B, respectively. The corresponding equilibrium quantities will be at Q_A (which is identical to $P_A U$) and Q_B ($P_B L$).

If trade is free between the two countries and transportation costs are negligible, there will be plywood trade between the two countries. The equilibrium price will be between P_A and P_B, where exports from Country A equal imports to Country B.

To see how the equilibrium price for the two countries is determined, we need to find the excess supply curve in Country A and excess demand curve in Country B. Note that when price equals P_A, the excess supply in Country A is zero. Thus, the excess supply curve in country A (X_S) intercepts the y-axis at P_A. When price is above P_A, there will be excess supply in country A, which is the difference between quantity supplied and quantity demanded, represented by the horizontal distance between S_A and D_A. Thus, if one knows the supply and demand functions in Country A, an excess supply curve can be drawn by algebraically or numerically varying the price above P_A.

Similarly, an excess demand curve (X_D) can be drawn for Country B. It intercepts the y-axis at P_B, signifying that excess demand is zero at the original equilibrium (without trade). By placing both the excess supply and excess demand curves at the centre of this figure, one will find that the point T, where these two curves cross each other, is the new equilibrium. At T, the price will be P_f, and the volume of trade will be Q_f (or P_fT), which is identical to WV or MK.

With the help of Figure 12.2, the economic consequence of trade in both countries becomes apparent. In Country A, the level of production will be increased from P_AU (Q_A) to P_fV, the producer surplus will increase by P_AUVP_f, and the consumer surplus will decline by P_AUWP_f. Consequently, plywood producers in Country A are better off, whereas consumers in Country A are worse off. The net social benefit will increase by UVW in country A.

In contrast, producer surplus will decline (by P_BLKP_f) and consumer surplus will increase (by P_BLMP_f) in Country B. The net social benefit will increase by MLK in Country B.

The total gains from trade in the two countries will therefore be UVW plus MLK. If we look at the excess demand and excess supply curves, this amount equals the area P_ATP_B. Trade enhances social benefits in both countries, and this is the fundamental reason why it has expanded greatly in the last century, despite some strong opposition to free trade from certain industries and segments of the labour force in some countries.

We need to point out that, to find out the equilibrium price in trade, both the excess demand (E_D) and excess supply (E_S) curves in Figure 12.2 need to be developed in either of the two currencies. This means that if E_D is measured in country A's currency, E_S needs to be converted to country A's currency using the exchange rate between the two currencies, and vice versa. This is important when we discuss the impact of exchange rates on trade.

More Sophisticated Regional and Global Trade Models

The foregoing discussion builds a theoretical foundation for more sophisticated trade models for forest products (or any other products). As we have seen, trade can enhance social benefits in both countries and, all else being equal, the equilibrium point for trade is where the net social benefit for both countries is greatest. Thus, the objective function of a forest products trade model for three or more countries is simply to maximize the net social benefits of these countries, after transportation costs, tariffs, and other trade barriers have been considered and incorporated into the model.

To build a regional or global forest products trade model requires enormous amounts of data (price, cost, quantity of various forest products) and rigorous estimation of supply and demand functions of various forest products in various countries. Further, a forest products trade model is often linked to forest resource supply and demand in these countries (outlined in Chapter 4), requiring additional delineation and careful considerations. Finally, to optimize or operate such a model requires linear programming or other sophisticated mathematical tools. Regardless of the intricacies involved in the forest products models that have been developed in the last few decades, all are based on maximizing the net social benefits of the countries involved, and are built on the constructs shown in Figure 12.2.

These forest products trade models have diverse uses. They are used for analyzing trade potentials and flows, trade policies such as log export bans and tariff escalations, resource-use policies such as harvesting regulations, changes in transportation (energy) costs, regional economic cooperation and integration efforts, and proposed carbon credits and taxes associated with climate change. Readers interested in global forest products trade modelling can begin with the Global Forest Products Model listed under "Further Reading" below.

Two Caveats

Having explained the conventional rationale for "gains from trade," we must add two caveats.

First, the demand and supply of products traded between the two countries, A and B, are not static but are influenced by trade flows. Once trade is facilitated, production in Country A will increase. Along with this comes increased employment and income. D_A is likely to shift outward because of the rise in employment and income, potentially reducing the level of export and trade benefits compared with a static demand. In Country B, production decreases, jobs are lost, and income falls. Assuming that those who lose jobs in Country B either stay unemployed for some time or can find only lower-paying jobs, D_B is likely to shift inward. Hence, the surplus gains from trade in Country B are likely to dissipate somewhat over time. These shifts in demand in both countries may reduce somewhat the level of trade and trade benefits predicted in the static model shown in Figure 12.2.

Second, if the high-cost producer is such because of environmental regulations and other efforts to protect the quality of life in Country B, while the low cost of production in Country A is due to the absence of such protections, then trade may cause overall net social benefit to drop as "clean" factories in Country B shut down and "dirty" factories in Country A expand. This is an example of "exporting the externalities," where the negative externalities generated by the dirty factories are exported from Country B to County A. Clean industry will simply not be able to compete in a global economy where not all producers are required to maintain the same standards. In this case, the resulting gains from trade will be an illusion. In recent years, some environmental groups (the greens) who support this argument have teamed up with labour unions (the blues) to form the "green/blue" alliance opposing free trade. More discussion on opposition to free trade will be presented under "The Political Economy of Trade Restrictions" below.

FACTORS INFLUENCING INTERNATIONAL FOREST PRODUCTS TRADE

Because bilateral trade involves two markets for a good, any factor influencing the supply of and demand for the traded good also influences its trade flow and price. On the demand side, these factors include income, preferences, and the prices of substitutes in both countries. On the supply side, they include comparative advantage in resource endowments (or factor costs, which determine production costs), marketing (advertisements), and shipping, handling, and insurance costs. Like income, most of these factors do not change quickly in the short term. Two factors, however, exchange rates and trade barriers (or lack thereof), can drastically change the balance of international trade in the short and long term.

Exchange Rates

An exchange rate is measured as a ratio of two currencies. The numerator can be either of the two currencies. Let us use US dollars (US$) and Canadian dollars (CDN$) as an example and assume that the current exchange rate between the two is 1 (CDN$/US$ = 1). If the US dollar appreciates 20%, then 1 US dollar will buy 1.20 Canadian dollars. On the other hand, 1 Canadian dollar will buy 0.833 US dollar, implying that the Canadian dollar has depreciated 16.67%.

How does this change in exchange rate affect bilateral trade between the two countries? This can be explained with reference to Figure 12.2, assuming that county A is Canada and county B is the United States.

In Figure 12.2, if the US excess demand (X_D) and the Canadian excess supply (X_S) were initially developed in US dollars when the exchange rate was 1, this 20% appreciation of the US dollar would not change or move the US excess demand curve. The Canadian excess supply curve, on the other hand, would shift downward or fall in parallel fashion by 16.67%, starting from point P_A. In other words, the new Canadian excess supply curve, valued in US dollars, would be 16.67% below the original excess supply curve. Thus, 1 Canadian dollar's worth of a good produced in Canada would now sell at 0.833 of a US dollar. This would increase the excess supply curve (X_S) and encourage Americans to buy more Canadian goods. In Figure 12.2, the latter represents a movement to the right along the excess demand curve (that is, quantity demanded increases) as price falls. Consequently, Canadian exports to the US would increase, and US exports to Canada would decrease. The change in equilibrium price, however, would be less than 16.67% measured in US dollars because the US excess demand curve is not perfectly inelastic.

It is worth noting that this 16.67% downward shift in the Canadian excess supply curve would have no impact on the Canadian domestic supply and demand curves, which are measured in Canadian dollars. It would only cause the original equilibrium price (P_A in Figure 12.2) measured in US dollars to fall, Canadian production (quantity supplied) to increase along the domestic supply curve, and Canadian exports to the US to expand. Also, quantity demanded in Canada would decrease, because more Canadian production would now be destined for export, driving up domestic prices.

Similarly, if both the US excess demand (X_D) and the Canadian excess supply (X_S) were initially developed in Canadian dollars, this 20% appreciation of the US dollar would not change or move the Canadian excess supply curve. The US excess demand curve, however, would shift upward by 20%, starting from point P_B. This means that US purchasing power, measured in Canadian dollars, would increase by 20%.

Whether the initial excess demand and excess supply is developed in either US or Canadian dollars, the market impact of this 20% appreciation of the US dollar would be the same. This conclusion is drawn from Figure 11.2 and related discussion. Interested readers may verify this by shifting either the excess supply or excess demand curve in parallel fashion (by the percentage given above) while holding the other curve constant. They should get the same new equilibrium price (and quantity) with the new exchange rate.

If the Canadian dollar were to appreciate, the trade impact would be just the opposite of what we have described.

A change in the exchange rate for a country's currency has both positive and negative impacts. A weaker US dollar, for example, encourages exports of American goods and discourages imports of foreign goods, and hence helps the US reduce its trade deficits. However, it raises the price of imported goods and foreign travel for American consumers and the cost of investing abroad for American investors.

It is unlikely that the performance of one sector can have much influence on the movement of exchange rates. Nonetheless, it is important to know what exchange rates can do to the profitability of a firm, the whole industry sector, and the country's balance of trade.

Exchange rates are critical for a country whose forest sector is largely export-oriented (such as Canada) or is heavily influenced by imports (such as the US, which imported some 13% of its forest products consumption in the mid-2000s). In the mid-2000s, when the US dollar was strong, American producers complained about the "invasion" of Brazilian plywood in their traditional markets in the southern US (particularly in the Miami and Atlanta markets) "without even a phone call." These producers were more at ease later, when the US dollar weakened and Brazilian plywood exports to the US declined drastically. Similarly, if four countries – the US, Canada, Germany, and Japan – produce similar types of logging equipment, variations in exchange rates have a heavy influence on where logging companies in these countries choose to buy their equipment.

The exchange rate between two currencies reflects the relative value of these currencies. Interested readers may refer to the balance of payments model, asset market model, and other models of exchange rates in international economics books.

Trade Barriers

Trade barriers are institutional measures that restrict trade, either overtly or, in many cases, covertly. They have historically played an important role in international trade by obstructing imports and exports of forest products. The many forms of trade barriers can be classified as:

- charges on imports or exports – tariffs, duties, export taxes, variable levies, tariff escalation
- quantitative limitations – quotas, export controls, log export bans, and embargoes
- regulations – industrial standards; packaging, labelling, and market regulations; customs procedures; health and sanitary regulations; licensing; and certification requirements
- government direct interventions – subsidies, countervailing duties, dumping and anti-dumping duties.

Sometimes two or more of these measures are combined to restrict trade. For example, a *tariff-rated quota* is a tariff applied after a certain level of imports is reached. Some measures, such as those connected with environmental issues, may not strictly fit in any one of these categories.

Formal trade restrictions have been the focus of the international multilateral trade negotiations held under the auspices of the General Agreement on Tariffs and Trade (GATT) between 1947 and 1994 and the World Trade Organization (WTO) thereafter. International trade has become freer as a result. For example, in the early 2000s, a decade after Canada and the United States entered into a free trade agreement, some 70-80% of all imports to the US and Canada entered duty-free, and the average duty rate on the rest was only 4-5%, meaning that the weighted average of duties on all goods imported by these two countries was only slightly over 1%. Nonetheless, some countries still impose high tariffs and other trade restrictive measures against some forest products imports.

Tariffs and Export Taxes

Tariffs are duties (taxes) applied to foreign imports, whereas an export tax is applied to domestic exports. Tariffs and export taxes with the same rate will have the same market impact in terms of protecting the domestic industry of the importing country. This can be readily seen in Figure 12.2 and with reference to Figure 11.2 and related discussion. Whether it is tariffs applied to imports (that is, on excess demand, E_D) or it is export taxes applied to exports (on the excess supply curve, E_S), the new equilibrium market price (and quantities imported and exported) will be identical when their rates are the same. The difference between them is that the government of the importing country gets the revenue from tariffs while the government of the exporting country gets the revenue from export taxes. Although it is unconstitutional for the government of the United States to impose a tax on exports, it has other means of controlling exports. In Canada and many other countries, export taxes can be applied, but some are implemented as a result of pressure from the importing country.

If an export tax is levied, the excess (export) supply curve of the exporting country will shift upward to the left if the export tax is based on export volume, or shift and rotate upward to the left if it is based on export value. In either case, the level of exports will decrease. On the whole, tariffs on forest products have declined in the past few decades, consistent with the overall trade expansion noted earlier.

A feature of many tariff schedules is tariff escalation, which occurs when tariff rates increase as the degree of processing increases. For example, logs are generally duty-free, sawn wood faces low tariffs, manufactured panels face higher rates, and more highly processed products face even higher rates. This affords more protection to domestic producers of highly processed products.

Quotas

A quota on a product limits the amount that can be imported. Quotas may take many forms, including an overall quota (regardless of a product's sources or origins), a quota allocated to specific countries of product origin, or a quota limiting eligibility for a particular tariff (that is, a tariff-rated quota). In general, quotas are allocated by both the importing country (to different sources of product origins, or countries) and the exporting source (to various firms within the same source/country).

Where they apply, quotas have a direct and often severe impact, and are generally regarded by economists as being more obstructive to economic efficiency than tariffs. They provide no revenue to the government, and they protect the domestic industry from competition more severely than a tariff does. If foreign producers are efficient enough, they can overcome a tariff, but they cannot exceed a quota. Finally, quotas cause the total supply curve, which consists of domestic and foreign sources, of the importing country to kink. As a result, when demand changes, market prices become more volatile than under a tariff or export tax, again at the expense of domestic consumers. Therefore, only the domestic industry – not domestic consumers or the governments in both countries – prefers quotas to tariffs, even though it has to endure more price volatility as a result.

Quotas may also be a preferred measure for some producers of the product in exporting countries if they can obtain quotas without export taxes or tariffs, and especially if they have market power in foreign markets because they can get a quota rent (or scarcity rent) by taking advantage of the high prices in protected foreign markets. The quota rent generated compensates part of their losses because of quota-induced restriction in export volume. Nonetheless, quotas, like all other trade restriction measures, limit the development of an exporting industry in the long term.

The extreme case of quantitative restriction is an outright prohibition on an export or import. Despite much criticism, logs cut on public lands in the US and Canada are subject to a general ban on exportation. In some other countries, log exports from private lands are also restricted.

Regulations

Trade-related regulations are very broad. A particularly troublesome area for exporters of forest products has to do with standards – both technical standards and those concerned with plant health, known as phytosanitary standards. Environmentally related technical standards – such as minimum recycled fibre content in newsprint, restrictions on wood panels that use formaldehyde glues, and a ban on trade in illegal timber – also distort trade. Most of these regulations have a good public policy purpose, so any trade losses must be weighed against public benefits.

Subsidies and Countervailing Duties

Private and especially government subsidies in the form of direct subsidization of producers, preferential treatment of producers who export, and preferential treatment of some producers or industries can result in trade distortions and injure an industry in another country. A subsidy that injures an industry in the importing country is subject to a retaliatory countervailing duty imposed by that country. Most countries have trade laws against predatory subsidies by foreign governments. The WTO and the North American Free Trade Agreement (NAFTA) have elaborate dispute settlement systems to adjudicate trade disputes that involve allegations of subsidy, but the three-decade-long softwood lumber dispute between the US and Canada demonstrates that such mechanisms do not always work well.

Dumping and Anti-Dumping Duties

Dumping occurs when imports are sold in a foreign market at unfairly low prices – lower than in the exporter's home market after all transportation costs are accounted for. Within this broad definition, various administrative methods and procedures have been used to determine whether dumping has occurred, which has resulted in inconsistent rulings, sometimes inconsistent with economic principles as well.

A positive finding of dumping and injury can lead to the imposition of an anti-dumping duty on foreign imports. Negotiations and litigation in the courts and under WTO or NAFTA rules and procedures often follow. The last 30 years have seen an increase in the number of anti-dumping cases filed in the US and the European Union.

THE POLITICAL ECONOMY OF TRADE RESTRICTIONS

If trade restrictions are so clearly harmful, why do some industrial executives, labour representatives, and politicians go to such great lengths to secure them? Possible explanations may be related to the political economy of trade restrictions.

Tariffs, quotas, and other measures used to manage trade are forms of government regulation of economic activities that can be issued in the interest of the general public or for special interest groups, as noted in Chapter 2. Some trade restrictions, such as the prohibition of trade in narcotics, benefit the public as a whole, but many are responses to special interest groups, mainly producers who press for protection against foreign competitors. In the latter case, the benefits of protection to producers will often fall short of the losses to consumers, with negative net gain to society as a whole even after all possible government revenues in tariffs and export taxes are considered.

Domestic producers understandably perceive foreign producers as a threat. If the foreign producers can produce the same products at lower cost, and have access to their market, domestic producers logically fear that they will suffer lower sales and profits. Jobs may be lost, plants may be shut down, and the capital they represent may be largely destroyed. Those who are hurt by imports often turn to their elected representatives for protection.

Politicians assess the benefits and costs of responding to demands from various interest groups and often make decisions based on their own self-interest, which includes gaining the support of interest groups – not only the threatened firms but also their employees, and the communities in which they operate. The beneficiaries of free trade are usually consumers, widely and thinly dispersed and without a coherent voice, while those seeking protection, though fewer, are more threatened, more vocal, and therefore more politically influential. Relatively small numbers of people, each of whom has suffered a large loss, have greater incentive for collective action and are better placed to organize themselves, and so their interest prevails in governmental decision making. Most recent empirical studies have found support for this interest group theory of government regulations to restrict trade.

This is not to imply that interest groups are immoral or that they always win at the expense of the public interest. The proliferation of political interest groups is perhaps a natural and benign consequence of economic development. They are regarded as essential and valuable participants in the democratic politics of a modern society, and without their participation policy would be made in far greater ignorance of what citizens actually wanted from their government. Fragmentation of authority – institutional and social pluralism – and a wide variety of competing interests using the same institutional machinery will prevent any single interest group from dominating. As we have noted, free trade has expanded in recent decades, helped by another interest group – the exporters who want to gain access to foreign markets.

It is undeniable, however, that the groups most visibly active in politics often do not represent a cross-section of economic and social interests. Some interests may be overrepresented even as others are underrepresented, as organizational resources, money, information, access to authority, skill, and bargaining power are distributed unequally among interest groups. This explains why some special interest groups can get their way in many economic policy issues, including trade restrictions, despite opposition from the general public.

FOREIGN DIRECT INVESTMENT IN THE FOREST INDUSTRY

With increasing trade in goods and services, some factors of production have become increasingly mobile globally. One such factor, capital, can move faster and more easily than others, such as labour and materials. Here we are less interested in the short-term movement of money around the world than in foreign direct investment (FDI), the movement of capital that is devoted to long-term production in a foreign country.

As defined by the International Monetary Fund, *foreign direct invest-ment* is an investment by a resident entity of one country in an enterprise in another country with the objective of obtaining a lasting interest in the enterprise and an effective voice in its management. For statistical purposes, an investment is classified as direct investment if the resident entity owns 10% or more of the shares of voting power of the foreign enterprise. As far as we know, most countries adopt this definition in their reporting of foreign direct investment, and firms with foreign direct investments are called *multinational enterprises.*

Foreign direct investment and new domestic capital expenditure differ in their definitions. New domestic capital expenditure means new equipment and machinery or new plants, thus representing new capital in the domestic economy. Although FDI often brings new capital expenditure to a country, it also includes investment that changes ownership from domestic owners to foreign investors. Like trade, FDI is a two-way street. Inward FDI means that foreign investors invest in a domestic economy; outward FDI implies investment in a foreign country.

Like trade in forest products, FDI in the forest industry has increased significantly in recent decades, as shown in Figures 12.3 and 12.4 for the US and Canada, respectively. Between 1983 and 2008, the US cumulative outward foreign direct investment position (accumulative stocks) in the forest industry increased from US$6.6 billion to US$10.4 billion, while cumulative inward investment rose by nearly three times, from

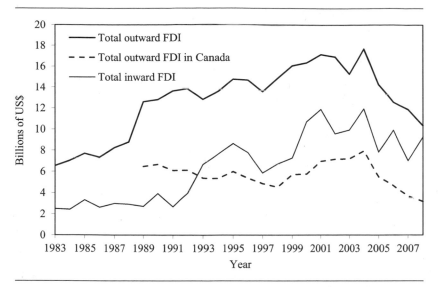

Figure 12.3   US outward and inward foreign direct investment in the forest industry in constant 2000 US$, 1983-2008. | *Source:* Nagubadi and Zhang 2011.

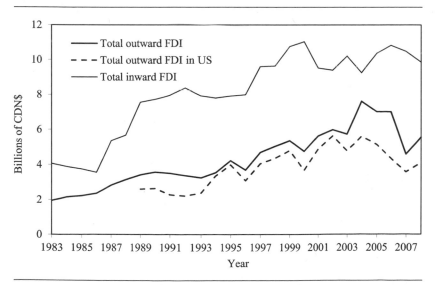

Figure 12.4   Canadian outward and inward foreign direct investment in the forest industry in constant 2000 CDN$, 1983-2008. | *Source:* Nagubadi and Zhang 2011.

US$2.5 billion to US$9.3 billion. For Canada, cumulative outward investment increased from CDN$1.9 billion to CDN$5.6 billion, while cumulative inward investment rose from CDN$4.1 billion to CDN$9.9 billion. Further, it is evident that Canada's forest industry received more inward FDI than its outward FDI, whereas the situation in the US saw just the opposite. Finally, and perhaps not surprisingly, each of these countries has the largest share of the other country's foreign direct investment in the forest industry. In recent years, however, other countries' share of outward FDI from the US and Canada has greatly increased.

Multinational enterprises can engage in international business through arrangements other than FDI, such as export and licensing. Why some choose FDI rather than export or licensing may be best explained by the ownership-location-internalization (OLI) paradigm, developed by economist John Dunning. The OLI paradigm is a blend of three different theories or factors of FDI. The so-called *OLI-factors* are three categories of advantages: *ownership advantages* (O), *locational advantages* (L), and *internalization advantages* (I).

- Ownership advantages answer the *why* question (why go abroad at all). To engage in any form of international activity, an enterprise must have one or more *firm-specific ownership advantages* that can be used for licensing or as inputs for expansion in domestic production and export, or, in the context of FDI, enable it to overcome the costs of operating in a foreign country. The ownership advantages, such as brand name, benefits of economies of scale, and technology, can be transferred within the enterprise at low cost. The advantages generate either higher revenues and/or lower costs, which can offset the costs of operating at a distance in a foreign location.
- Locational advantages address the *where* question (where to locate). Locational advantages are required for FDI but not for export and licensing. The motivation for moving offshore is the ability to use the ownership advantage in conjunction with factors (such as labour, land, or resources) in a foreign country. Through these factors, the enterprise makes profits on its ownership advantages. The choice of investment location depends on a complex calculation that includes economic, social, and political factors.
- Internalization advantages address the *how* question (how to go abroad). The enterprise chooses *internalization* (that is, the internal route, through wholly owned subsidiaries or controlled joint ventures) where the market does not exist or functions poorly so that the transaction costs of the external route (markets) are high. Through

FDI, the enterprise substitutes its own internal market and reaps some efficiency gains. For example, a firm can go abroad by simply exporting its products to foreign markets, but uncertainty, search costs, and tariff barriers are additional costs that will deter such trade. Similarly, the firm could license a foreigner to distribute its product, but the firm must worry about opportunistic behaviour on the part of the licensee. In these cases, FDI would be a logical choice.

FDI can be supply-driven or demand-driven, or both. Resource-seeking FDI looks for destinations that have abundant resources to produce goods and services at a lower cost than elsewhere. Market-seeking FDI looks for destinations with large existing and potential markets. Arguably, US outward foreign direct investment in the forest industry is geared more towards resource seeking, since its main destinations are forest resource–rich countries such as Canada and Brazil, whereas Canadian outward FDI is more market seeking, since its main destinations are the US and other developed countries.

FDI can have significant implications for the economies of both the host country and the home country. For the host country, FDI transfers not only financial resources but also technology, managerial skills, and employment. It also brings in various sales and procurement networks to expand business opportunities and trade, and it increases competitive pressure on other domestic firms to improve technical and allocative efficiency. Conversely, as foreign capital comes and goes in response to global markets and conditions in the host country, it can create a boom-and-bust local economy. For the home country, FDI brings higher returns to capital, enables efficient use of domestic resources, and often facilitates exports of goods and services. On the other hand, outward FDI may be negatively correlated with domestic capital expenditure, affecting domestic employment.

Trade in forest products and foreign direct investment in the forest industry promote forest-based economic development, affect the rate of forest resource depletion and renewal, and thus impact global forest sustainability. Chapter 13 discusses the economic and environmental aspects of global forest resources.

REVIEW QUESTIONS

1 What is comparative advantage? What is absolute advantage? What is specialization?

2 What are the measuring sticks for comparative advantage and absolute advantage, respectively?

3   How are the terms of bilateral trade determined?

4   On what basis is a forest products trade model that involves more than two countries and many forest products built?

5   What is the gain of trade? How do exchange rates affect the flow of international trade?

6   What is a countervailing duty? Under what conditions is a public or private subsidy subject to countervailing action in international trade?

7   What is dumping?

8   Describe the role of the green/blue alliance in restricting international trade.

9   In what ways are foreign direct investment and domestic capital expenditures similar? In what ways are they different?

FURTHER READING

Adams, Darius A., and Richard W. Haynes. 1980. The 1980 softwood timber assessment market model: The structure, projections and policy simulations. *Forest Science* Monograph 22.

Bourke, I.J., and Jeanette Leitch. 1998. *Trade Restrictions and Their Impact on International Trade in Forest Products.* Rome: Food and Agriculture Organization of the United Nations.

Buongiorno, Joseph, Shushuai Zhu, Dali Zhang, James Turner, and David Tomberlin. 2003. *Global Forest Products Model: Structure, Estimation, and Applications.* San Diego: Academic Press.

Food and Agriculture Organization of the United Nations (FAO). 2010. FAOSTAT, http://faostat.fao.org/site/626/default.aspx#ancor.

Dunning, John H. 2001. The Eclectic (OLI) Paradigm of international production: Past, present and future. *International Journal of the Economics of Business* 8 (2): 173-90.

Nagubadi, R., and Daowei Zhang. 2011. Bilateral foreign direct investment in forest industry between the US and Canada. *Forest Policy Economics* 13(5): 338-44.

Zhang, Daowei. 1997. Inward and outward foreign investment: The case of US forest industry. *Forest Products Journal* 47 (5): 29-35.

-. 2007. *The Softwood Lumber War: Politics, Economics, and the Long US-Canada Trade Dispute.* Washington, DC: RFF Press.

Chapter 13   Global Forest Resources and the Environment

In this final chapter, we look at the current status of global forest resources and economic issues related to global forest management. Our focus is on the allocation of forest resources at a global scale, especially the role of forest resources in economic development and environmental protection. Although economic principles applicable in a mixed capitalistic system may also be used in other systems, we stress that each country has its unique history, culture, and political governance structure and that global issues are inherently more difficult to tackle than domestic issues. Nonetheless, forest resources and forest management in one country impact the cultural and economic well-being of not only its people but also people in other countries and the global environment. Thus, we must examine forest resource management from a global perspective.

Global concerns about forest resource use and management include the role of forests in the economic development of less-developed countries, in protecting non-market environmental resources such as biodiversity, which are often global public goods, and in combating climate change. These roles are obviously linked, but the latter two deserve special attention. We begin with an overview of global forest resources, then continue with the economics of forest-based industrialization, deforestation, and the associated decline in forest-related environmental goods and services. Finally, we note several emerging issues in global forest resources and the environment.

GLOBAL FOREST RESOURCES
For statistical purposes, a forest is often defined as a collection of trees that covers at least 10% or 20% of the land. Forests can be considered as

either closed (with dense canopy) or open. Closed forests generally occur in moist temperate and some tropical climates, and open forests are usually found in drier climates such as Sub-Saharan Africa. According to the Food and Agriculture Organization of the United Nations (FAO), closed forests account for some 85% of global forests, open forests account for 12%, and the remainder are forest fallow, where tree cover is returning on lands that have been subjected to shifting agricultural practices and abandonment.

Table 13.1 provides a summary of the world's forests. In 2005, the world's 3.9 billion hectares of forests covered around 30% of the total global land base. Forest areas are unevenly distributed among countries, however, especially in per capita terms. The Russian Federation, Brazil, Canada, the United States, and China have the largest total forestland areas in the world, with 809, 478, 310, 303, and 197 million hectares, respectively, but China's per capita forestland area is far less than those of the other four countries and far below the global average. Many other countries, such as Cambodia, the two Congos (the Republic of the Congo and the Democratic Republic of the Congo), Costa Rica, Ecuador, Ghana, and Peru, although small in total land base and forest area, have rich and important forest resources. Whereas the global average for forestland per capita is around 0.62 hectare, some 64 countries with a combined population of 2.0 billion have less than 0.1 hectare of forestland per capita.

Of the world's total forestland area, tropical forests comprise 47%, subtropical forests 9%, temperate forests 11%, and boreal forests 33%. Usually, tropical and subtropical forests are composed of broad-leaved (mostly hardwood) species, while boreal forests are composed mostly of coniferous (softwood) species. In developed countries, mostly in the Northern Hemisphere, more forest products are made from softwood species than from hardwood species, but many hardwood species have important aesthetic and environmental service values and functions, if not commercial value.

Forests can also be classified based on their characteristics and ownership. As of 2005, natural forests, which regenerate on their own or become established with assisted natural regeneration with native species, accounted for the vast majority (95%) of the world's forests. Natural forests include primary forests, modified natural forests, and semi-natural forests. Planted forests, established through planting or seeding, accounted for the remaining 5%. Planted forests have increased significantly and have accounted for a large share of global timber supply in recent decades. Public ownership dominates in most countries, with rights to use and to manage often assigned to private entities.

TABLE 13.1 Characteristics of forestland in different regions of the world

Region	Country	Population in 2004 (millions)	Total forest area in 2005 (1,000 ha)	Ownership of forestland in 2000			Change in forestland (2000-05)	
				Public (%)	Private (%)	Other (%)	Rate (1,000 ha/yr)	%
Africa		868	635,412	97.6	1.8	0.6	-4,040	-0.6
Asia		3,838	571,577	94.4	5.0	0.6	1,003	0.2
	China	1,327	197,290	100.0	0.0	0.0	4,058	2.2
	Indonesia	218	88,495	100.0	0.0	0.0	-1,871	-2.0
	Japan	128	24,868	41.9	58.1	0.0	-2	n.s.
Oceania		33	206,254	61.3	23.7	15.0	-356	-0.2
	Australia	20	163,678	72.0	27.1	0.9	-193	-0.1
Europe		723	1,001,394	89.9	10.0	0.1	661	0.1
	Germany	83	11,076	52.8	47.2	0.0	0	0.0
	Russia	143	808,790	100.0	0.0	0.0	-96	n.s.
	Sweden	9	27,528	19.7	80.3	0.0	11	n.s.
North America		429	677,464	66.7	29.3	4.0	-101	n.s.
	Canada	32	310,134	92.1	7.9	0.0	0	0.0
	United States	294	303,089	42.4	57.6	0.0	159	0.1
	Mexico	104	64,238	58.8	-	41.2	-260	-0.4
Central America		39	22,411	42.5	56.1	1.4	-285	-1.2
South America		365	831,540	75.9	17.3	6.9	-4,251	-0.5
	Brazil	179	477,698	-	-	-	-3,103	-0.6
	Ecuador	13	10,853	77.1	0.0	22.9	-198	-1.7
World		6,335	3,952,025	84.4	13.3	2.4	-7,317	-0.2

Note: n.s. = not significant.
Source: FAO 2006.

TABLE 13.2   Countries with largest net losses and net gains in forest area, 2000-05

	Country	Annual change (1,000 ha/yr)
Largest net losses	Brazil	−3,103
	Indonesia	−1,871
	Sudan	−589
	Myanmar	−466
	Zambia	−445
	Tanzania	−412
	Nigeria	−410
	Democratic Republic of the Congo	−319
	Zimbabwe	−313
	Venezuela	−288
	Total	−8,216
Largest net gains	China	4,058
	Spain	296
	Vietnam	241
	United States	159
	Italy	106
	Chile	57
	Cuba	56
	Bulgaria	50
	France	41
	Portugal	40
	Total	5,104

Source: FAO 2006.

Most countries with a higher deforestation rate than the global average are less developed and are located in the tropics. Table 13.2 shows that all 10 countries with the largest net loss in forest area from 2000 to 2005 are tropical developing countries, and that most of the 10 countries with the largest net gains in forest area are either developed countries or countries with high rates of economic growth in recent decades.

POPULATION, ECONOMIC GROWTH, AND THE ENVIRONMENT

Forest resources, and the capital contained therein, have often been used to stimulate economic growth and development in forest-rich countries. Economic development is seen as necessary if growing populations are to maintain and raise living standards. Thus, the use and management of global forest resources have much to do with population growth, economic development, and the environment.

Global Population Trends, Scarcity, and Growth

The world population, of less than 1 billion in 1800, had climbed to 2.5 billion by 1950, 6 billion by 1999, and nearly 6.7 billion by 2008. With a larger base and an expected lower growth rate, global population is widely expected to reach 7 billion in 2011 and 9.3 billion by 2050. Although population growth provides additional labour for economic development, it creates pressure on global natural resources, including forests. Data from FAO have shown a continued decline in global forest area in the last few decades, although the rate of loss has slowed and the forest area has stabilized or increased in some countries.

A related issue is the rise in per capita incomes and standards of living. Although it is difficult to compare consumption possibilities and levels of different generations, it is estimated that the average consumption of the billion people on the planet two centuries ago can be valued at $600 to $700 per year in today's US dollar values. By the turn of the last century, this had increased to well over $1,000 a year. It was over $2,000 a year in 1950 and now stands at some $8,000 per year. This shows a continued increase in real global per capita income.

Rising populations and living standards demand continued economic growth, but is this possible in the long run with limited land and natural resources? Economists, philosophers, and demographers have engaged in a perennial debate concerning scarcity of resources and its implications for economic growth.

In the 18th century, Thomas Malthus developed a highly influential theory explaining how an expanding population on a fixed land base would always be reduced to bare subsistence and held in check by starvation and disease. Earlier we drew attention to the conservation movement of the early 20th century, driven by concern that resources were being exploited too fast to sustain the growing American economy. Depletion of the great white pine forests of eastern Canada was seen to portend a "timber famine." After the Second World War, there were renewed anxieties about an impending scarcity of natural resources, especially in the US. More recently, anxieties have arisen over continuing access to oil supplies. Nevertheless, long-term economic growth in most countries, and in the world, has continued.

Some argue that while land and some natural resources are limited in supply, growth can be sustained through expansion of both the human workforce and manufactured capital, but without more natural resources, the gains from employing more of these other factors of production may become progressively smaller. This principle of diminishing returns

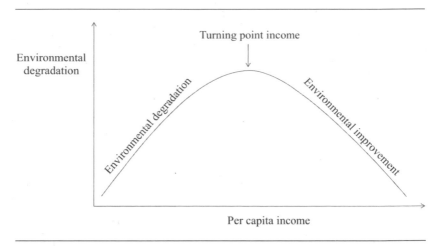

Figure 13.1   The Environmental Kuznets Curve.

This may be too optimistic a prognosis in several respects, however. First, there may be a *composition effect* (defined as compositional changes in pollution associated with the rise in income in a county or with variations in income among countries) in the income/pollution relationship. Wealthier societies often concentrate their own economic activities in the high-technology and service industries, leaving or "exporting" dirtier industries to less-developed countries. Second, not all pollutants can be expected to follow the inverted-U relationship of the Environmental Kuznets Curve, at least not yet. Carbon dioxide and other greenhouse gases in the atmosphere have not abated despite a rapid rise in global per capita income in the last two centuries. Third, cumulative pollutants such as greenhouse gases show that the accumulation process cannot be quickly reversed. Finally, losses in biodiversity may be completely irreversible. No one is certain how many species now live on earth and how many become extinct each year, but some biologists suggest that the current rate of species extinction exceeds that shown in the geological record by a factor of 1,000.

FOREST-BASED INDUSTRIALIZATION AND TROPICAL DEFORESTATION

Forest-Based Industrialization
In Chapter 9, we discussed the long-term trend in the forest sector, which is essentially a forest-based industrialization process in a mixed capitalistic economy. As J.R. Vincent and C.S. Binkley (1992, 93-94) write,

"Forest-based industrialization begins with an unexploited old-growth forest. The principles of forest ecology dictate that timber growth in such forests is nil, as mortality just balances any net photosynthetic activity. Under such circumstances, any level of harvest will exceed growth, and the inventory of timber will decline. Ecological capital contained in the form of timber is converted to economic capital, which can be used in the development process. The area of forest will also decline as land is converted to agriculture and other uses yielding higher economic returns."

Most industrialized countries have experienced this type of initial decline in forest inventory and forestland area. What is different among countries is the ensuing adjustment process. Harvests greater than growth and the occurrence of deforestation do not necessarily imply that exploitation is bad or unsustainable. More important are the adjustments in economic and ecological systems that are induced by the decline in timber inventories.

Vincent and Binkley further note that "timber (stumpage) prices play a key role in signalling increased scarcity and in inducing the ensuing adjustments." If property rights are well defined and stumpage markets operate efficiently, rising stumpage prices create financial incentives to sustain a stable and "permanent forest area, albeit one that is smaller than the country's original endowment and one that may be too small if non-market values are not fully considered. Timber growth will increase as inventories decline, possibly increasing to the point at which growth equals harvest and the system tracks a sustainable path through time." Simultaneously, the timber-processing sector will respond to the higher price of timber by becoming more efficient in the use of wood.

Many forest-based industrialization projects – initiated by a country's government, sometimes with help from international aid agencies – have failed, however, and the industrialization process in some countries has been far from sustainable. Moreover, some aid agencies have had concerns regarding the impact of these projects on biodiversity and other environmental values. As a result, some international organizations have recently refrained from financing such projects.

Problems in the past stem from inappropriate macroeconomic and sector-specific policies, especially in some less-developed countries. Some governments have fostered an unstable macroeconomic environment and insecure property rights arrangements. Others have tried to keep the price of wood artificially low to encourage investment in (often inefficient) domestic processing industries. The results, note Vincent and Binkley, have been depletion of forests, inadequate management of natural forests, limited establishment of plantations, and underutilized natural or old-growth forests. The principles of forest ecology dictate that timber growth in such forests is nil or close to zero, as mortality just balances any net photosynthetic activity. Under such circumstances, any level of harvest will exceed growth, and the inventory of timber will decline. Ecological capital contained in the form of timber is converted to economic capital, which can be used in the economic development process. The area of forest will also decline as land is converted to agriculture and other uses that yield higher economic returns.

Most industrialized countries have experienced this type of initial decline in forest inventory and forestland area. What is different among countries is the ensuing adjustment process. Harvests greater than growth and the occurrence of deforestation do not necessarily imply that exploitation is bad or unsustainable. More important are the adjustments in economic and ecological systems that are induced by the decline in timber inventories.

Timber (stumpage) prices play a key role in signalling increased scarcity and in inducing the ensuing adjustments. If property rights are well defined and stumpage markets operate efficiently, rising stumpage prices create financial incentives to sustain a stable and permanent forest area, albeit one that is smaller than the country's original endowment and one that may be too small if non-market values are not fully considered. Timber growth will increase as inventories decline, possibly increasing to the point at which growth equals harvest and the system tracks a sustainable path through time. Simultaneously, the timber-processing sector will respond to the higher price of timber by becoming more efficient in the use of wood.

Many forest-based industrialization projects – initiated by a country's government, sometimes with help from international aid agencies – have failed, however, and the industrialization process in some countries has been far from sustainable. Moreover, some aid agencies have had concerns regarding the impact of these projects on biodiversity and other environmental values. As a result, some international organizations have recently refrained from financing such projects.

Many problems in the past can be traced to inappropriate macroeconomic and sector-specific policies, especially in some less-developed countries. Some governments have tended to foster an unstable macroeconomic environment and insecure property rights arrangements. Others have tried to keep the price of wood artificially low to encourage investment in (often inefficient) domestic processing industries. The results have been depletion of forests, inadequate management of natural forests, limited establishment of plantations, and underutilized processing capacity. These policies often create a boom-and-bust cycle in forest-based economic activities, employment, and income, and do not encourage sustainable forest-based economic development. More recently, environmental concerns, ranging from the depletion of old-growth forests, to wasteful deforestation and adverse consequences for the biodiversity of forest plantations, have hindered forest-based industrialization efforts in some countries.

Tropical Forests: Private Goods and Global Commons
As noted at the beginning of this chapter, the 10 countries with the largest losses in forest area are all in the tropics. Brazil led the world in terms of total deforested area between 2000 and 2005, losing 15,515,000 hectares (38,012,000 acres) of forest in five years, roughly the size of the state of Georgia in the United States. In addition to absolute size, we can look at deforestation in terms of the percentage of a country's forests that was cleared over time. The 10 countries with the largest annual net rates of forest depletion over the same period are: Comoros (north of Madagascar, -7.4%), Burundi (central Africa, -5.2%), Togo (-4.5%), Mauritania (-3.4%), Nigeria (-3.3%), Afghanistan (-3.1%), Honduras (-3.1%), Benin (-2.5%), Uganda (-2.2%), and the Philippines (-2.1%). Again, other than war-torn Afghanistan, all are in the tropics.

As it is elsewhere, some tropical deforestation is the result of conversion of forestland to more highly valued uses, representing a useful and needed conversion of ecological capital to economic capital. Excessive or wasteful tropical deforestation has profound, sometimes devastating, economic, social, and environmental consequences, however, including the persistent poverty of indigenous and other people, social conflict, extinction of plants and animals, and climate change.

Thus, a great deal of research effort has been devoted to understanding the dynamic of deforestation and its associated land-use changes in the tropics since the 1970s. The direct, or proximate, causes of tropical deforestation are agricultural expansion, wood extraction (either commercial logging or timber harvesting for family uses, which can be part of a well-intentioned forest-based industrialization activity), infrastructure expansion, and urbanization. Usually it is the result of multiple processes that work simultaneously or sequentially to cause deforestation.

Figure 13.2 is a simple illustration of the deforestation process, highlighting the direct and underlying causes of tropical deforestation. It shows that deforestation is a complex mixture of players and processes.

For example, poverty and population growth may drive some people to migrate to forest frontiers, where they engage in slash-and-burn

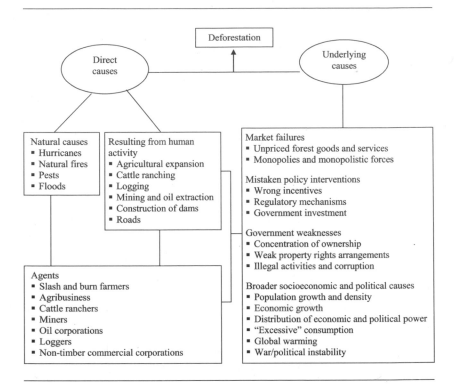

Figure 13.2   Direct and underlying causes of tropical deforestation.

forest clearing for subsistence. State policies that encourage economic development, such as road and railway expansion projects – which may cause significant, unintentional deforestation on their own – further expedite this process. Agricultural subsidies and tax breaks, land title registration (which requires demonstrated uses and occupation), and poorly designed timber concessions have also encouraged forest clearing. Political instability and poor governance introduce uncertainty and make the definition and enforcement of property rights problematic. Global economic factors such as a country's foreign debt, expanding global markets for rainforest timber and pulpwood, or low domestic costs of land, labour, and fuel can encourage deforestation instead of more sustainable land use.

Access to technology may either enhance or diminish deforestation. The availability of technologies that enable "industrial-scale" agriculture can spur rapid forest clearing. Inefficient technologies in the logging industry increase collateral damage in surrounding forests and wood waste, making subsequent deforestation more likely, whereas wood-saving technologies in forest products manufacturing help reduce deforestation.

So far, we have treated tropical forests as private goods that are available for governments, businesses, or local populations to explore and use. Tropical forests are also a global public good as they harbour more species than any other ecosystem and contain a significant amount of carbon. Undisturbed tropical forests may be nearly neutral with respect to carbon, but deforestation and degradation have made the tropics into a source of carbon going into the atmosphere, and have the potential to turn them into an even greater source in coming decades.

Indeed, global concerns regarding tropical forests are mostly related to the decline of the public good values affected by deforestation. Tropical forests are owned by public or private agencies in various countries that usually do not get much compensation for their non-market values. Unless there is an effective mechanism to compensate the owners for these values, the forests will be undervalued and perhaps exploited too quickly.

THE ROLE OF FORESTS IN MITIGATING GLOBAL CLIMATE CHANGE

Global climate change affects economic activities, human welfare, the environment, and survival of all living organisms, and certainly has not always supported life as we know it today. Global warming is a major form of climate change.

The term "global warming" refers to the increase in global average temperature that has been observed over the last hundred years or more. There is little dispute that the global average temperature has been on the rise during this period. It is also true that mankind's burning of fossil fuels (mostly coal, petroleum, and natural gas) releases carbon dioxide (CO_2) and other greenhouse gases into the atmosphere and that, as of 2008, the concentration of CO_2, the most important of these gases, was around 40-45% higher than it was before the Industrial Revolution.

Scientists have not reached a consensus, however, on how much human activities have contributed to global warming.[1] There is even more debate about what can or should be done to control global warming, how much each country should contribute towards the reduction of greenhouse gas emissions, and what policy instrument should be used to control emissions.

What does global warming have to do with forests and forest economics? Quite a bit. Forests, like other ecosystems, are affected by global warming and other forms of climate change, be it a rising sea level that threatens coastal forests, rising temperature that helps expand the temperate forest zone northward, or changes in temperature and rainfall patterns, fire and pest regimes, or forest productivity. In some places, the impacts may be negative; in others, they may be positive. Much research is now being conducted to assess the economic impacts of climate change–induced forest productivity changes and changes in fire/pest regimes. In fact, global warming and other forms of climate change could become one of the most important influences on management of the world's forests.

Forests also influence climate and the climate change process, by absorbing carbon in wood, leaves, and soil and releasing it into the atmosphere when these decay or are burned or cleared. Thus, forests can play significant roles as carbon stores, as sources of carbon emissions, and as carbon sinks in the whole global carbon cycle, and are important in discussions of global warming.

Deforestation has been seen as contributing to the build-up of carbon in the atmosphere in the last two centuries. The Intergovernmental Panel on Climate Change (IPCC) estimates that removal of forests and conversion of forestland to other uses have contributed close to 20% of the overall amount of greenhouse gases entering the atmosphere. Forest degradation also contributes significantly to emissions. As we have seen, deforestation continues in some countries, especially tropical countries where forests and forestland contain relatively high levels of carbon per hectare. In 2008, the United Nations established a collaborative program known as REDD+ (Reducing Emissions from Deforestation and Forest Degradation and the Role of Conservation, Sustainable Management of Forests and Enhancement of Forest Carbon Stocks in Developing Countries), which is aimed at generating the requisite resources to significantly reduce global emissions from deforestation and forest degradation. Afforestation, on the other hand, absorbs atmospheric carbon, and wood products that are expected to last a long time can be seen as storage for carbon.

Carbon Markets

Because the burning of fossil fuel is seen as the single largest contributor of anthropogenic greenhouse gas emissions, several measures have been proposed to reduce the use of fossil fuels. These include support for technological development and transfer to reduce the energy use associated with global economic activities, regulations restricting energy consumption (such as those mandating that vehicles meet certain fuel efficiency standards), and carbon taxation and emissions charges. Increased use of renewable energy, discussed in the next subsection, can also help reduce greenhouse gas emissions. Finally, a cap-and-trade system for carbon emissions offers a market mechanism to help achieve emissions reduction targets.

Under a cap-and-trade system, the total amount of emissions is set as a mandatory *cap* that is allocated to individual firms through *emission permits;* firms have flexibility in how they comply. In this way, the regulatory authority decides on the total quantity of emissions to be allowed and issues corresponding permits. Whether the initial distribution is by public auction or some other allocation system, the *trading* of permits eventually sets a price for permits. Firms will limit their emissions to the level at which the marginal cost of control is equal to the price at which they can buy and sell permits, and marginal costs will thus be equal among emitters.

The logic behind the cap-and-trade program is that some companies can reduce their emissions below their allowed amount of emissions less expensively or more easily than others. These more efficient companies can sell their extra permits to companies that are not able to make reductions as easily. This creates a system that guarantees a set level of overall reductions while rewarding the most efficient companies and ensuring that the cap can be met at the lowest possible cost to the economy.

Examples of successful cap-and-trade programs include the Acid Rain Program implemented in the Great Lakes Region of the US and Canada and the NO_x Budget Trading Program in the US Northeast. Empirical studies have found that these emission trading programs generally resulted in significant cost savings relative to regulatory or command-and-control approaches.

As for carbon markets, a trading program may include both *carbon allowances* (the share of the cap) and *carbon offsets.* The latter would require a source (such as a power producer) that produces new or increased emissions of carbon dioxide to offset those emissions by obtaining equivalent reductions of CO_2 emissions from other sources, such as carbon sinks. As trees sequester carbon, forests can become carbon sinks. Thus, a forestry project that can sequester carbon may be used to offset carbon emissions from a carbon emitter that uses fossil fuels.

Accordingly, various voluntary carbon registration standards and one voluntary market (the Chicago Climate Exchange) have emerged in North America in recent years, even though a carbon emission trading program has not yet received government sanction and no carbon allowance (cap) has been set in the United States or Canada. Eligible forestry carbon sequestration offset projects that are traded on the Chicago Climate Exchange and other climate exchanges in the world include afforestation, reforestation, and sustainably managed forests. Often, the values or prices of carbon traded on the markets are based on carbon credits. One carbon credit is equal to one ton of CO_2 or, in some markets, CO_2-equivalent gases.

From its beginning in 2003 through 2009, the cash price of carbon credits traded in the Chicago Climate Exchange varied from US$0.10 to US$7.40 and averaged US$2.40 per metric ton of CO_2 or equivalent. We can illustrate the opportunities these developments present for forest landowners by referring to an Alabama example.

In 2000, the net annual growth of pine plantations in Alabama, on average, was 6.8 cubic metres per hectare per year. This translates into 3.4 tons of carbon biomass per hectare per year (as 1 cubic metre of growing stock is equal to 0.5 tons of carbon biomass in North America). A ton of carbon is equal to 3.66 tons of CO_2. Thus, a typical pine plantation in Alabama will store 12.5 (3.4 × 3.66) tons of CO_2 per hectare per year. Using the average market price of US$2.40 per ton of CO_2, owners of pine plantations in Alabama could sell their carbon credits for $30 per hectare per year (or $12 per acre per year). From this, landowners would have to deduct the transaction costs associated with the production and sale of carbon credits, which may be significant insofar as they must be certified and their forest practices verified. As a result, their net revenues are not likely to be lower. This is good for landowners as an additional source of income, but carbon credits alone may not be able to entice them to make much change in their forest management decisions. This could change, however, if the carbon credits were to reach the much higher European prices.

This explains why only a handful of forest carbon offset projects in North America have been traded in the voluntary market through the Chicago Climate Exchange: the price of carbon is quite low and there is limited demand for offsets. Perhaps little progress will be made until a mandatory emissions regulation system (the cap) is in place and carbon has a much higher value than we see in North America today. Many forest carbon offset projects have been implemented in other parts of the world, particularly in South America and Asia, through the Clean Development Mechanism of the Kyoto Protocol, although the effectiveness and efficiency of these projects have yet to be systematically studied.

In addition, concerns about forestry offsets in particular relate to three key principles of the United Nations Framework Convention on Climate Change: *additionality*, *leakage*, and *permanence*. The principle of *additionality* ensures that carbon offset credits for a project are awarded only for the additional carbon sequestered beyond what would have occurred without the project. The baseline from which the carbon sequestration benefits for a project are calculated is known as *Business as Usual* (BAU). The BAU level of carbon sequestration for forests is notoriously difficult to calculate due to the uncontrollable events that can occur. Thus, it will be necessary to develop clear and mutually acceptable methods for determining such levels.

The principle of *leakage* ensures that the beneficial impacts of a forestry project are not negated by impacts that are induced by the project outside the project boundary. For example, many believe that the reduction in the US federal timber harvest in the 1990s directly caused an increase in harvests on private lands in the US, and public and private lands outside of the US, particularly in Canada. Thus, projects that effectively reduce US harvests for the purpose of selling offset credits may result in increased harvests elsewhere.

The principle of *permanence* requires that, to qualify for an offset, the reduction in greenhouse gas emissions resulting from a project must be permanent – that is, it cannot just delay the emission. The principle is appropriate for direct emitters but is perhaps difficult to achieve in forestry projects, which are long-term and uncertain. This principle will probably limit the number of forestry projects that qualify.

Wood-Based Bioenergy

Forests can also contribute to reducing greenhouse gases through renewable biomass for energy production that replaces fossil fuel emissions. Wood-based energy is by and large carbon-neutral. Even though harvesting and transporting wood may use fossil fuels and cause some emissions, wood-based energy emits less greenhouse gases than grass and non-cellulosic biomass (corn ethanol and soybean biodiesel) per unit weight of energy produced.

The problem is that the cost of producing bioenergy from wood is higher than the cost of energy from alternatives. A recent study shows that the cost of producing electricity from logging residue is substantially higher than from coal largely because the energy content of wood per unit weight (BTU) is about one-half that of coal, and transportation and handling costs for timber are much higher than for coal. The difference in cost would translate to a carbon sequestration cost of $57 per ton of CO_2. In other words, this type of project would be carried out only if CO_2 values exceeded $57 per ton; otherwise, more cost-effective mitigation projects would be chosen.

In the United States, however, recent government policies encourage wood-based bioenergy production. The 2007 US Energy Independence and Security Act requires a huge increase in the use of renewable fuels, from some 7 billion gallons per year in 2007 to 36 billion gallons per year by 2022. Much of this increase will come from biofuels from non-food feedstock (including cellulosic woody biomass), which would jump from 600 million gallons to 21 billion gallons. The US government has provided subsidies for start-up bioenergy plants that use woods, as well as funding for research and development of technologies that would convert woody biomass to ethanol, methanol, and other chemicals. Should a technological breakthrough reduce the cost of producing cellulosic biofuels and woody biomass–based cellulosic ethanol plants actually succeed financially without government grants and subsidies, then there will be a new wood-based bioenergy market for timber. This new market may begin with an increased use and recovery of logging residues and end up with more forests that are planted in fast-growing species and harvested in short rotations for the sole purpose of producing wood-based bioenergy.

EMERGING ISSUES IN FOREST ECONOMICS

In this section, we note three emerging and important topics in forest economics: ecosystem services, forest health, and the need to understand governance, policy making, and institutions that support sustainable forestry. Although these issues are not new, the expansion of human population and economic activities has brought them to the attention of the public and the forest economics profession.

Ecosystem Services

The city of New York is endowed with abundant water for its over 8 million residents. Most of the water from reservoirs built in upstate watersheds flow through tunnels. To meet federal water quality standards for clear tap water, the city has been dumping 16 tons of chemicals a day, on the average, into the water supply.

In the late 1990s, city and federal officials worried that the fabled water coming from the largest unfiltered system in the country, was becoming cloudy, and the sediment interfered with chlorination to eliminate contaminants. If the city could not find another solution, it would be required to build a huge filtration plant at a cost of about US$6-8 billion, plus hundreds of millions of dollars in operating costs each year.

The city chose a different route. City officials realized that the increase in muddiness was mostly caused by increasing runoff from land development in upstate watersheds, so they sought to restrict development of forest and agricultural land throughout the relevant watersheds. They considered pushing for statewide land-use regulation, but because such restrictions were not popular, they settled on outright land purchases and, for the most part, paying landowners to sign conservation easements. In essence, the city bought some of the "sticks" in the bundle of property rights (discussed in Chapter 10) from forest landowners. Between 1997 and 2006, the city was able to protect over 70,000 acres at a cost of $168 million, and would continue to do so in the foreseeable future. This investment in land purchase and conservation easements, which was estimated at $1-1.5 billion eventually, can be seen as an investment in natural capital. In this way, the city has avoided building and operating a much more costly filtration plant.

This story sheds some light on an emerging economics and policy issue in forest resources and the environment: ecosystem services.

Humans receive vital goods and services from both managed and natural ecosystems. Thus, *ecosystem services,* broadly defined, cover all benefits that humans receive from managed and natural ecosystems, including products such as clean drinking water and processes such as the decomposition of wastes. The United Nations 2004 Millennium Ecosystem Assessment (MA) grouped ecosystem services into four broad categories: *provisioning,* such as the production of food and water; *regulating,* such as the control of climate and disease; *supporting,* such as nutrient cycles and crop pollination; and *cultural,* such as spiritual and recreational benefits.

The forestry profession has long recognized the joint production of beneficial forest outputs, such as timber, forage, water, recreation, and habitat for species. The recent call for ecosystem management instead of stand- or forest-level management has popularized the more encompassing term of "ecosystem services," however, largely to advocate ecosystem protection. The difference between the current notion of ecosystem services and the traditional concept of multiple-use forest management discussed in Chapter 6 is only the emphasis on ecosystems as an organizing structure of benefits.

Some ecosystem services are traded in markets. Until fairly recently, much of the work in forest economics and other fields of economics emphasized the role of nature as a storehouse that provides raw materials for the production of economic goods. Less thought was given to environmental quality, wilderness, or other ecosystem services. As economists expand their view of the value of natural systems, they have begun to value the myriad services created by ecosystems. So far, only the most preliminary attempts have been made at valuing most of them.

The challenges in valuing ecosystem goods and services lie in the facts that most ecosystem services are not marketed, that the often unknown "production functions" for ecosystem services could be complicated in both space and time, and that the public-good nature of most ecosystem services makes it difficult for policy makers to devise incentives to encourage their supply. In order to maximize the total value of ecosystem services, we may need to unbundle them into tradable or ordinary goods and services, collective goods, and public goods.

The New York watershed protection project is a good example of how a seemingly public good becomes a collective good that is traded in a voluntary market. The city paid landowners to conserve and restore some polluted upstate watersheds that had previously provided the city with the ecosystem service of water purification. This resulted in an improvement in water quality to levels that met government standards, a saving of over $5 billion for the city, and additional income for rural landowners in upstate watersheds.

Such a voluntary market for certain types of ecosystem services is akin to the types of private arrangements negotiated between individuals who create or are affected by externalities, as noted in Chapter 10. When property rights are well defined and the net gains from trading are sufficiently high for individuals to overcome the transaction costs, successful negotiations are likely to ensue. Voluntary negotiations will also be more likely when the benefits of such arrangements are more certain, measurable, and enforceable.

On the other hand, private bargaining will not occur when the transaction costs are prohibitively high or the benefits of such arrangements are less certain, less measurable, and not enforceable. This is perhaps the typical situation for most ecosystem services. At the extreme, for example, is carbon emissions and climate change, a problem so big, with so many actors, and with conceivable impacts so uncertain in their exact spatial configuration and timing that people have great difficulty grasping how it might affect them personally and what to do about it. The struggle to reach a political agreement among nations at the 2009 Copenhagen Climate Summit demonstrated such difficulties.

Because of this, many traditional and nongovernmental organizations are stepping up to play the role of intermediary, which could lower transaction costs involved in private ecosystem service arrangements where voluntary markets for many ecosystem services still have not emerged or have not functioned well. However, many successful cases of voluntary markets for ecosystem services have government involvement or are created by government regulations in the first place.

The aforementioned Acid Rain Program is a successful case of government cap and private trade. The emerging market for wetland mitigation, where landowners sell mitigation credits to developers, utility producers, and municipalities, was created by the US policy of no-net-loss in wetland. In 1997, a forest landowner in South Carolina was able to sell the right to create additional habitat for, and increase a cluster (family) of, endangered Red-cockaded Woodpeckers on her land for $200,000 to a developer who wanted to take a pair of the birds on a piece of land that he intended to develop. In this case, "take" means to legally eliminate the habitat of the pair, which could cause their deaths. The forest landowner was paid to offset the loss of endangered species on the developer's land. Such a transaction is possible only because, although the US Endangered Species Act prohibits the "taking" of any endangered species, new government regulations allow the taking of endangered species if the loss is perhaps inevitable (because isolated endangered birds cannot survive long) and if a mitigation plan and an offset are provided.

As noted earlier, governments have other means of regulating ecosystem services. In theory, markets can result in the same level of ecosystem services being offered by regulating or taxing ecologically damaging activities or subsidizing mitigations, or by public ownership of land. Hence, when voluntary markets for many ecosystem services do not emerge, governments can step in. Forest economists can help governments make informed decision by valuing various ecosystem services, individually or in aggregate.

Forest Health: Wildfire, Insect, and Disease Control, and Invasive Species Management

Although the management of wildfires, insect and disease outbreaks, and invasive species is not new to the forestry profession, the intensity and frequency of some of the wildfire and insect outbreaks in North America and elsewhere in recent years have been catastrophic and have caused great concern. A 1997-98 forest fire in South Asia burned at least 1.85 million acres (750,000 hectares) of forests in Indonesia, Papua New Guinea, and Malaysia. The 2010 summer forest fires in Russia also seized the attention of the global forestry and environmental community. Large forest fires in the US and Canada have been in the news almost every summer in recent decades. The mountain pine beetle outbreak that began around the middle of the last decade in the US Pacific Northwest and western Canada (mostly in British Columbia) has killed millions of acres of pine trees and is expected to result in the loss of more than 1 billion cubic metres of softwood timber inventory in both countries when it is over. This loss of timber amounts to 4-5% of total recoverable softwood inventory (21 billion cubic metres) in Canada or about 3% of total recoverable softwood inventory in both countries (35 billion cubic metres).

The increased intensity and frequency of wildfire and insect out-breaks is partly related to climate change, but often they are caused by human activities and public policy. For example, the fire prevention policy implemented in the US since the 1920s has probably increased the risk of large, catastrophic wildfires. Growing human populations in forested areas that are labelled as wildland-urban interface (WUI) have also contributed to the risk of fire outbreak and made fire management more difficult. The spread of invasive species is closely related to trade, travel, and tourism.

The core theory in wildfire management, insect and disease control, and prevention of the spread of invasive species is known as the *least cost-plus-loss* theory, which explains the trade-off between damage and management or damage aversion efforts. For example, fire management policy has typically been guided by three concepts: providing adequate protection, minimizing damage, and thus minimizing fire management costs plus damage. This means that the optimal amount of fire management effort (or fire budget) is the amount that can be expected to minimize the total of management cost and property loss.

Management cost includes the cost of fire preparedness, prevention, and suppression. Property losses are more difficult to estimate, however, because of uncertainty and the difficulty of quantifying non-market forest products and services.

Understanding Governance and Building Policy and Institutional Supports for Sustainable Forestry

Sustainable forestry depends on strong policy and institutions that nurture and foster appropriate incentives for forest landowners and forest users. Wasteful deforestation in some countries is attributed to poor governance and weak institutional arrangements in land tenure and forest policy. Thus, understanding the role of governance and building policy and institutions that foster sustainable forestry are an important emerging issue for global forest resource conservation and management.

Governance, defined as the traditions and institutions by which authority in a country is exercised, has multiple dimensions, including public participation and government accountability, political stability and integrity, public security, government effectiveness, regulatory quality, and the rule of law. These characteristics collectively define the quality of governance, which is found to have an impact on forest management.

Governance is often seen as a subject for political scientists. Foresters and forest economists tend to shy away from it. They cannot totally detach themselves from it, however, just as they cannot avoid laws and regulations. Forest certification, for example, is a new governance mechanism that is intended to supplement the traditional government-led governance mechanism in forest management. In this case, nongovernmental organizations have a say on how forests are managed.

Foresters and forest economists perhaps have a better appreciation for policy and institutional building. Policy and institutional failures that hinder sustainable forestry are as common in developed countries as in developing countries. For example, revenues from timber sales from US national forests have not been able to cover the cost of timber sales, reforestation, and road construction for decades. These so-called below-cost timber sales are allowed to take place and persist partly because timber sales expenses are too high and some of these expenses (such as building permanent roads instead of temporary logging roads) should be attributed to multiple uses (including motorized recreation, for example). Some economists, however, argue that more significant is the subtle arrangement in the property rights structure of US national forests: their manager – the US Forest Service – is under no legal obligation to make a profit, and has little incentive to keep timber sales expenses down. Indeed, the current US statutory, regulatory, and administrative laws do not permit the US Forest Service to bring them down.

To show how the US Forest Service loses money in its timber sales, we turn to a comparative study of public timber sales in the state of Montana, where national forests and state forests are intermingled in location, have similar topography and timber quality, and have access to the same markets. Both state and national forests are managed for multiple uses and comply with the same environmental regulations. Environmental audits have shown that state forests are managed no worse, and sometimes better, than national forests for environmental services. Timber sales from state forests generated $13.3 million in income over the

1988-92 period, whereas sales from Montana's 10 national forests yielded a loss of nearly $42 million. Remarkably, state forests generated this income by harvesting only 8% (191 million board feet) of the quantity harvested by the US Forest Service in Montana over this period (23 billion board feet).

This disparity did not come about because the US Forest Service is inherently inefficient compared with the state forest agency. Where the two agencies differ markedly, however, is in the stated purpose of the use of their lands. State forests are mandated by law to generate income from timber and other uses, to fund public schools. National forests have no such mandate, so the costs of timber sales in national forests are much higher than in state forests in Montana.

A similar situation can be found in British Columbia, where silvicultural investment and reforestation performance differs markedly between companies that hold their timber-harvesting rights under Tree Farm Licences and those that hold Forest Licences. Both Tree Farm Licences and Forest Licences are long-term forest tenures and their holders are required to comply with the same forest practices and reforestation regulations – that is, trees must reach a free-to-grow status within a few years after timber harvesting. The only difference between these two forms of tenure is that Tree Farm Licences are more secure, applying to a defined area, with long terms of 25 years renewable every 10 years, whereas Forest Licences provide a right to cut a volume of timber and have terms of 15 years renewable every 5 years. Because Forest Licences are volume-based, their holders, unlike the holders of Tree Farm Licences, are not likely to return to the sites they have harvested and reforested. Further, because both are subject to a government takeback of 5% each time they are renewed, the more frequent renewal of Forest Licences causes their holders to lose more than the holders of Tree Farm Licences. These two subtle differences in property rights arrangements imply that the holders of Tree Farm Licences have a stronger incentive to reforest thoroughly the land they log, whereas the holders of Forest Licences are more likely to be satisfied with meeting the minimum regulatory requirements.

Our last example relates to land tenure reform in the Brazilian Amazon. Until 2006, Brazil lacked a mechanism to regulate forest management on the public land that covers some 80% of its Amazon region, where illegal occupation, illegal logging, and violence were prevalent and sustainable forest management was lacking. In 2006, the country passed a new Public Forest Management Law, which regulates the management of public forests for sustainable use and conservation. The law

sets out the approach to be taken in allocating new forms of timber concessions in public forests – those forests located on federal lands – for
sustainable production involving the private sector, communities, and
other potential stakeholders. Under the law, forest concessions are allocated for 40 years through a bidding process, (although communities
and nongovernmental organizations are exempt from normal bidding
requirements). Concession fees are set on a case-by-case basis and take
into consideration the characteristics of the forest, its location, and other
aspects. Under the law, 20% of all revenue from land use goes to the
newly created Brazilian Forest Service and the Brazilian Institute of
Environment and Renewable Resources.

The Brazilian government hopes that forests under concessions will
be managed for multiple uses and better protected from encroachment
and clearance for agriculture, the main cause of deforestation in the
Amazon. The initial allocation for the first 10 years will be 11 million
hectares, or 3% of the Amazon region. These innovations in forest tenure in Brazil illustrate the growing recognition of the importance of forest policy and institutions in achieving sustainable forest management.

In short, policy and institution building are critical to global sustainable forestry. In the last few decades, we have witnessed calls for reforms
in forest tenure in some developed countries, as well as actual land and
tenure reforms in China, Brazil, Vietnam, Eastern European countries,
and others. Even with varying political systems and economic and cultural backgrounds, different countries can benefit by learning from each
other. Armed with an understanding of the theory of property rights and
institutions and with solid empirical evidence of their practical impacts,
forest economists can help governments design policy and institutions
that foster sustainable forestry and protect the environment.

NOTE

1 A consensus is not necessarily required in this case, as consensus is
mainly a political concept. Further, it may be too late to act by the time
a consensus on climate change is reached among scientists and among
the leaders of all countries. A report of "consensus among scientists"
on climate change can create a controversy of its own. Certain news
media have reported that some 3,750 scientists from more than 130
countries have endorsed the 2007 report by the United Nations
Intergovernmental Panel on Climate Change (IPCC), which states that
anthropogenic global warming represents a threat to the planet. The
great majority of reviewers and authors of the IPCC report worked on
chapters that dealt with historical or technical issues, however. Only

the authors and those who reviewed favourably Chapter 9 of the report ("Understanding and Attributing Climate Change") explicitly endorsed the IPCC's conclusions on anthropogenic climate change. Chapter 9 had 53 authors and 7 favourable reviewers, or 1 scientist for every 200 from the more than 130 countries.

REVIEW QUESTIONS

1 What is the approximate percentage of global forest cover? Where are countries with the highest deforestation rates located?

2 What is the Environmental Kuznets Curve? What does it imply?

3 Describe how the voluntary market under the cap-and-trade system works.

4 Find out how much a typical forest landowner in your area can get from selling carbon credits by growing trees. Discuss the management actions the landowner has to take in order to get these credits.

5 Describe the three key principles of carbon offsets and their applications to forestry-related carbon offset projects.

6 What policy instruments can be used to encourage the production and use of wood-based bioenergy?

7 What are the barriers to the creation of a voluntary market for ecosystem services?

FURTHER READING

Banerjee, Onil, Alexander J. Macpherson, and Janaki Alavalapati. 2009. Toward a policy of sustainable forest management in Brazil: A historical analysis. *Journal of Environment and Development* 18 (2): 130-53.

Chichilnisky, Graciela, and Geoffrey Heal. 1998. Economic returns from the Biosphere. *Nature* 391: 629-30.

Food and Agriculture Organization of the United Nations (FAO). 2006. *Global Forest Resources Assessment 2005.* FAO Forestry Paper 147. Rome: FAO.

Gan, Jianbang. 2007. Supply of biomass, bioenergy, and carbon mitigation: Method and application. *Energy Policy* 35 (12): 6003-9.

Geist, Helmut J., and Eric F. Lambin. 2001. *What Drives Tropical Deforestation? A Meta-Analysis of Proximate and Underlying Causes of Deforestation Based on Subnational Case Study Evidence.* Louvain-la-Neuve, Belgium: LUCC International Project Office.

Leal, Donald R. 1995. *Turning a Profit on Public Forests.* PERC Policy Series PS-4. Bozeman, MT: Political Economic Research Center.

Lindsey, Rebecca. 2007. Tropical Deforestation. http://earthobservatory. nasa.gov.

Mannes, Thomas. 2009. Forest management and climate change mitigation: Good policy requires careful thought. *Journal of Forestry* 107 (3): 119-24.

Simpson, R. David, Michael A. Toman, and Robert U. Ayres. 2005. "Introduction: The 'New Scarcity.'" In *Scarcity and Growth Revisited: Natural Resources and the Environment in the New Millennium,* edited by R. David Simpson, Michael A. Toman, and Robert U. Ayres. Chapter 1. Washington, DC: RFF Press.

United Nations Millennium Ecosystem Assessment. 2005. *Ecosystems and Human Well-Being: Synthesis.* Washington, DC: Island Press.

Vincent, J.R., and C.S. Binkley. 1992. Forest-based industrialization: A dynamic perspective. In *Managing the World's Forests: Looking for Balance between Conservation and Development.* Washington, DC: World Bank.

Zhang, Daowei, and Peter H. Pearse. 1996. Differences in silvicultural investment under various types of forest tenure in British Columbia. *Forest Science* 44 (4): 442-49.

–. 1997. The influence of the form of tenure on reforestation in British Columbia. *Forest Ecology and Management* 98: 239-50.

Index

AAC (Allowable Annual Cut), 232-33.
    See also maximum sustained yield
ability to pay principle (taxation), 303
access fees (for recreational sites),
    145, 151-56
Acid Rain Program, 367
ad valorem tax 215, 307, 309-11. *See
    also* property taxes
advantage: absolute, 337-39; com-
    parative, 337-39; internalization,
    352; locational, 352; ownership,
    352
aesthetic values (of forest), 141,
    163-66, 241
afforestation, 204
aggregated approach (of forest
    products demand), 114
Agricultural Conservation Programs
    (of the US), 266
Alabama, 81, 85-86, 251, 368
allocation of resources, xv, 10-11, 15,
    40, 44-50
allowable cut effect, 236-37, 265
annuity: finite, 58-59; perpetual, 58
appraisals: pre-merchantable timber
    84-92; stumpage, 83-84, 316-17;
    tax, 307-10
area control, 232
average rate of growth (in stumpage
    value), 194(f), 195, 210-11

Bandwagon effect, 161
bare land value, 60, 83, 89-90, 198;
    tax on, 214, 307-10, 327
barriers, trade. *See* trade barriers
barriers to entry, 41, 44-46, 118
below-cost timber sales (in the US),
    375-76
Benefit/Cost Analysis, 63-70
benefit/cost ratio, 65-66, 68-70
benefit principle (taxation), 303
benefits conferred (of property
    rights), 284
bequest value, 6-7
Brazil: deforestation, 357(t), 358(t),
    363; as a destination for foreign
    direct investment, 353; Eucalyptus
    plantation, 251; forest resources,
    356, 357(t); plywood, 345; Public
    Forest Management Law, 294,
    376-77; transition to normal
    forest, 231
British Columbia, 6, 110, 200-1(t),
    251, 374, 376
Brundtland Commission, 19

Canadian Council of Forest Ministers,
    133
capacity: environmental, 360; of a
    recreation site, 162;
cap and trade, 366-68